Japan Environmental Council (Ed.)
The State of the Environment in Asia
2005/2006

Japan Environmental Council (Ed.)

# The State of the Environment in Asia

## 2005/2006

 Springer

Japan Environmental Council (JEC)

AWAJI Takehisa (Editor-in-Chief)
Professor
Law School
Rikkyo University
3-34-1 Nishiikebukuro, Toshima-ku, Tokyo 171-8501, Japan

TERANISHI Shun'ichi (Editor-in-Chief)
Professor
Graduate School of Economics
Hitotsubashi University
2-1 Naka, Kunitachi, Tokyo 186-8601, Japan

Rick DAVIS (Translator)
Ashigawa, Japan

Cover: Terraced paddies in South Slawesi, Indonesia
Photo: Inoue Makoto

Library of Congress Control Number: 2005921888

ISBN 4-431-25028-X Springer-Verlag Tokyo Berlin Heidelberg New York

Springer is a part of Springer Science+Business Media
springeronline.com
© Springer-Verlag Tokyo 2005


Typesetting: SNP Best-set Typesetter Ltd., Hong Kong


Printed on acid-free paper

# Preface to the English-Language Version

The Japan Environmental Council (JEC) published the English-Language Version of the first book of this series, *The State of the Environment in Asia 1999/2000*, from Springer-Verlag, Tokyo in November 1999, followed by the second book, *The State of the Environment in Asia 2002/2003*, also from Springer-Verlag, Tokyo in November 2002. This book is the third one.

The Preface and the Foreword to the English-Language Version in the 1999/2000 volume provide a somewhat detailed discussion on the background of this book and the course of events leading to its publication. Readers interested in knowing why JEC began work on this series are encouraged to read them.

This book is an English-Language Version of *Ajia Kankyo Hakusho 2003/04* ("Asian Environment White Paper 2003/04") published in October 2003 by Toyo Keizai, Inc. During the transition to English, maximum efforts have been made to add new information and update the data, and some editorial changes have also been made.

The basic message we convey through this third English-Language Volume is how people in Asia can work together to build and develop Asia-wide networks for environmental cooperation towards launching the era of global environmental governance from Asia and guaranteeing fairness through the involvement principle in the 21st century. International environmental cooperation is a primary task in the 21st century, and it is our sincere hope that this series will benefit its development and serve as a substantial platform that everyone can use to facilitate this effort.

I would like to express my gratitude to all the people who lent their cooperation and support in the editing, production, and publication of this third English-Language Volume. The following people deserve a special word of appreciation. First, I want to thank the people at Springer-Verlag, Tokyo for publishing this series. Thanks are also due to the Japan Fund for Global Environment for a grant that went a long way toward the production of this book. Second, we owe a major debt of gratitude to the translator, Mr. Rick Davis, for not only tackling the translation, but also more or less single-handedly shouldering the task of editing. Without his dedicated effort, this book would not have been possible.

Work on this volume also benefited significantly from the generous cooperation of the many writers listed on pages IX–XII of the book. Their patience and diligence in cooperating with Rick during the translation process were of immeasurable help in getting this book finished. And last but not least, this project owes much to the people who worked behind the scenes performing the administrative chores associated with editing and production: Dr. Yamashita Hidetoshi, Mr. Hayashi Kiminori, Mr. Yamakawa Toshikazu, Ms. Choi Sunyong, and Mr. Fujiya Takeshi for laboring at the task of making the figures and tables and for creating the index. All of them deserve applause for their unstinting efforts at these toilsome jobs.

Secretary-General of the Japan Environment Council
TERANISHI Shun'ichi
January 2005

## Addendum

In late December 2004 when this English-language version was in the editing stage, countries on the Indian Ocean suffered unprecedented disaster from the severe earthquake off the coast of Sumatra and the giant tsunamis that quickly followed.

On behalf of the editors and writers of this volume, I wish to respectfully express our condolences for the hundreds of thousands of victims, and offer our heartfelt prayers that in 2005 humanity can chart a new course toward peace, public security, and environmental progress throughout Asia and the world.

# Preface

This book is the third in the NGO-oriented series *Ajia Kankyo Hakusho* ("Asia Environment White Paper"). In December 1997 we published the Japanese original of the first book in this series, and the second appeared in October 2000. Our greatest encouragement over this period of time was the enthusiastic reception for the series from a broad spectrum of people. So far the first book has been through seven printings, and the second through four. Further, the English-language books based on these two volumes—*The State of the Environment in Asia 1999/2000* and *The State of the Environment in Asia 2002/2003* (published by Springer-Verlag, Tokyo)—have both had two printings, and are attracting interest not only in Asia, but also among international institutions, NGOs in the West, and other parties around the world.

This third volume has the same three-part arrangement as the first two in the series, but also includes some new experimental editorial innovations throughout the book.

First, Asia by Theme consciously addresses some crucial themes that must be taken into account when considering Asia's environmental problems in the 21st century, and in an anticipatory fashion it somewhat boldly proposes the challenges Asia will face in this century. Its four chapters cover military impacts on the environment, the environment and trade, agriculture/food and the environment, and the biodiversity of forests and rice paddies. As this mix of themes shows, this part describes the basic challenges that, from a mid- and long-term perspective, Asia must address within a framework of mutual cooperation.

Second, Asia by Country and Region takes a new tack by organizing problems and posing challenges from the unique perspective of "regions" that transcend individual countries. This volume is the first to incorporate such an approach, and although it deals with only three regions—Northeast Asia, the Mekong region, and Inner Asia—frameworks based on the regional perspective will be increasingly indispensable for envisioning solutions for environmental problems not only in Asia, but also internationally and globally.

Third, and as the Introduction observes, this volume does its best to offer specific suggestions and concrete proposals for future action. Our basic message in the first volume of this series was "Conservation of the global environment starts in Asia," and that in the second was a slogan proposing how to realize the first: "New advances in Asian environmental cooperation in the 21 century." The present volume is more substantial because it includes a large number of suggestions and action proposals that go further toward fleshing out the message and slogan. It is our sincere hope that this third installment in the series will find an even larger audience and wider use as a guidebook for environmental initiatives in Asia and across the globe.

As the list of Editors, Writers, Collaborators, and Assistants shows, overall editing of this volume was achieved through the close collaboration and teamwork of leading and young researchers including Inoue Makoto, Kojima Michikazu, Oshima Ken'ichi, and Yamashita Hidetoshi. What's more, the roster of people who worked on each part is an even more diverse lineup than before. With enthusiasm for preserving Asia's environment, and feeling encouraged by the new broadening of valuable networks comprising many researchers, NGO people, and others who lend their unstinting cooperation, we have set to work on preparations for the fourth book in this series. We hope that our dedicated efforts on this series will persuade everyone to continue benefiting us with their advice, support, and cooperation.

Finally, we wish to acknowledge that part of the expenses for the editing and production of this book was covered by a 2002 grant ("Continuance of the *Ajia Kankyou Hakusho* and Development of the Asian Environmental Cooperation Network," Awaji Takehisa, representative) from the Japan Environment Corporation's Japan Fund for Global Environment, and a 2002 science research grant ("Policy Research on the Possibilities for Conversion to Ecological Economic Systems in Asia," Teranishi Shun'ichi, representative) from the Japan Society for the Promotion of Science.

August 2003

For the editorial committee
AWAJI Takehisa
TERANISHI Shun'ichi

# Editors, Writers, Collaborators, and Assistants

Note: Names of East Asians are written according to East Asian custom with surnames first.

## Editorial Advisors

HARADA Masazumi, Kumamoto Gakuen University
ISHI Hiroyuki, Hokkaido University
MIYAMOTO Ken'ichi, Osaka City University, Prof. Emeritus
OKAMOTO Masami, Nihon University
SHIBATA Tokue, Tokyo Keizai University, Prof. Emeritus
UI Jun, University of Okinawa, Prof. Emeritus
UZAWA Hirobumi, The University of Tokyo, Prof. Emeritus

## Editorial Committee Members

AOKI Yuko, Yokohama National University
ASUKA Jusen, Tohoku University
AWAJI Takehisa, Rikkyo University
CHEN Li-Chun, Yamaguchi University
INOUE Makoto, The University of Tokyo
ISONO Yayoi, Tokyo Keizai University
ISOZAKI Hiroji, Meijigakuin University
IWASA Kazuyuki, Kochi University
JUNG Sung-Chun, Korea Institute for International Economic Policy
KANAZAWA Kentaro, Kobe College
KOJIMA Michikazu, Institute of Developing Economies
MATSUMOTO Satoru, Mekong Watch
MORI Akihisa, Kyoto University
NAKANO Ari, Waseda University
OSHIMA Ken'ichi, Ritsumeikan University
OTA Kazuhiro, Kobe University
OTSUKA Kenji, Institute of Developing Economies
SAKUMOTO Naoyuki, Institute of Developing Economies
TERANISHI Shun'ichi, Hitotsubashi University
YAMASHITA Hidetoshi, Hitotsubashi University
YOSHIDA Fumikazu, Hokkaido University

## Writers

AIKAWA Yasushi, Tottori University of Environmental Studies
AKIMOTO Yuki, Attorney in the United States
AOKI Yuko, Yokohama National University
ASAZUMA Yutaka, Hokkai Gakuen University
ASHINO Yuriko, Japan Family Planning Association
ASUKA Jusen, Tohoku University
AWAJI Takahisa, Rikkyo University
BAO Zhiming, Central University for Nationalities, China
CHANG Jung-ouk, Matsuyama University
CHEN Li-Chun, Yamaguchi University
CHEON Kyung-ah, Ritsumeikan University, Graduate School
ENDO Gen, Daito Bunka University
FUJITA Masanori, Osaka University
HARADA Kazuhiro, Institute for Global Environmental Strategies
HATA Akio, Osaka City University
HAYAKAWA Mitsutoshi, Citizens' Alliance for Saving the Atmosphere and the
    Earth
HAYASHI Kiminori, Hitotsubashi University, Graduate School
HAYASHI Tadashi, Takasaki City University of Economics
HISANO Shuji, Kyoto University
INOUE Makoto, The University of Tokyo
ISAYAMA Kenji, Ritsumeikan University, Graduate School
ISHIDA Norio, People's Institute of Environment
ISONO Yayoi, Tokyo Keizai University
ISOZAKI Hiroji, Meijigakuin University
IWASA Kazuyuki, Kochi University
IZAWA Arata, TRAFFIC East Asia-Japan
JUNG Sung-Chun, Korea Institute for International Economic Policy
KAMO Yoshiaki, National University of Mongolia
KANAZAWA Kentaro, Kobe College
KATSURAGI Kenji, Fukuoka Institute of Technology
KAWAKAMI Tsuyoshi, ILO, Bangkok
KAWASAKA Kyoko, Citizens' Alliance for Saving the Atmosphere and the
    Earth
KIYONO Hisako, TRAFFIC East Asia-Japan
KOJIMA Michikazu, Institute of Developing Economies
KOYAMA Shin'ya, University of Hyogo
KUSUMI Ariyoshi, Chukyo University
LONG Shi-Xiang, Toyama University
MATSUMOTO Satoru, Mekong Watch
MATSUMOTO Yasuko, Kyoto University
MINATO Kunio, Kobe University, Graduate School
MORI Akihisa, Kyoto University

NA Sungin, Hiroshima Shudo University
NAKAMURA Yuriko, Waseda University, Graduate School
NAKANO Ari, Waseda University
NODA Koji, Hitotsubashi University, Graduate School
OKADA Tomokazu, Tokyo Metropolitan Government
OKUBO Noriko, Osaka University
OSHIMA Ken'ichi, Ritsumeikan University
OTA Kazuhiro, Kobe University
OTSUKA Kenji, Institute of Developing Economies
SAKAMOTO Masayuki, Japan Wildlife Conservation Society
SAKUMOTO Naoyuki, Institute of Developing Economies
SAWANO Nobuhiro, Seiryo Women's Junior College
SHIBASAKI Shigemitsu, The University of Tokyo
SUGIMOTO Daizo, Embassy of Japan in New Delhi
SUNITA Narain, Centre for Science and Environment, India
TACHIBANA Satoshi, Forestry and Forest Products Research Institute
TAKAMI Kunio, Green Earth Network
TAKAMURA Yukari, Ryukoku University
TANI Yoichi, Solidarity Network Asia and Minamata
TATEDA Masafumi, College of Technology, Toyama Prefectural University
TERANISHI Shun'ichi, Hitotsubashi University
THENG Lee Chong, Universiti Putra Malaysia
TSUJITA Yuko, Institute of Developing Economies
UETA Kazuhiro, Kyoto University
UEZONO Masatake, Shimane University
XU Kezhu, Center for Legal Assistance to Pollution Victims, China University
    of Political Science & Law
YAMANE Masanobu, Kanagawa Prefecture Natural Environment Conservation
    Center, Research Division
YAMASHITA Haruko, Meikai University
YAMASHITA Hidetoshi, Hitotsubashi University
YOKEMOTO Masafumi, Tokyo Keizai University
YOSHIDA Aya, The University of Tokyo, Graduate School

## Collaborators (Comments at APNEC and other conferences, comments, and other cooperation)

Alexander LACSON, People's Taskforce for Bases Clean-up, the Philippines
ASANO Masahiro, Ryukoku University
CHIAU Wen-Yan, National Sun Yat-Sen University, Taiwan
CHOI Sunyong, Hitotsubashi University, Graduate School
CHOU Loke Ming, National University of Singapore
CHUN Man Kyu, the Maehyang-ri Task Force for the Closing of the Bombing
    Range

Corazon Valdez FABROS, Nuclear Free & Independent Pacific Movement/ Pacific Concerns Resource Center, People's Taskforce for Bases Clean-up, the Philippines

DUNU Roy, Hazards Center, India

FUJIYA Takeshi, Hitotsubashi University, Graduate School

ICHINOSE Toshiaki, National Institute for Environmental Studies

IKE Michihiko, Osaka University

ISHIGAKI Tomomoto, National Institute for Environmental Studies

JIN Jian, Interpreter

KIM Bok Nyeo, Interpreter

KIM Hye-ok, Ritsumeikan University, Graduate School

KIM Jung Wk, Seoul National University

MEZAKI Shigekazu, Nanzan University

MORIZUMI Takashi, Photojournalist

O'lola Ann Zamora-OLIB, People's Taskforce for Bases Clean-up, the Philippines

PHAM Hung Viet, Vietnam National University, Hanoi

Rick DAVIS, Translator

Surya DHUNGEL, Lawyer, Nepal

TABUCHI Hitoshi, Shinano Mutsumi High School

USAMI Yoshifumi, Osaka Prefectural University

WANG Canfa, Center for Legal Assistance to Pollution Victims, China University of Politics & Law

WANG Xi, Shanghai Jiao Tong University

YAMAKAWA Toshikazu, Hitotsubashi University, Graduate School

# Contents

## Part III   Data and Commentary

# Introduction: Launching the Era of Global Environmental Governance from Asia
## Guaranteeing Fairness Through the Involvement Principle

## 1. Economic Globalization and Environmental Deterioration

Rows of high-rise buildings line broad avenues, motorbikes and automobiles dash through the cities while honking their horns furiously, traffic congestion stalls hordes of vehicles, and people choke on clouds of exhaust gases. And amid all of this, the vibrant throngs of people. On the surface at least, Asia brims with vitality and is sizzling with activity. Asian economies suffered a recession triggered by the 1997 currency and financial crisis, yet recovered faster than experts anticipated.

But strangely enough this led to revision of the Washington consensus, which was the underlying doctrine for North-South international economic cooperation in the post-Cold War world. The Washington consensus comprised policy recommendations agreed upon by the US government and international institutions at an international conference held in 1989 at the Institute of International Economics in Washington. Its 10 policy areas included fiscal discipline, liberalizing interest rates, competitive exchange rates, trade liberalization, liberalization of foreign direct investment, privatization, deregulation, and guaranteeing property rights. In the 1990s globalization driven by multinational corporations proceeded in Asia as the International Monetary Fund (IMF) and World Bank encouraged development under the Washington consensus.

Meanwhile, poverty worsened and gaps between regions and nations widened as environmental damage proceeded and ecosystems declined. We described these events in the first two volumes of this series, whose Asia by Theme sections discussed accelerated industrialization and explosive urbanization, growing motorization, pollution and health damage, the conservation and use of biodiversity, mining development and pollution, the transfers of wastes around the globe, energy policy, and marine pollution and conservation. We also argued that local governments play an important role in environmental conservation.

There is no longer any doubt that globalization can endanger the environment and increase poverty. The United Nations Development Programme (UNDP) clearly points out the negative aspects of globalization, such as widening income disparities, social destabilization, and environmental damage, and to control them it proposes institutionalizing mechanisms built around human rights and consideration for people.[1]

Faced with this reality, the World Bank and IMF had no choice but to effectively revise the Washington consensus, and they are shifting the emphasis of their strategy from the quest for economic growth to initiatives aimed at accentuating poverty reduction, reviving the positive role of governments, and forming partnerships with civil society.

## 2.    Relativizing the Vicious Circle of Environmental Damage and Poverty

In this new strategy one glimpses a perception that sees an interlocking relationship between poverty and the environment, in which reducing poverty allows us to achieve environmental conservation at the same time. Underlying this thinking is the specious theory of the vicious circle of environmental damage and poverty, which holds that poverty causes environmental degradation, which in turn increases poverty. If true, development is given priority until poverty disappears, and conservation comes later.

The UN World Summit on Sustainable Development held in Johannesburg, South Africa in 2002 was a test to determine how international society saw the relationship between the environment and development. Its purposes were to assess the achievements realized after the 1992 Earth Summit and to discuss initiatives for sustainable development. The Johannesburg Summit's accomplishments were the Johannesburg Declaration on Sustainable Development and the Plan of Implementation, which is a set of guidelines for initiatives, but they merely reaffirm the substance of other international agreements to date.

It is important that this summit showed an awareness of the need to integrate the three elements of economic growth, social development, and environmental conservation to bring about sustainable development, showing that the environment was considered just one of three elements. A clear difference from the Earth Summit was that, because the environment was not seen as undergirding development, the environment took a back seat and development assumed the leading role. This view of the situation put the elimination of poverty at the top of the agenda.

But does this vicious circle theory really hold water? Let's examine some evidence to the contrary. Many readers will recall how the widespread Indonesian forest fires of 1997 and 1998 brought about smoke damage in Malaysia, Singapore, Thailand, and other nearby countries. Not only were airports closed due to poor visibility, the smoke also caused aircraft crashes with casualties, and physically injured many people in the region, such as by damaging their respiratory systems and eyes.

When the fires first became an issue, Indonesia's government claimed that farmers practicing swidden agriculture were to blame, that poverty had forced them to burn the surrounding forest for cropland, and that they do not understand the value of the forests (the environment) because they are uneducated. It was the classic argument that poverty destroys the environment. But as studies by international institutions and NGOs proceeded, people began to see the government's explanation as mistaken, and suspected that the cause lay elsewhere.

In time there was agreement that in fact the major causes were the felling and burning of trees to create oil palm plantations or plantation forests in Kalimantan and Sumatra. Companies favored burning off the natural forests because it allowed them to clear land cheaply, and local people helped because they could earn cash income. It was a case of development destroying the environment.

Residents of the fire-affected areas had their own opinions. One of the authors (Inoue) visited East Kalimantan in February 1998, and in an indigenous village previously studied, experienced the overpowering threat of a fire that had reduced orchards to ashes and was bearing down on dwellings. That fire had originated in a nearby coconut palm plantation. People were futilely battling the fire by thrashing the flames with tree branches.[2] When people who depend on the forest to make a living lose it to fire, and when fire burns the orchards that provide their cash income, those people become fire refugees. It is a case of environmental damage causing poverty.

This example illustrates the process by which development leads to environmental damage, which in turn brings about poverty, and shows that the vicious circle theory does not always hold up. Putting development before conservation to eliminate poverty is not always the right answer.

# 3.    Glocal Perspectives Generate Commitment

It is important that the environment and development be seen from a distance, which calls for a global perspective. Many reports published by international institutions and discussions at international conferences are based on just such a perspective. But that is like watching a bombing on television because the people who live with environmental problems and the attendant hardships are invisible to us. No matter how much information one may have, it is very difficult for people not on the scene to make the commitment spoken of by Amartya Sen. Commitment means having a moral concern for people one doesn't know, and resolving to take action to stop infringements on their rights. We need information generated from the perspective of locales where environmental problems are happening, or it is impossible to have many people easily make commitments.

An example would be tropical forest policy, where the views of experts on forest management (foresters, corporate silviculturalists, scientists) have been considered authoritative. These experts have published the results of their scientific analyses in journals, and published their results as annual reports, but it is only a handful of other experts who read them. Certainly no citizen would arrive at a commitment after reading arcane papers written in language understood among only a small number of experts, but many people would be motivated to make a cross-border commitment if more effort were made to frame problems from the viewpoint of people living in the locale, and to make the information more understandable. Global citizens take action due to their concern for the situation in a certain place. This "think locally, act globally" stance, along with the preexisting "think globally, act locally," is associated with a "glocal" (global + local) viewpoint in the sense that people think about local problems in connection with global problems.

# 4. Involvement Supports Local Environmental Governance

It is hoped that providing information based on a glocal viewpoint will result in international commitments, and improve the capabilities of local people. "A person's 'capability' refers to the alternative combinations of functionings that are feasible for her to achieve. Capability is thus a kind of freedom: the substantive freedom to achieve alternative functioning combinations (or, less formally put, the freedom to achieve various lifestyles)."[3] Local environmental governance (arrangements for using and managing local environments) results from commitments by not only local people but also outsiders including foreigners. Governance is not governments alone, for the use and protection of local environments happens through collaboration among local people, businesses, citizens, NGOs, NPOs, and administrative authorities.

Collaboration through governance assumes the existence of a mature civil society with a global awareness of human rights and the environment. Although we will not try to define "citizen" here, we will say that citizens are not all alike because some people have zero interest in their own local environments, while others will even go abroad to work on conservation. Thus brandishing abstract terms like "citizen" leads to a superficial discussion.

For that reason we offer a principle derived from actual conditions in the tropical forest itself. The type and extent of involvement with the forest differ according to the individual and group. People who live in the forest and get their food there are quite deeply involved. City dwellers do not depend on the forest for their livelihoods, but they visit the forest for recreation, or donate money for its conservation. They perforce have a shallower involvement than forest dwellers do. People in developed countries have a shallow involvement through their interest in tropical forests abroad and through their economic support for NGOs working there. This is one kind of commitment.

If we ignore this depth of involvement and allow everyone to participate equally in the process of developing forest policy, those with only shallow involvement will be the majority and have a larger voice, and their views will likely become forest policy. A typical example of this the designation of national parks and other protected areas, the consequence being that people deeply involved with the forest are chased out.

On the other hand, narrow localism holding that forests belong only to locals is no longer convincing. Needed now is open localism, in which local people play an important role while at the same time using and managing local forests and other resources in partnership with involved outsiders.

In this light, we would like to define "involvement" as the principle of recognizing the most diverse possible range of actors involved in local environmental governance, and then granting them a say that is measured by the depth of their involvement. This "involvement principle" allows people in other countries to be involved in local environmental governance in their own ways. Some people will go abroad and work side by side with local people, while others will offer specialized knowledge from their home countries. Some people cannot get away from their jobs, but cooperate by financing NGOs whose work they support.[4]

Needless to say, the involvement principle has never been used in the actual process of consensus-building. If one were to design a consensus-building forum based on this principle, it would necessitate the troublesome process of creating indicators for each actor's degree of involvement with the local environment in question, and weighting each actor's vote according to the indicator. And sometimes governments will refuse commitments by foreigners with the usual claim of interference in their internal affairs. Yet those who have worked in the forest are intuitively convinced by the involvement principle.

The involvement principle encourages a certain approach: Instead of turning the other way with the excuse that it's another country or doesn't concern me, stand and make a commitment; there are many ways to make a commitment, but outsiders should always lend help to local people from the sidelines; and the best possible help should be provided if the locals are victims, but we should be careful of meddling that ignores local people's rights.

## 5.   The Fairness Debate and the Involvement Principle in International Negotiations

Many readers probably feel that the involvement principle sounds good but is unrealistic, or that it is mere idealistic twaddle because the venues of international negotiations, where governments do the talking, are the scenes of harsh confrontations over fairness between the developed and developing countries.

A typical example is international negotiations over global warming. It is the developed countries that produced a volume of greenhouse gases perhaps large enough to disrupt the ecosystem. For example, Japan's 1999 $CO_2$ emissions were the world's third-largest at over 315 million tons carbon equivalent. The emissions of this one country are far higher than the approximate 217 million tons of the entire African continent.

However, the developed countries' GHG emissions are so high that their efforts alone are insufficient to cope. Leaving future generations with a sustainable environment requires global reduction efforts that include the developing countries, to whom this is a quite unfair owing to the huge difference in per capita emissions between the two groups. Compare, for example, the per capita emissions of Japan and China in 1999, which were 2.49 TCE in Japan but only 0.61 TCE in China. These data show that first developed countries must reduce their own GHG emissions, and that it is quite unfair to saddle developing countries with this requirement until developed countries have achieved their own reductions. It is from this standpoint that developing countries strongly oppose initiating discussions for long-term GHG cuts beyond the Kyoto Protocol. How are long-term emission reductions to be achieved in view of this highly unfair relationship between the two groups? Future international climate change negotiations are likely to feature heated discussion on how to assure fairness.

In fact, the involvement principle is useful in resolving such clashes because people always take a stance as individual people or citizens. The principle sets forth a code of conduct to which people must subscribe if they are to be truly free individuals unfettered by national interests or by vested interests as residents of developed countries. The involvement principle lends legitimacy to the involvement of outsiders in the problems of other regions. Many people recognize that as long as outsiders respect local people and do not impose their own values, the involvement of outsiders is legitimate.

Our task, based on such an accord, is to create an involvement zone (a kind of social sphere) that transcends national borders. It is none other than "interregionalism,"[5] a job akin to putting the vital finishing touch on a work of art. The conservation and management of the public interest on a global level can be achieved only by developing and building up mechanisms to conserve and manage the public interest on a regional level based on open relationships between regions.

# 6.    Proposals for Action

The Japan Environmental Council wants to learn from the lessons of the 20th century—the century of war and environmental destruction—and make the 21st century into one of peace and environmental progress, but the world is full of ethnic conflicts and religious disputes—not to mention the US invasion of Iraq—and the grave environmental damage caused by these uses of force. War is the worst environmental scourge, and for that reason the Asia by Theme part features a chapter on it. Any serious consideration of environmental problems cannot fail to consider military impacts. Although we are not experts on military issues or international politics, we feel the time has come to examine military actions from an environmental perspective.

To tackle the problems Asia faces, below we list specific actions that are derived from the Asia by Theme and Asia by Country and Region parts of this book. They are arranged into three categories depending on the actors.

## 6-1 Actions by Asia as a Region

(1) Build a conservation-oriented framework to reduce and restrain military activities. Especially in geographical areas that need urgent action, create international funding systems or other institutions to help people take back healthy and safe living.

(2) Create arrangements to share information obtained through trade, such as customs inspections and plant quarantines, and use the information to protect endangered species, properly manage hazardous wastes, and limit the inappropriate use of agricultural pesticides.

(3) The WTO system limits protection of domestic agriculture and keeps domestic policy from developing sufficiently. Sustainable development of food production makes it essential that trade agreements include provisions for eliminating starvation and that all countries have "food sovereignty" which allows them a certain self-sufficiency rate.

(4) Enhance monitoring of agribusiness, which is the force behind the internationalization of agriculture and food issues, and institute international growth management of agriculture- and food-related trade and investment.

(5) Prepare a corporate code of conduct and an environmental convention to internationally control the indiscriminate development of agribusiness, and set up an Asia-wide system, in parallel with national regulations, that would allow returns on capital investments to be used for environmental conservation in producing areas.

(6) Set up systems to manage the Sea of Japan, the Mekong watershed, and other internationally shared resources on the national, local, and NGO levels.

(7) Asian governments and citizens, including NGOs, should cooperate in beefing up laws and arrangements that regulate genetically modified crops and foods.

## 6-2 Actions by Governments

(1) Asian governments should strictly control environmental pollution and damage caused by military exercises and bases.

(2) Trade and investment are progressively liberalized under free trade agreements and economic partnership agreements, but governments should also assess those agreements' environmental impacts, and more vigorously implement environmental policies to keep trade and investment liberalization from damaging the environment.

(3) Asian governments should regulate activities that damage the environment in agricultural areas, and reconsider the way they go about development.

(4) Asian governments should work out strategies to support the development and dissemination of environmentally sensitive agricultural technologies. Some specific examples are: (1) Teach farmers how to properly use pesticides, and register and manage all pesticides; (2) take measures to restrict or ban hazardous pesticides; and (3) promote and disseminate small-scale, environmentally compatible agricultural technologies.

(5) Asian governments should adopt an ecosystem approach that integrally manages land, water, biological resources, and other resources and encourages their sustainable use.

(6) Japan's government should encourage more governments to ratify the Ramsar Convention and other conventions for conserving Asia's wetlands and rice paddies.

(7) Japan and other countries with laws that allow a certain level of information disclosure and citizen participation should provide countries lacking such laws with material and personnel support for such systems and their development.

(8) Asian governments should immediately regenerate polluted areas, such as those with contaminated soil.

## 6-3  Actions by Citizens (NGOs) in Asian Countries

(1) Facilitate networking among environmental NGOs to establish an Asian Environmental Cooperation Organization (AECO).

(2) Asian environmental NGOs should cooperate with NGOs working on the military base issue from a peace perspective, and approach the same issue from an environmental perspective.

(3) NGOs should conduct studies to determine the extent of environmental damage at the former Subic Bay Naval Base and Clark Air Base in the Philippines, and make the results useful in understanding base-caused harm throughout Asia.

(4) In countries hosting foreign military forces, NGOs should demand revision of their status of forces agreements, with emphasis on restoration of bases and their environs to their original condition, and the application of local environmental laws.

(5) NGOs should monitor illegal logging, the endangered species trade, and the recycling of wastes in Asian countries.

(6) NGOs, farmers' organizations, labor unions, cooperatives, and other organizations that comprise chiefly producers and consumers, who make up the two sides of the agricultural and food system, should collaborate in raising questions and using their influence to have desirable policies developed.

(7) Japanese consumers have a particularly heavy dependence on Asia's agriculture and food trade, and because they are a party to the overconsumption and excess food imports that accelerate environmental damage, they need to reassess their lifestyles from the ground up.

(8) Agricultural producers throughout Asia should have an awareness of them-selves as those who practice environmentally compatible agriculture and local resource management while making use of modern environmental science and traditional knowledge.

(9) When implementing projects that claim to include transparency and the participation of native peoples and local stakeholders, Asia's NGOs should monitor the extent of actual participation and encourage real local partici-pation in the management of forests and other local resources.

(10) Asia's NGOs should work on rectifying Asia's quickly expanding afforesta-tion because the top-down, large-scale afforestation practiced until now is standardized and ignores local conditions, a situation which has deprived people of their land and created social unrest.

(11) Asia's NGOs should educate the populace and governments about pre-serving wetlands (rice paddies), and they should perform functions such as the interface between governments and local people, and monitoring preservation for the government.

Inoue Makoto, Kojima Michikazu, Oshima Ken'ichi

# Essay:   Climatic Disasters in Bangladesh[6]

Bangladesh is known as a country of frequent climatic disasters that can be categorized into floods caused by water from the mountains in the north, and cyclone damage along the Bay of Bengal coast in the south. Scientists anticipate that they will worsen as climate change proceeds.

Flooding happens often because the amount of water flowing into the country from Nepal and India is greater than Bangladesh's rainfall. Nepal's rivers carry more water because the country's glaciers are melting due to global warming, and because its forests are being logged. India is building mammoth dams, whose gates are opened and closed to control the water amount on India's side. Bangladesh must therefore control flooding through cooperation with neighboring countries sharing the same rivers.

On top of this, Bangladesh has another problem arising from its own geography. Ninety percent of the country's land is plain no higher than 9 m above sea level, and it has a large delta that with an elevation of between 1 and 2 m. Bangladesh is therefore highly sus-ceptible to the impact of sea level rise.

Flooding in recent years has submerged three-fourths of the country. Flood control is difficult because in some places river channels shift as much as 800 m in a year, and erosion frequently breaks levees. The difficulty of predicting erosion leads to the construction of inappropriate levees, actually hindering the drainage of certain areas. In such situations local people break the levee themselves in a "public cut," which can further hobble the levee's ability to control floods.

Depending on levee construction and other artificial measures to control all flooding makes it hard to cope with unexpectedly large floods and storm surges, and can cause a vicious circle. Emphasis should be placed on managing the inflow from neighboring countries through international cooperation, and emphasis should be shifted to stopping large-scale disasters while allowing floods of a certain size.

Bangladesh has such frequent cyclones that it suffers 75% of the world's total cyclone damage, and it is getting worse because global warming and extreme weather events such as those due to El Niño are causing more and larger cyclones. Poor Bangladeshis are concentrated in the high-risk area along the southern coast, where they have experienced many powerful cyclones and storm surges. Cyclone-driven storm surges can be as high as 10 to 12 m. The clearing of coastal mangrove forests to make way for shrimp ponds and other development eliminates this natural breakwater and weakens water retention capacity. Yet the poor have nowhere else to live despite their knowledge of the danger.

The Department of Forest conducts planned mangrove afforestation along the coast to cope with cyclones and storm surges. Bangladesh's government, the Red Crescent Society, and other agencies have been augmenting prediction and warning equipment, building cyclone shelters, and taking other measures. Despite the various means of prediction and warning, they have limited effect owing to the low literacy rate (49% for men, 26% for women) and low radio ownership of 20 to 30%.

Cyclone shelters are little help in places where river channels move widely. Shelters sometimes disappear below sea level in just a few years after being built. Some of the older ferroconcrete shelters have deteriorated so badly that their steel bars are exposed. The Japan International Cooperation Agency and other aid organizations are building shelters in different locations, but maintenance is inadequate. Yet, some areas are making progress by improving their warning systems and stepping up community disaster prevention education. Because bigger and more numerous cyclones are expected, it will be necessary to enhance not only physical components such as cyclone shelters, but also education and other institutional components.

Isayama Kenji, Oshima Ken'ichi

# Part I

# Asia by Theme

# Chapter 1
# Military Impacts on the Environment: Working Toward the Century of Peace and Environmental Progress

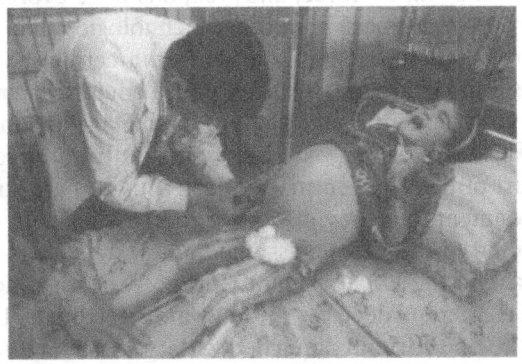

A Baghdad girl with bone marrow cancer. She was exposed to radiation from depleted uranium munitions used in the Gulf War, and doctors think her internal organs are enlarged because of heavy metal poisoning.
Photo: Morizumi Takashi, April 1998

Gunnery targets at Nong-do, Kooni Fire Range, Republic of Korea. Already two-thirds of Nong-do has disappeared because of firing practice. Training continues except for Saturdays and Sundays.
Photo: Oshima Ken'ichi

## 1.   From the Century of War and Environmental Destruction to the Century of Peace and Environmental Progress

War and environmental destruction characterized the 20th century. There were two world wars and other frequent wars and conflicts around the globe. Although progress in mega-science and technology brought economic growth, military technologies advanced to such extremes that they can instantly wipe out urban environments with histories of several thousand years, and even the global environment with its history of hundreds of millions of years. Ultimately, militaries are meant to do one's will by force, and weapons have evolved to an ultimate state for that purpose. Advances in military technologies changed the character of war itself, which used to be simply a fight on a certain battlefield, but now makes all citizens into victims and destroys entire countries. Once a war starts, it leads to drastic environmental destruction, damaging not only cities, cultural assets, and other historical stock, but also wiping out whole ecosystems.

Here at the start of the 21st century military activities are still highly destructive to the environment, but they also have the new element of "never-ending war." Already military actions have assumed a new dimension. Triggered by the 9/11 terrorist attack in the US, President George W. Bush conceived retaliation against terrorism as war and declared all-out war on terrorism. Less than a month after 9/11 the US and Britain launched a large-scale military operation in Afghanistan, and in 2003 the US and Britain struck Iraq.

The attack on Iraq was not even a preemptive strike, but a "preventive war" launched despite the lack of direct danger to the US.[1] Since 9/11 the US military has constantly been on a war footing under the pretext of preventing terrorism and containing it through war. Military activities could become a constant part of everyday life, over and beyond the traditional definition of war, and there is a rapidly increasing danger of military environmental damage on a daily basis.[2]

Protecting the global environment necessitates reining in militaries and their environmentally destructive activities because sustainability and military action are incompatible. Achieving a sustainable society requires that the 21st century become the century of peace and environmental progress. It is imperative that the world limit and reduce military activities as soon as possible. Humanity stands at the crossroads of a crucial decision: Will this be another century of war and environmental destruction, or the century of peace and environmental progress?

World citizens are already taking action. During the 20th century the possibility of environmental catastrophe increased, but it was also the first century when movements to oppose and restrain military activities arose around the world. As the Iraq War loomed, the groundwork laid by the peace movement over the last century blossomed into the biggest anti-war demonstrations since the Vietnam War, and possibly the largest ever expression of anti-war sentiment in terms of a simultaneous world upwelling.[3]

One feature of the movement against the Iraq War was opposition to the use of depleted uranium, which was also opposed on environmental grounds. It stands to reason that the peace and environmental movements would come together because of the devastating military impact on the environment. Because war both kills people and destroys the environment, the Iraq War should serve as the occasion for the peace and environmental movements to integrate themselves and their policies.

Military activities damage the environment in four ways: base construction, base operations, war preparations (training and maneuvers), and battle. This chapter will examine each of these aspects with regard to environmental problems and how citizens are addressing them. Nearly all the problems dealt with here are caused by the US military because that information is comparatively easy to come by, and because of the US military's overwhelming preponderance throughout the world. Grounds for this second criterion are America's colossal military expenditures (43% of total world military spending), which surpass even the combined expenditures of the second-place (Japan) through 15th-place (Israel) countries,[4] the size of its forces around the world, and its technological superiority.

# 2.    Military Base Construction

## 2-1    Base Size and Impacts

Military bases need a variety of facilities including ammunition depots, and garages and hangars for tanks and aircraft. They combine ports and airfields, and have facilities such as training areas nearby. Since military installations occupy a considerable land area, information on how much land is used for military activities worldwide is vital to getting the full picture on military-caused environmental problems. Unfortunately, even these most basic statistics are not released in the military sector. Even yearly defense reports are published only by Japan and the United States. In fact the large majority of countries keep their military information totally secret. It is especially hard to get a complete picture of military activities in Asian countries. The Republic of Korea (South Korea) formerly published a yearly defense report, but it is now biennial. The Democratic People's Republic of Korea, Russia, Pakistan, Myanmar, and the Philippines do not publish defense reports, while China, India, Taiwan, Thailand, and Indonesia publish them irregularly.[5] This makes it impossible even to determine base sizes.

*State of the World* 1991[6] estimates that in 1981 peacetime military use of land was 1% in developed nations and 0.5 to 1.0% worldwide. Assuming 1%, that nearly equals the entire land area of Indonesia. Land used for military purposes should also include the grounds of defense industry companies, minefields, and other uses, but there is no way to estimate these.[7] The US dedicates 27,050,000 acres, or about 1.2% of its land, to military installations.[8] The US also has many bases on foreign soil, 21.3% of which are in Asia. Large bases in the Philippines were returned in 1991 and 1992, making the current breakdown 14.6% in Japan and 6.7% in South Korea.

Although total base area in Asia is unknown, the environmental impact is huge. As bases are often located in undeveloped areas, they sometimes have fatal impacts on valuable ecosystems and rare species.

## 2-2   Relocation of Futenma Air Station (Japan)

Asia's biggest issue regarding environmental damage by base construction is the plan to build a sea-based facility offshore from Henoko in Nago City, Okinawa to relocate Futenma Marine Corps Air Station. The current plan[9] calls for building a 184-ha base over a reclaimed coral reef (Fig. 1, Photo 1). If built, this base

FIG. 1. Proposed Site for Offshore Facility to Replace Futenma Air Station

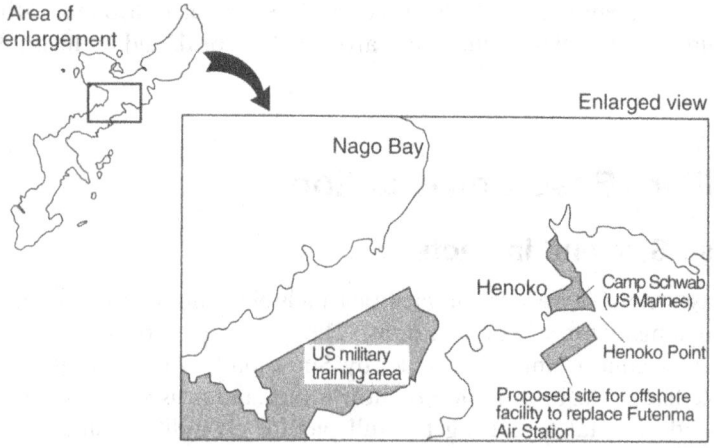

Source: Prepared from documents including "Basic Plan for the relocation of Marine Corps Air Station Futenma," July 29, 2002.

PHOTO 1. Coast in the Henoko Area (1988 photo by Mezaki Shigekazu)

would likely bring noise and marine pollution to the area because of its close proximity to Henoko and the nearby island of Hirashima. Inside the reef is a large, high-quality seaweed bed that serves as a dugong feeding area. Dugongs are sensitive to environmental changes and are designated an endangered species in the Fisheries Agency's Red Data Book. Fifty or fewer dugongs live in Okinawan waters, and they are protected under Japan's Wildlife Preservation and Game Act. If the base is built despite opposition, it could mean the extinction of this population.

Okinawa had been demanding the unconditional return of Futenma Air Station, but the Japanese and US governments agreed that Japan would provide a replacement site for the facility on the eastern coast of Okinawa island, and the site offshore from Henoko was chosen.

Henoko has the highest classification as an area deserving protection under Okinawa Prefecture's Nature Conservation Guidelines. Even in Okinawa, which hosts 75% of US forces in Japan, no new bases have been created over the last 40 years, but if Japan builds this new base it would be one of the worst blows to the environment in Asia. Nago City's mayor stipulates that a 15-year limit on use is a condition for accepting the base, but this is not legally binding, and once the base is built, environmental damage by the US military will worsen.

In response to requests from the World Wide Fund for Nature Japan and other environmental organizations, the World Conservation Union in October 2000 recommended that the Japanese and US governments make conservation efforts at Henoko. This recommendation urges Japan's government to carry out an EIA on the facility as soon as possible, and to implement dugong protection measures as soon as possible to stop the population's decline and help it recover. It also asked for the US government's cooperation in conducting the EIA.[10]

The July 29, 2002 Basic Plan for the relocation of Marine Corps Air Station Futenma stipulates among environmental measures that an EIA shall be conducted for base construction, and that appropriate measures shall be taken to minimize base impacts. Japan's EIA system was often criticized for not questioning the need or appropriateness of development projects, but the 1999 Environmental Impact Assessment Law effected many improvements and made planning-stage assessments possible. Under the law, local citizens must be shown the scoping document and asked their opinions before the draft EIS is released.[11] Therefore the EIA called for by the basic plan must not assume from the outset that the base will be built.

Since April 2003, however, the Japanese government through the Naha Defense Facilities Administration Bureau has been defying citizen opposition in its attempts to conduct boring at 63 sites off the coast of Henoko, calling it a technical study. In view of the heavy environmental impact of this boring, this study itself would normally be subject to an EIA. If the government goes ahead with the boring against the citizens' wishes, its heavy impacts on the environment would affect the local ecosystem before the EIA for base construction is performed, making it impossible for that EIA to be accurate. This study should not be performed because it would render the EIA system meaningless.

The EIA planned in this instance must be properly done, by holding explanatory meetings and public hearings, and then incorporating community opinions from the scoping document stage. The problem arises in the many impediments in EIA implementation: For example, not only are the scoping documents, draft EIS, and EIS very long documents, the inspection period is very short, and copying is restricted.[12] The system must be improved to allow citizens to participate adequately. Without such changes, conducting a proper EIA will be impossible, and end up being a perfunctory procedure for the construction of this base.

A scientific survey of the ocean by Henoko is proceeding with the cooperation of local citizens, environmental NGOs, and other parties. If improvements are effected in the EIA system's implementation, and if citizen views are incorporated with a scientific basis, it is quite possible the plan could be changed or withdrawn as in the case of Fujimae tideland in Nagoya.

# 3.    Base Pollution and Environmental Damage

## 3-1    Information Lacking About On-base Pollution

### The Need to Manage Dangerous Substances

Military bases store and handle many environmental pollutants in large amounts, including explosives; fuels; lubricants; cleaners; insulation material; nuclear, biological, and chemical weapons; heavy metals, and herbicides. Just as civilian factories are required to manage such substances in accordance with environmental regulations, military bases should assess the dangers of the chemical substances they use and properly manage those substances at all times.

Nevertheless, information about such substances on military bases is often secret, which is so even for nuclear weapons, the most dangerous. Because Japan professes its three non-nuclear principles, nuclear weapons have supposedly never been brought onto Japanese soil. Prior consultations are to precede the introduction of nuclear weapons, and the Japanese government's official position is that since prior consultations have never been held, the US has never brought nuclear weapons into Japan. Yet, it is an open secret that nuclear weapons have been brought into Japan.[13] But we have no idea about the extent of nuclear weapons contamination due to secrecy and the inability to investigate military installations.

Information is also unavailable about chemical and biological weapons. Even local governments hosting bases have no idea if such weapons exist on local bases, and if they are properly managed. In fact, such information is unavailable even for other substances such as ordinary chemicals, fuels, waste oil, and lubricants. In Japan the only way such information about US military bases comes to light is when an accident happens.

In the mid-1990s the US Marines Okinawa Environmental Branch began a pioneering initiative that involves enhancing its control over PCBs and certain other substances, and creating opportunities for regular discussions with Okinawa Prefecture,[14] but the US military as a whole does not systematically make such efforts. In many cases it is only after a base has been returned that people discover what chemicals had been used there. Even in Okinawa, which hosts many US military bases, it was only after the 1995 return of the Onna Communications Site that attention was focused on the hazardous chemicals that bases have. An analysis of the site's septic tank sludge, which was to be used as fertilizer, revealed that it was contaminated with high concentrations of mercury, cadmium, arsenic, PCBs, and other substances. Prior to this, there had been hardly any discussion on the possibilities of contamination from hazardous substances on US bases.[15]

## Yokosuka Base Contamination

There is likely serious contamination on and around Yokosuka Naval Base. Aircraft carriers are not mere ships, but mobile bases that carry many weapons and large amounts of chemicals. Bases that serve as home ports for carriers present an especially serious danger of contamination.

For example, in 1988 serious heavy metal contamination was discovered at Yokosuka's berth 12, which is used by aircraft carriers. Yokosuka is the home port of the Seventh Fleet's Carrier Group Five, and is the US military's only overseas carrier home port. A 1991 US Government Accounting Office report contained material on the heavy metal contamination at Yokosuka, and a 1993–94 study by the US Navy found that the groundwater near berth 12 had lead contamination that was 250 times Japan's environmental quality standard (EQS). A Japanese government study similarly found that the berth's soil, groundwater, and seabed sludge contained heavy metals and organochlorines in high concentrations. The US Navy merely surfaced this area with asphalt and did not decontaminate it.

In 1997 contaminated soil washed into the ocean when work began on expanding berth 12, and lead concentration detected in the water nearby was 4.8 times Japan's EQS. Studies of gobies and other marine life have been conducted by the Kanagawa Prefecture Health Insurance Doctors Association. The 1999 study found that seven of 22 (31.8%) gobies caught had spinal abnormalities. In 2000 a higher percentage of gobies had congenital spinal deformities, and in 2001 there were even gobies with tumors.[16] However, in December 2002 the Yokohama Defense Facilities Administration Bureau gave the green light for berth 12 expansion construction without conducting a comprehensive environmental study, and with Yokosuka City approval the construction began in August 2003.

In addition to the berth 12 area, there could be widespread contamination on Yokosuka Naval Base. In June 2000 an investigation of the soil on a family housing construction site in the base's residential area discovered mercury, arsenic, and lead far in excess of Japanese EQS, but the source and cause of the contamination are totally unknown because the site had never been used for ship repair or other pollution-generating tasks, suggesting that the entire base might be similarly contaminated.

By 2017 the two remaining conventional diesel-powered aircraft carriers that call Yokosuka their home port, *Kitty Hawk* and *John F. Kennedy*, will have been decommissioned and replaced with nuclear-powered carriers. Facilities and maintenance work on nuclear-powered ships will increase the danger of radioactive contamination. Local environmental NGOs are calling for a halt to berth 12 expansion work because of the hazards its presents, and they are asking the US government to conduct a comprehensive survey of contamination on the base, to perform remediation, and to abandon making Yokosuka into a home port for nuclear-powered carriers.

## 3-2    Base Contamination in the Philippines[17]

### Health Damage

Grave health damage in the Philippines, which was the first place in Asia where entire US military bases were returned, illustrates the slipshod way in which hazardous substances were managed on those bases because such damage is occurring in the areas of two former large US bases, Clark Air Base (returned in 1991) and Subic Bay Naval Base (1992). Both sites are undergoing the transformation from military to civilian use, and many companies from foreign countries have sited there.

The Philippine environmental NGO People's Taskforce for Bases Clean-up is visiting homes and using other means to investigate the health damage among people living near the former bases. As of August 2004 there were 2,460 pollution victims in all around the bases, with a large variety of ailments including leukemia, cancers, renal disorders, and respiratory disorders (Table 1).

### Former Clark Air Base

A case of particular note in the Philippines is the health damage suffered by evacuees taken into a temporary evacuation center set up in the Clark Air Base Communications Center when Mt. Pinatubo erupted. Their maladies were typical of those appearing in people exposed to pollutants on a base by living there.

For drinking and bathing, evacuees were using groundwater from shallow wells that were easily polluted by contaminated soil on the base. They were not permitted the everyday use of water from the deeper and safer wells used by the US military. From the time they entered the shelter, the evacuees complained about the smell, taste, and color of the groundwater, and they found oily substances floating on water they used. Health problems appeared immediately after they moved into the center, but it is not known how many people suffered ill effects because no official records remain.

TABLE 1. Documented Cases of Toxic Waste Victims around Former US Bases in the Philippines (as of April 30, 2004)

| Former Clark Air Base | Living | Dead | Total |
|---|---|---|---|
| Central nervous system disorders / cerebral palsy: all children ages 1–7 | 39 | 0 | 39 |
| Congenital heart disease, heart ailments w/lung or kidney problems | 26 | 28 | 54 |
| Leukemia, signs of leukemia | 11 | 117 | 128 |
| Skin diseases and various skin disorders | 68 | 8 | 76 |
| Kidney disease | 28 | 8 | 36 |
| Lung disease, TB | 34 | 11 | 45 |
| Cancer of breast, throat, uterus, liver, urinary bladder, etc. | 29 | 40 | 69 |
| Stomach problems | 8 | 4 | 12 |
| Spontaneous abortions, still births | 16 | 5 | 21 |
| Asthma | 26 | 5 | 31 |
| Sudden Death | 0 | 9 | 9 |
| TB meningitis | 2 | 0 | 2 |
| Enlargement of testes | 0 | 1 | 1 |
| Congenital strawberry hemangioma | 1 | 0 | 1 |
| Total | 288 | 236 | 524 |

| Former Subic Bay Naval Base | Living | Dead | Total |
|---|---|---|---|
| Central nervous system disorders / cerebral palsy: all children ages 1–7 | 15 | 1 | 16 |
| Congenital heart disease among children ages 1–14 | 19 | 4 | 23 |
| Cardiac aliments and/or abnormal heart functions among adults | 11 | 6 | 17 |
| Leukemia, signs of leukemia including severe anemia | 47 | 252 | 299 |
| Skin disease and various skin disorders | 71 | 5 | 76 |
| Kidney disease | 17 | 7 | 24 |
| Asbestosis, lung disease (among former workers at shipyard repair facility) | 833 | 487 | 1,320 |
| Cancer of breast, throat, uterus, liver, urinary bladder, etc. | 15 | 42 | 57 |
| Tumors in various parts of the body | 14 | 4 | 18 |
| Spontaneous abortions, still births | 6 | 5 | 11 |
| Weak lungs including asthma | 52 | 5 | 57 |
| Hydrocephalus | 6 | 4 | 10 |
| TB meningitis | 3 | 0 | 3 |
| Enlargement of testes | 0 | 2 | 2 |
| Congenital strawberry hemangioma | 3 | 0 | 3 |
| Total | 1,112 | 824 | 1,936 |

Source: People's Taskforce for Bases Clean-up <www.yonip.com/main/victims.html>.
Note: This table tallies only documented cases; the actual numbers of cases are believed to be much higher.

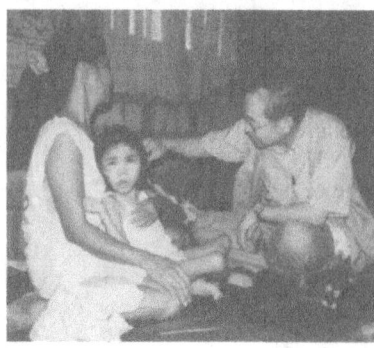

PHOTO 2. Dr. Harada Masazumi examining a child thought to be a victim of base pollution. Women who lived in the Clark Air Base Communications Center have given birth to many children having cerebral palsy (August 2002 photo by Oshima Ken'ichi).

The only documentation indicating the extent of the damage consists of data kept by Mandy Rivera, a community leader at the evacuation center.[18] Rivera carried out a health survey of the 500 households living in the center in 1994, and found that 144 people had been affected by cancer, leukemia, deformities, miscarriages, still births, heart disease, and kidney disorders. The Senate committee performed a follow-up study which found that, as of May 2000, 76 of those 144 people had already died. About 20,000 households used the center over the nine years from its creation in 1991 to its closing in 1999. Calculations based on the number of households, number of people having health problems, and the percentage of people who died as found in Rivera's study indicate the possibility that nearly 7,000 people suffered pollution-caused health damage, and that almost half of them have already died. Another distinguishing feature is that, in addition to those people directly affected, many of the children born to women who had stayed in the center have cerebral palsy. Some of the children have symptoms resembling those of congenital Minamata disease (Photo 2).

## Worker Illness at Subic Bay Naval Base

Heavy pollution also affects base workers. Health impacts among base workers in the Philippines were apparent even before the bases were returned, and the damage is especially serious among people who worked at ship repair facilities because they handled asbestos often. In 1993 over 1,000 workers at Subic Bay Naval Base filed a lawsuit in the US against asbestos-related companies, and settlements have been reached with some of the defendant companies. As of this writing about 400 plaintiffs are still involved in litigation. Similarly affected workers at the ship repair facility in Yokosuka, Japan have filed lawsuits.

## Base Waste Disposal Sites

Serious health damage has appeared around on- and off-base waste disposal sites. In the Philippines countless sites were created over several decades to dispose of the colossal amounts of waste generated by US military bases, and wastes were improperly disposed at many of them. An example would be the recycling area of the on-base Public Works Center at Subic Bay, where workers under the direction of the military recovered metal, paper, cloth, and other recyclable materials from wastes. One male worker was employed at the recycling area from 1973 to 1979, working without any protective gear. After 1979 he performed the same work at on-base disposal sites until base closure. He has asthma plus lung and stomach ailments, and his wife, who performed the same job, died from cancer at age 46. A physician attributed her death to hazardous wastes. The worker attests to seeing asbestos and chemicals at the recycling area and disposal sites, and said the military told him not to approach buried drums.

Health damage also appears among scavengers who live near off-base disposal sites and make their livings by recovering recyclable and useful items from wastes. A case in point is a girl without a cardiac septum born to a woman scavenger living at Pag-Asa, a disposal site for wastes from the Subic Bay base.

## US General Accounting Office Report

The 1947 Military Bases Agreement between the US and the Philippines gives the US responsibility for withdrawal-related costs, and for that reason the Senate Appropriations Defense Subcommittee asked the Government Accounting Office (GAO) for a study on the government's financial responsibility to determine the extent of environmental damage, and the obligation for remediation and environmental recovery. Study results were released in a report on January 22, 1992.[19]

This report was the first admission by the US itself of pollution on bases in the Philippines, and it made the following revelations.

- Neither Clark nor Subic Bay abided by US domestic environmental standards, and had contaminated locations and facilities.
- Neither base had leak prevention equipment installed on underground storage tanks, and because their firefighting facilities lacked drainage equipment, fuel and chemicals used in firefighting contaminated soil and groundwater. At Subic Bay there was runoff into the bay.
- Subic Bay Naval Base treated only 25% of its daily 5 million gallons of effluent.
- Lead and other heavy metals emitted from the ship repair facility sandblasting site were released directly into the bay or landfilled.
- PCBs emitted by the Subic Bay base power plant exceeded US domestic standards.

This report suggests strongly that the US military itself was well aware of the pollution but thought the cost of base remediation would be colossal. If the military had performed an extensive survey of pollution when closing these bases and given the results to the Philippine government, it would have been possible to prevent the health damage especially to the Pinatubo evacuees on Clark Air Base. The US government has a grave responsibility for this.

# 4.  Remediation of Former Bases and Compensation for Damage

## 4-1  Compensation for Damage in the Philippines

### Little Progress in Remediation and Relief for Victims

The US government has neither performed environmental remediation nor compensated victims for the environmental and health damage discussed in the previous section. It cites Article 17 of the Military Bases Agreement as its grounds for not restoring sites to their original state or compensating damage, because the article explicitly states that when bases are returned America has no obligation to restore the land to its original state. This is the basis for the US claim that the Philippine government has waived its right to demand compensation.

But the previously mentioned Philippine Senate Committee report says that when the bases agreement was signed in 1947 there was no awareness of pollution impacts on human life and health, for which reason the Philippines has not waived its right to demand the remediation of facilities. It continues to claim that despite the explicit terms absolving the US of compensation and restoration obligations, the agreement did not unconditionally give America the right to bury wastes, damage human health, and otherwise cause severe environmental impacts, and thus the health damage to local people obligates the US to remediate the environment and compensate damage.[20]

The Philippine Senate Committee recommended that the Foreign Ministry strengthen its demands to the US government and also file suit in the International Court of Justice, but owing to interests affecting both nations the Philippine government has taken no such action. Meanwhile, there has been no remediation in most polluted areas, and victims have received no compensation at all.

### Base Pollution Lawsuits

Pollution victims are trying to achieve a breakthrough by working with the environmental NGOs, lawyers, and scientists who support them. With the help of an environmental NGO, people who had suffered health damage because of the Clark and Subic Bay bases, or in some cases their families and survivors, filed lawsuits in August 2000 against the US and Philippine governments demanding compensation and remediation. Although starting out with only 97 plaintiffs, the number subsequently grew to over 300 victims on the complaint, with the total number of victims coming to over 1,000. Total compensation demanded comes to more than US$100 billion (Table 2).

TABLE 2. Compensation Demanded for Pollution

| | | Subic Bay Naval Base | Clark Air Base |
|---|---|---|---|
| Claims against US govt | Actual damages | 3.5 million pesos | 25.2 million pesos |
| | Non-pecuniary damages | $25 billion | $25 billion |
| | Punative damages | $25 billion | $25 billion |
| Claims against Philippine govt | Actual damages | 3.5 million pesos | 25.2 million pesos |
| | Non-pecuniary damages | 12.5 billion pesos | 12.5 billion pesos |
| | Punative damages | 12.5 billion pesos | 12.5 billion pesos |
| Total | About $100.8 billion | | |

Source: Based on petition by attorney Alexander L. Lacson.

In July 2001 Angeles District Court dismissed the lawsuit by victims of Clark Air Base on the grounds that under international law a lawsuit filed in the Philippines cannot have the US government as a defendant. As of November 2002 there has been no ruling in the lawsuit by Subic Bay victims, but victims are planning to file a single appeal by plaintiffs of both original suits. In addition to these domestic lawsuits, victims and environmental NGOs launched initiatives making use of international networks. With an American support group called Filipino-American Coalition for Environmental Solutions, they demanded that environmental studies be performed at former base sites in the Philippines under the Superfund Law, and filed a lawsuit in the US in December 2002.[21] This lawsuit aims to use comprehensive environmental studies to demonstrate that the former Clark and Subic Bay bases are polluted, and to develop into a movement seeking redress for the victims.

## 4-2    Status of Forces Agreements with Japan and South Korea

### Exemption from Requirements for Restoration and Compensation

Japan and South Korea have status of forces agreements (SOFAs) with the US that, like the Military Bases Agreement with the Philippines, exempt the US from obligations to restore areas to their original conditions or to provide compensation (Box 1).

As long as these provisions remain, the US will likely use them to refuse compensation or remediation demands. Worse, it could even tempt the military to dump contaminants on a base just before returning it. An April 1993 investigation at the former Subic Bay Naval Base by a World Health Organization team found that the US military had abandoned large quantities of hazardous chemicals including insecticides, herbicides, and PCBs, as well as fuel, explosives, and other items.[22]

**Box 1** Article 4.1 of the Status of U.S. Forces Agreement in Japan (the Republic of Korea)

> The United States (The Government of the United States) is not obliged, when it returns facilities and areas to Japan (the Government of the Republic of Korea) on the expiration of this Agreement or at an earlier date, to restore the facilities and areas to the condition in which they were at the time they became available to the United States armed forces, or to compensate Japan (the Government of the Republic of Korea) in lieu of such restoration.

The same thing could happen in Japan and South Korea, whose citizens need to take the Philippines lawsuits seriously. In that sense the SOFA issue is of crucial environmental significance.

## South Korea's SOFA Revision Movement

In recent years South Korea has had the biggest movement in Asia demanding SOFA revision, and it has made significant progress. This movement was triggered by a July 2000 incident in which a large amount of formaldehyde from the US military base at Yongsan in Seoul was dumped in the Han River. The South Korean environmental NGO Green Korea United discovered the dumping and conducted a major protest action using the media, which mobilized public opinion against US military pollution and brought about the first official apology by US Forces Korea to the people at a press conference.[23] This sequence of events led to a SOFA revision at the end of 2000 and to the January 18, 2001 signing of a Memorandum of Special Understandings on Environmental Protection.

The Memorandum includes the following provisions: Environmental governing standards are based on the environmental standards of either country, whichever are more protective. Setting of new rules and standards is considered every two years. Access to facilities and areas follows procedures established by the Joint Committee. The two governments consult on environmental risks. US forces develop plans and programs to minimize environmental impacts, procure the required budget, and take prompt action to deal with imminent contamination-caused dangers to health. Regular Joint Committee meetings will discuss environmental issues related to facilities and areas.

A similar document, called the Joint Statement of Environmental Principles, was released on September 11, 2000 in Japan.[24] Although US forces are not obligated to observe Japan's environmental laws, under this joint statement the US military nominally abides by the environmental standards of either Japan or the US, whichever are more protective. Unlike the Memorandum with South Korea, however, this statement has no provisions for environmental budgeting by US forces, and it is not clear whether it is legally binding.

**Box 2**   Main Provisions of the Joint Environmental Information Exchange and
Access Procedures

---

(1)   When an environmental accident occurs on a US military base, South
Korea's government is notified by telephone, and subsequently in
writing within 48 hours.
(2)   The SOFA environmental panel forms a joint investigation team to
survey the contaminated area.
(3)   When an environmental accident occurs or when a base is returned,
procedures are initiated to allow South Korean public officials access.
(4)   Through discussions with the Ministry of Environment, US Forces
Korea shall revise its environmental governing standards to comply
with South Korean law.

---

On January 18, 2002 the US-South Korea SOFA Joint Committee established
the Joint Environmental Information Exchange and Access Procedures, which
was more in-depth than the Memorandum (Box 2).

The protest movement against crimes by US military personnel reached a new
high following the June 2002 death of two 14-year-old Korean girls run over by
a US armored vehicle. In response, new US-South Korea agreements were
reached in March 2003 at a special meeting of the SOFA Joint Committee, and
one of them pertained to environmental contamination.

According to a Ministry of Foreign Affairs and Trade news release, an agreement on contamination surveys and remediation to be conducted when US bases
are returned stipulates that both South Korea and the US are to perform environmental surveys when a base is returned, and that if contamination is found,
the US is to remediate the site. Detailed procedures are provided for the surveys,
talks on the extent of remediation, and how remediation steps would be carried
out. Although the agreement's text itself has not been released, it is a landmark
agreement in that it requires the US military to perform remediation, and the
citizens' movement deserves credit for it. The agreement not only is a step toward
revision of the US-South Korea SOFA, but also will likely have a significant
impact in Japan, which hosts many US bases.

In the future the agreement must be publicly released, and the environmental
surveys and remediation must be performed within a framework allowing the
participation of local communities because in South Korea citizens have pointed
out many problems in the Joint Environmental Information Exchange and
Access Procedures. Specifically, this document does not expressly require the
payment of remediation expenses or compensation for pollution damage, while
talks and base access only begin when the US military recognizes that a serious
problem exists.

# 5.   The Environment and War Preparations

## 5-1   Lack of Consideration for the Environment and Society

Because militaries ultimately exist to be used, they are constantly engaged in military exercises, training, and other preparations for war to maximize their battlefield results. Owing to the imperative for victory, there is no avoiding the use of weapons just because they are environmentally destructive, and speed cannot be limited just because of the noise. Hence training takes precedence over the environment. Live-fire gunnery training, firing and bombing training, and takeoff and landing practice damage the environment and the health of nearby people.

## 5-2   Noise

Aircraft landings and takeoffs at military air bases blast nearby residents with earsplitting noise and deprive them of their health. People living near bases suffer physical damage such as sleep disturbance, hearing loss, and ringing in the ears, psychological and emotional damage, and impediments to everyday activities such as conversation and thought. They report that at times their houses shake, and they cannot hear the television or radio, or talk on the telephone.[25] Author Oshima interviewed a person living in Gunsan, South Korea whose house had developed cracks from noise and vibration. The loudest sound recorded near the US base at Kadena in Okinawa was 127 dB, which is extreme considering that hearing loss occurs at about 130 dB.[26] Such deafening sound occurs near places with air bases or aircraft carrier ports, including, in Japan, Kadena, Ginowan, Yokota, Atsugi, Yokosuka, Misawa, and Komatsu, and in South Korea, Uijeongbu, Josan, Gunsan, and Dae-gu.

Yokota Air Base in Japan, which is home to the HQ Fifth Air Force under the Pacific Command, is the only foreign military air base in the world that is in a national capital. The surrounding area is densely populated, with people in the surrounding nine cities and one town adversely affected by the base. People living near US military bases in Japan have filed a considerable number of lawsuits against base noise.[27] At Yokota between 1976 and 1994 there were three lawsuits with over 700 plaintiffs, who demanded (1) an injunction on flights between 9:00 p.m. and 7:00 a.m., (2) compensation for past damage, and (3) compensation for future damage. Five court decisions recognized the claim for compensation for past damage, but turned down the demand for an injunction on late-night and early-morning flights. For this reason local citizens, hoping to amass 10,000 plaintiffs, started the New Yokota Base Noise Pollution Lawsuit against the Japanese and US governments to win compensation and an injunction. Three lawsuits were filed in 1996, 1997, and 1998. In April 2002 the Supreme Court dismissed the first suit's claims against the US government as not allowed by law, and on May 30 the second and third lawsuits were dismissed by the Tokyo District Court for the same reason. On that same day the district court ruled on claims against the Japanese government and ordered the government to pay a total of 2.4 billion yen in damages to 4,763 of 5,917 plaintiffs. The district court decisions were appealed the following month and are now pending. In South Korea citizens are preparing noise lawsuits at places including Maehyang-ri, Gunsan, Pyeongtaek, and Chuncheon.

## 5-3   Environmental Damage by Military Exercises

### Bombing Range at Maehyang-ri

Training areas, especially bombing ranges that are bombing targets for fighter jets, are in a constant battlefield state and suffer the environmental consequences. A case in point is the US military's Kooni Fire Range near the village of Maehyang-ri. In 1955 a 2,281-ha bombing range was created offshore near the village, and in 1968 the military created a 96-ha firing range, but a firing range for US forces actually began service in 1951. Even now US fighters from Okinawa, Guam, Thailand, and Japan conduct bombing practice here, adjacent to farmland and fisheries, and near human dwellings. Nong-do Island serves as the target for missiles and other weapons fired by fighters. The former target, an island in the same area called Gwi-do, completely disappeared. Two-thirds of Nong-do has disappeared, and the island bears no resemblance to its former self (see photo at start of this chapter). Bombing practice also harms Maehyang-ri villagers, who have suffered fatalities, harm from noise (hearing difficulty, inability to use the telephone, stress, interrupted sleep), and economic losses (damage to dwellings by explosion vibrations, and reduced income due to the disruption of fisheries, livestock abandonment, and restrictions on farmland access).

### Opposition Movement

In February 1998 14 Maehyang-ri victims filed a lawsuit against the government demanding compensation for damage caused by US military bases. It was the first such lawsuit in South Korea. On April 11, 2001 the Seoul Civil District Court, and on January 9, 2001 the Seoul Civil Appellate Court, handed down decisions in favor of the citizens. The case is now pending in the Supreme Court. This lawsuit showed that noise in the Maehyang-ri vicinity exceeded both the USEPA standards and South Korea's environmental noise standards.

This issue induced a significant citizens' movement throughout South Korea, which was triggered by a May 8, 2000 bombing error in Maehyang-ri that caused serious damage when a US Air Force A-10 fighter-bomber dropped six MK82 bombs offshore of the village to lighten itself because of engine trouble. Damage caused by this incident included 170 broken windows, cracks in walls, 13 people with injuries, and 42 milk cows suffering miscarriages.[28]

Citizen pressure made South Korea's Ministry of National Defense announce changes in training, and although practice on the firing range has stopped, bombing practice on Nong-do continues except on Saturdays and Sundays.

Citizen efforts have had a significant effect on South Korean society. Especially noteworthy is that US Forces Korea now conduct environmental contamination investigations. Pressing the government on its responsibilities has also had a strong impact on SOFA revisions. Further, the Maehyang-ri court decision influenced not only the US military, but also South Korea's Ministry of Defense, which has plans to enact a special law on noise pollution by 2005, and phase it into effect starting in 2006.

Nevertheless, many problems remain. One serious shortcoming is that the court decision limits compensation for noise damage to the last three years, and there is no consideration at all of coastal ecosystem disruption or soil contamination, or of unexploded ordnance (UXO) clearance and disposal, problems which will have the largest impacts on future generations. What is more, despite the strong opposition to training at Maehyang-ri, the April 2002 US-South Korea Agreement on a Combined Land Management Plan does not call for the return of Maehyang-ri.

## 5-4    Depleted Uranium

Radioactive depleted uranium (DU) ordnance is sometimes used on practice ranges. The US Marines used 1,520 rounds of DU armor-piercing incendiary shells on Torishima Island in Okinawa on December 5 and 6, 1995, and January 24, 1996. DU munitions are not used for training inside the US, and they are as tightly controlled as nuclear weapons. The use of this ammunition on Torishima violates Japan's Law Concerning Regulation of Nuclear Raw Materials, Nuclear Fuel Materials, and Nuclear Reactors, and the US military admitted this.

It wasn't until a year later in January 1997 that the US military notified Japan's government about the use of DU ammunition, but the government withheld the information from Okinawa Prefecture, which found out about it after a February 10 *Washington Times* article on DU ammunition that discussed the Torishima incident. Investigations were later performed by the US five times and by Japan twice, but they reported no contamination other than near the recovered shells.[29] Nevertheless, fewer than 200 of the 1,520 rounds were recovered, while the rest are likely buried deeply in the soil where they will cause long-term contamination.

## 5-5    Training Area Restoration

Restoration of retired training areas requires a very long time. For example, until 1997 US forces conducted artillery live-fire training over Prefectural Route 104 in Okinawa, and even now the problem of UXO clearance remains totally unresolved. The approximately 40,000 shells fired in practice present the national and prefectural governments with long-term problems of heavy-metal contamination and UXO clearance. No matter what kind of ordnance is fired, the users of training areas should be responsible for site remediation and environmental restoration, and this is yet another reason for SOFA revisions.

# 6.  Battlefield Environmental Damage

## 6-1  The Difficulty of Environmental Restoration

Actual use of weapons in battle releases chemicals, heavy metals, and other substances that contaminate the battlefield, and fighting itself destroys historical assets and damages the local ecosystem. Warfare brings all military impacts to bear on the environment at the same time.

About 14 million tons of bombs were dropped on Vietnam during the Vietnam War, gouging between 10 and 15 million large craters. This alone allowed soil erosion and had a devastating impact on the ecosystem. Many landmines and UXO also remain.[30]

Okinawa was the only part of Japan to experience ground fighting in World War II. About 5%, or 10,000 tons, of the estimated 200,000 tons of bombs dropped on Okinawa are thought to be unexploded. Immediately after the war those bombs were still near the surface of the ground and therefore rather easy to discover, and by the time Okinawa reverted to Japanese control in 1972 Okinawans themselves had cleared about 3,000 tons, and the US military about 2,500 tons. Up to 2000 Japan's Self-Defense Forces cleared 1,501 tons, but Okinawa Prefecture estimates that discounting 500 tons that will never be found, there are still about 2,500 tons of unexploded bombs buried on the islands. In 1989 the prefecture commenced a wide-area project that clears UXO while examining land divisions. Okinawa Prefecture and the national government fund the project, which will require another 40 to 50 years to complete.[31] Yet, these four or five decades are only the time needed for clearing and disposing of unexploded bombs; the much longer time needed for remediation is not included.

These examples illustrate how recovery from war-caused environmental damage requires long years and entails much difficulty for environmental restoration and remediation.

## 6-2  Weapons Meant to Destroy the Environment

### Defoliant in Vietnam

Asia has seen the use of weapons specifically meant to destroy the environment. Typical of them is the defoliant used in the Vietnam War to destroy the tropical forest and eradicate the enemy forces hiding there. The environmental impact of defoliant has even spawned the term "ecocide." At least 2 million hectares of forest have been seriously affected.

Even now, nearly 30 years after the Vietnam War, trees that used to grow in affected areas still show no signs of coming back naturally. Overgrown with weeds and having meager fauna, those areas bear little resemblance to their prewar state. Vietnam has programs that plant trees for the recovery of forests and the ecosystem, and although some progress has been achieved, such as bringing back mangrove forests and cranes, the job still requires massive funding and many years.[32] This environmental damage has also taken a severe human toll, as in the many conjoined twins and deformed children born since the war.

## Determining the Full Extent of Defoliant Damage

To investigate long-term defoliant effects, Vietnam's Ministry of Public Health and other organizations created Vietnam's National Committee for the Investigation of the Consequences of the Chemicals Used During the Vietnam War (10–80 Commission) at Hanoi Medical School in October 1980. Results of the extensive studies and research have been reported on several occasions including the first international symposium on the long-term effects of wartime defoliant on humans and nature, held in Ho Chi Minh City in 1983, and at the second international symposium, held in Hanoi in 1993. These reports showed that dioxins have seriously affected the ecosystem and people of southern Vietnam. Since then there has been much research mainly in Vietnam by researchers from Western countries and Japan on the extent of damage, residual dioxins, and other aspects. Their research covers a broad spectrum including the areas and extent of defoliant application, measuring dioxin concentrations in the ecosystem, residual concentrations in the human body, human health impacts, genotoxicity, and forest damage and recovery.

The defoliant Agent Orange damaged the health of soldiers from the US, South Korea, Australia, and other countries that sent forces to Vietnam, not to mention the serious impacts on the health of Vietnamese soldiers. Years after the conflict ended, the world is finally learning about the congenital deformities and other kinds of damage appearing even among their children and grandchildren. Defoliant might have genetic effects reaching the third and fourth generations, requiring continued attention to these and other future impacts.

As part of the focus on dioxins in recent years, reports delivered at dioxin-related international conventions on defoliant impacts in Vietnam and other Indochinese countries have heightened international concern. In February 2002 the US and Vietnamese governments held a conference in Hanoi, where they inaugurated a US-Vietnam program to investigate defoliant impacts, and that July an international conference was held in Sweden. These are some of the international initiatives for studies and research now in the beginning stages.

Recently high contamination has been found in and around the bases created by the US in Vietnam during the war. Of particular interest is dioxin contamination of the former bases at Bien Hoa near Ho Chi Minh City and Da Nang in central Vietnam. Although the environmental dioxin concentration has fallen over the four decades since the 1960s, there is serious dioxin-caused damage around former bases where defoliants are thought to have been stored or abandoned. A study performed in 2000 and 2001 by Dr. Arnold Schecter at the University of Texas found an average of 413 ppm, or 206 times the usual level, of dioxins in the blood of 43 people living near the former base at Bien Hoa. Many children with congenital deformities have been born in the area.

In July 2002 Vietnam's Vice President Nguyen Thi Binh, head of the Vietnam Red Cross Agent Orange Victims Fund, stated that he intended to seek compensation from the US for children with congenital defects and other Agent Orange victims.

## 6-3  Health Damage from Depleted Uranium Munitions

Some weapons are used for their overwhelming effect despite knowledge of their devastating environmental consequences, and DU munitions are a typical example. The density of DU shells allows them to pierce tank armor, but because they burn fiercely on impact, the heat fries a tank's occupants and the DU turns into an aerosol that spreads throughout the area. DU has two kinds of toxicity: it is both a heavy metal and radioactive. Once in the body it has a variety of deleterious impacts, and even the soldiers who use DU munitions on the battlefield cannot avoid its effects.

DU ordnance found its first major use in the 1991 Gulf War, where the multinational force used 290 tons. Later military operations used 3 tons in Bosnia-Herzegovina and 9 tons in the Kosovo conflict.[33] There are strong suspicions that DU ordnance was used in the attack on Afghanistan in 2000. Much was also used in the 2003 attack on Iraq, including within Baghdad.[34]

Since the Gulf War, environmental NGOs, UNEP, and other organizations have performed environmental studies on the impacts of DU munitions,[35] but because of limited information availability, political restrictions, and other factors, there have been no comprehensive studies at all on where DU munitions have been used, the amounts used, and what health impacts have resulted. This makes it hard to determine the full effects, including the establishment of cause-and-effect relationships.

In Iraq a few journalists, environmental NGOs, and anti-war organizations have found many people who appear to have suffered health damage from DU munitions used in the Gulf War (Photo 3).[36] The victims have received no compensation at all. Comprehensive environmental studies, compensation for victims, and an international framework for banning the use of DU are sorely needed.

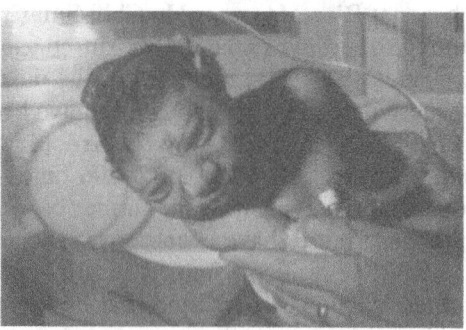

PHOTO 3.  Baby with anencephalia. In post-Gulf War Iraq the rate of children born with serious congenital defects is very high at 3% (January 2003 photo taken by Morizumi Takashi in Baghdad).

# 7.    Peace and Environmental Preservation in the 21st Century

## 7-1    Military Force Cannot Defend Peace and the Environment

Advancing globalization since the 1990s has made it impossible for individual nation-states to attain policy goals on their own. International society must therefore cooperate to accomplish common tasks, which is especially true for peace and environmental preservation. Yet the National Security Strategy released by the Bush Administration in September 2002 concerns itself almost totally with military solutions and enunciates a policy under which the US will use force to contain terrorism even if it must do so alone.

Technological innovation now allows even small groups to have weapons of mass destruction, and it is quite possible that no matter how much military force the US brings to bear, small groups will proliferate and continue to mount challenges. Dependence on military force will actually make the world less stable, and there is a limit to what such a strategy can accomplish. Continued "preventive war" actions against countless terrorist groups will rapidly increase defense costs and lead the US down the road to economic crisis. Military actions destroy not only the environment, but also economies. From an environmental viewpoint, it is clearly an illusion that military force can achieve peace and security.

## 7-2    Environmental Disarmament

In view of the mutual exclusivity of military activities and environmental preservation, we need to take a new look at military activities from an environmental standpoint. Initiatives for the environment share much with those for peace. Assuming military activities are to be restrained and reduced from an environmental standpoint, more effort will be needed for initiatives with emphasis on the following areas.

First is freedom of information. It is often the case that information on the environmental impacts of military activities is kept secret, and the consequence is that damage already done gets worse. In many places there is hardly any progress in determining the state of military-caused pollution and health damage, and people living in areas where such damage could already have occurred are sometimes unaware of the situation. Because governments are always loath to release information, it is crucial that citizens increase pressure on them, gather and organize information, and actively release it to the public. Local governments and citizens must obtain the right to access military facilities.

Second is citizen involvement in two ways. One is the involvement of citizens, especially local people who have suffered harm, in environmental measures. While the need for citizen involvement is advocated with respect to other environmental issues, in Asia there is hardly any citizen involvement, especially in military-related issues. In Western countries there are arrangements for citizen participation when determining land use after a base is closed. Environmental NGOs and experts can also be involved.[37] Institutions for citizen participation in environmental decision-making have critical importance to protecting the environment, and military-related environmental issues must be no exception.

Another type of citizen involvement is to become aware that people living near military installations bear the brunt of military-caused environmental problems, and have outside citizens become involved in some way. The environment is totally ignored for reasons of "national security," but national security cannot be divorced from environmental concerns. Especially in Japan the military burden strongly tends to be imposed on people in certain geographical areas. South Korea's movement for SOFA revision developed into such a large groundswell because the presence of US military bases throughout the country makes it hard for anyone to see this as someone else's problem.[38] This national movement is making headway, at least on the environmental front. In Japan about 80% of US bases are imposed on Okinawa, making them relatively sparse in the rest of the country, which breeds a tendency for the Japanese—except for Okinawans and a few others—to see this as someone else's problem. The Japanese need to shake themselves out of complacency to deal with the military environmental issue head-on.

Third, even military polluters should be required to pay for the pollution they cause. Military activities have been equated with national security and accorded priority under all policies, thereby insulating them totally from one of the principles of environmental conservation, the polluter pays principle. There is no justification whatsoever for unconditionally making military activities an exception to this principle. Germany and other countries in fact require foreign military forces in their countries to perform remediation and pay compensation for environmental damage, and to allow base inspections. In view of the severe environmental damage in Asia, all military activities should be subject to remediation and compensation requirements as in Germany. Asia must discard the Cold War ideology that military activities trump all policy.

It is not the "greening" of military activities that we should work for in this century. Although efforts such as the pioneering initiatives of US Marines Okinawa Environment Branch as described in section 3-1 are welcome, the greening of military activities is fundamentally impossible. Faced as we are with the environmental impacts of the military, human society's ultimate goal should be environmental disarmament—the drastic downsizing of military activities from an environmental stance, and the Asian citizens' movements described here are playing an important role in the historic initiative to bring about environmental disarmament and make this the century of peace and environmental progress.

Oshima Ken'ichi, Yokemoto Masafumi, Tani Yoichi, Cheon Kyung-ah,
Hayashi Kiminori, Na Sungin

## Essay 1:   Landmines and Unexploded Ordnance in Cambodia

Landmines and unexploded ordnance (UXO) pose two threats. First, they remain in the soil for many years and can wound or kill people long after fighting ends. Second, they indiscriminately kill farmers and other people unconnected to wars, as well as children born after wars are over. In 1991 the parties in Cambodia's civil war signed a peace treaty, but even now between 4 and 6 million mines remain on at least 2,000 square km of the country, continuing to claim victims who are often members of the general public (Tables 1 and 2).

Rural Cambodians have for many years coexisted with the mines by exchanging information on mine locations and staying away from them. But mines and UXO have a close connection with poverty because under the current land ownership system people can have the land they clear themselves and because there is much good timber in minefields.

TABLE 1. Accidents Caused by Landmines and Unexploded Ordnance in Cambodia

| Year | No. of accidents | Deaths/month | Deaths/day |
|------|------------------|--------------|------------|
| 1996 | 3,046 | 254 | 8 |
| 1997 | 1,810 | 151 | 5 |
| 1998 | 1,929 | 161 | 5 |
| 1999 | 1,049 | 87 | 3 |
| 2000 | 802 | 67 | 2 |

Source: Prepared from: International Campaign to Ban Landmines, *Landmine Monitor Report 2001*, August 2001.

TABLE 2. Percentages of Military and Civilian Victims (%)

| Year | Military | Civilians |
|------|----------|-----------|
| 1997 | 49 | 51 |
| 1998 | 41 | 59 |
| 1999 | 14 | 86 |
| 2000 | 7 | 93 |

Source: Same as Table 1.

For these and other reasons, not a few people venture into minefields to open the land or fell trees. Other people unearth and dismantle mines or UXO themselves and attempt to sell them to ordnance clearing personnel for cash income. Farmers poignantly explain their predicament by noting that they are more afraid of starvation than mines.

As mines were usually laid in certain places, farmers who were formerly soldiers have knowledge of where they are, but field work always involves danger from unexploded shells and bombs because there is no guessing where they might be. While many people have knowledge about mines, few people know how to deal with UXO. Mines and UXO might be found in schoolyards, or they might be set off by the vibration from boring a well.

Generally when someone discovers mines or UXO, they are to contact the Cambodian Mine Action Center (CMAC) or foreign NGOs that clear mines, but sometimes people will tell only others whom they can trust and say nothing to public agencies or NGOs, or they will make no report if a victim is not from their own village. Several factors underlie this problem, such as local relationships and a lack of trust in government agencies and foreigners. In a village near Siem Reap, former soldier Lo A-kila clears mines himself for no fee. Villagers do report the discovery of mines and UXO to him and other people who are truly working for the village.

Cambodia currently has no nationwide normative legal document on arms control, and hence administrative authorities in each area arbitrarily decide on permission for exploding or storing mines and UXO. Some people in power demand bribes to grant permits, or arrest individuals who have discovered and cleared mines on the charge of the "illegal possession of arms," and then demand high bail. The interests who profit from mines and UXO are hampering the activities of NGOs and individuals working to clear them.

Most of the foreign aid for mine clearing is embezzled before it gets to the minefields. While it is claimed that clearing one anti-personnel mine costs several hundred dollars, the fact is that it costs hardly anything for an experienced deminer to dismantle one by hand. Parties monopolizing the rights to use foreign aid for mine clearance fear that such aid would dry up if the real cost were known, and therefore apply various kinds of pressure on those who clear mines for their villages. To keep the foreign aid coming in, CMAC performs "mine clearance" in places known to be free of mines.

An unfortunate fact is that landmines and UXO are "resources" for Cambodia to obtain foreign aid. They are the biggest vested interest, and a source of corruption. Administrative authorities need to be watched so as to protect the autonomous efforts of citizens like Mr. A-kila. This would perhaps necessitate the creation of a permanent monitoring system from outside Cambodia.

Nakano Ari

❖                    ❖                    ❖

## Essay 2:   Japanese Chemical Weapons Abandoned in China

The Chemical Weapons Convention, which was ratified by Japan in September 1995 and took effect in April 1997, requires signatories to dispose of all chemical weapons abandoned in the territories of other signatory countries, and as a rule to complete that task within 10 years of the convention's effective date.

From the time the former Japanese army invaded China until the end of World War II, it took many chemical weapons into China, where they were used or abandoned. After the war China's government collected the abandoned chemical weapons (ACWs) and buried them in a number of locations, most of which are in Jilin Province, not a few of them near human dwellings. There are 14 currently known locations where ACWs are buried or stored, and the total quantity is about 700,000 rounds according to Japan's estimate. This number is exceptionally large compared with the quantity of non-stockpiled old chemical weapons being disposed in Western countries.

The ACWs include a variety of dangerous chemical agents such as blistering agents like mustard and lewisite, the pulmonary agent phosgene, and the sneeze gas diphenyl cyanarsine, while the weapons containing the agents also come in various types including mortar rounds, howitzer shells, aerial bombs, and smoke bombs. Even now many of these are buried in large collections with explosives still inside. Some of them are leaking their chemical agents, many of which include arsenic.

Japan is to shoulder the entire cost of disposal. While the project's cost is not yet known, it will likely be considerable judging by disposal costs in other countries. Steps in the plan consist of narrowing down the field of possible disposal technologies, conducting experiments, finalizing the technologies, setting up a pilot plant in China, building the disposal plant, and carrying out disposal. Japanese government sources say it will be virtually impossible to complete the job by the 2007 deadline under the convention.

In China there are still many instances in which people are injured by accidentally coming into contact with ACWs on farmland or at construction sites. Such incidents stand as the detrimental legacy of Japan's war against China.

On August 4, 2003 an unfortunate accident occurred when five abandoned metal chemical-weapon canisters were discovered underground on a construction site in the city of Qiqihar, Heilongjiang Province. One of the canisters was damaged on site, and an oil-like substance soaked the soil. A worker bought the canisters without knowing they were chemical weapons, cut them open, and sold them to a recycling center. Mustard gas escaped from the canisters, and some contaminated soil from the site was transported to several locations. Forty-three people were hospitalized and one died.[39] Victims are demanding that the Japanese government pay seven categories of compensation including livelihood assistance, child support, and solatia.

In a lawsuit by Chinese victims and their survivors seeking compensation from the Japanese government for having been harmed by poison gas and other chemical weapons abandoned by the former Japanese army in China at the close of WWII, on September 29, 2003 Tokyo District Court handed down a decision completely recognizing the plaintiffs' demands and ordering the government to pay a solatium. For disposal of chemical weapons abandoned in China by the former Japanese army, on August 21, 2004 Japan's government revealed plans for the construction of excavation and disposal facilities at Harba Peak in China's Jilin Province. Owing to the danger of exploding artillery shells and working contact with chemical agents, buried shells are first covered with an excavation facility. Remote-control robots then unearth the shells, which are temporarily kept in a storage facility. Further, at a site of 700 to 800 thousand square meters and located about 4 km from the excavation site, two incineration plants would be built to dispose of the of the chemical weapons. Total construction cost is estimated at about 300 billion yen. Plans call for the excavation facility to be completed at the end of 2006, and the disposal facility at the end of 2007.

In Japan people have fading memories of the war, and concern about ACWs is not high among the Japanese. In fact, many Japanese are unaware of this issue. News reports about the August 2003 accident were fewer in Japan than in China, a sign of the widening gulf between the two nations.

Asuka Jusen

# Chapter 2
# Trade and the Environment: Promoting Environmentally Friendly Trade

Mixed metal scrap imported by China from Japan. November 2002, Zhejiang Province, China.
Photo: Kojima Michikazu

A market for foodstuffs and traditional Chinese medicines in Guangzhou, China.
Photo: TRAFFIC Japan

# 1.   Introduction

Efforts toward creating bilateral or multilateral free trade agreements (FTAs) are broadening in Asia. The Association of Southeast Asian Nations (ASEAN) has been lowering tariffs and enacting other measures to facilitate even more economic integration. In January 2002 Japan and Singapore signed the Japan Singapore Economic Agreement for a New Age Partnership. Negotiations and joint research are underway to bring about FTAs for ASEAN with China, South Korea, and Japan, and between Japan and South Korea. A World Trade Organization (WTO) ministers' meeting in November 2001 decided to start new multilateral negotiations.

Meanwhile, environmental and labor organizations are critical of progress in trade liberalization talks owing to their concerns that free trade will aggravate environmental problems and hamper implementation of preventive environmental policies by governments. Trade controls are needed because many of the items traded internationally have significant environmental impacts.

A number of multilateral environmental agreements allow trade in certain items, such as rare species and hazardous wastes, among signatories only if certain conditions are met. Many Asian countries are signatories to such agreements, under which they have instituted environmental trade measures (Table 1). In addition to those agreements in the table, measures are likewise to be taken under other environmental agreements which entered into force in 2004. Two of these are the Rotterdam Convention on PIC and the Stockholm Convention on Persistent Organic Pollutants.

Two currents affect trade and the environment: one attempts to facilitate free trade, while the other attempts to limit trade owing to environmental concerns. Many countries are members of the WTO and other international organizations that pursue free trade, but they have also ratified multilateral environmental agreements that restrict trade. As there is no international accord on the relationship between WTO rules and environmental agreements, delegates have started discussing it in multilateral talks.

This chapter reviews the state of Asian trade in rare plant and animal species, marine resources, forest resources, and wastes, and on that basis discusses how Asian countries must cooperate on trade and the environment, and what approaches they should take. Multilateral environmental agreements and corresponding measures have been adopted for rare plants and animals and for wastes, but although marine and forest resources are partially restricted by the Convention on International Trade in Endangered Species of Flora and Fauna (CITES) and other agreements, trade measures are inadequate. Section 2 is an overview of trade and its environmental impacts in each of these sectors, and Section 3 addresses the issues that Asian countries must together address in the pursuit of environmentally friendly trade.

TABLE 1. Ratification and Accession of Asian Countries to Organizations and International Environmental Agreements

| Country/region | WTO (Started 1995) | ASEAN (Started 1967) | APEC (Started 1989) | CITES (Opened 1973, effective 1975) | Montreal Protocol (Opened 1987, effective 1989) | Basel Convention (Adopted 1989, effective 1992) |
|---|---|---|---|---|---|---|
| Japan | 1995 | — | 1989 | 1980 | 1988 | 1993 |
| ROK | 1995 | — | 1989 | 1993 | 1992 | 1994 |
| DPRK | — | — | — | — | 1995 | — |
| Russia | — | — | 1998 | 1992 | 1988 | — |
| China | 2001 | — | 1991 | 1981 | 1991 | 1991 |
| Taiwan | 2001 | — | 1991 | — | — | — |
| Philippines | 1995 | 1967 | 1989 | 1981 | 1991 | 1993 |
| Indonesia | 1995 | 1967 | 1989 | 1978 | 1992 | 1993 |
| Singapore | 1995 | 1967 | 1989 | 1986 | 1989 | 1996 |
| Malaysia | 1995 | 1967 | 1989 | 1977 | 1989 | 1993 |
| Vietnam | — | 1995 | 1998 | 1994 | 1994 | 1995 |
| Laos | — | 1997 | — | — | 1998 | — |
| Cambodia | — | 1999 | — | 1997 | 2001 | 2001 |
| Thailand | 1995 | 1967 | 1989 | 1983 | 1989 | 1997 |
| Myanmar | 1995 | 1997 | — | — | — | — |
| Bangladesh | 1995 | — | — | 1981 | 1990 | 1993 |
| Sri Lanka | 1995 | — | — | 1979 | 1989 | 1992 |
| India | 1995 | — | — | 1976 | 1992 | 1992 |

Source: Prepared from statistics of international agencies and convention secretariat websites.

# 2. The Environmental Impacts of Trade

## 2-1 Asian Trade in Threatened Flora and Fauna

Many East Asian countries have achieved rapid economic growth over the last 30 years. While this has raised the standard of living, it has also increased environmental impacts, destroying and desertifying habitats through changes in land use, and losing biodiversity due to pollution and other factors. Direct use of wildlife resources is also on the rise.

Asians use wildlife in many products including pharmaceuticals, food, and ornaments. Especially to people in developing countries, wildlife is a vital resource for sustaining their livelihoods. The annual turnover of the huge global wildlife trade market is estimated at billions of dollars.

Based on declared import values, it is estimated that in the early 1990s timber alone accounted for 65% of the annual value of the global international wildlife trade, followed by fisheries food products (25%) and non-timber forest products

(7%). Other wildlife commodities, such as live animals, animal products for clothing and ornaments, medicinal products, wild meat, and live ornamental plants, accounted for the remaining 3%.

## Environmental Impacts of the Wildlife Trade

Habitat destruction is the greatest threat to wildlife, but another major factor causing species extinctions, and also threatening whole ecosystems, is the excessive capture or harvesting of certain animals and plants. Further, extinctions bring economic crises to communities whose livelihoods depend on wildlife. In many areas of biodiversity-rich Asian countries, food supplies and traditional health care are dependent on wildlife. Seen over the long term, unsustainable use of flora and fauna adversely affects communities and their economies.

## Wildlife Use

*Medicines.* At all times and places people have used wildlife as the ingredients for traditional medicines. In East Asia at least 1,000 plants and animals have been so used, including tigers, musk deer, rhinoceroses, bears, costus root (*Saussureae radix*), and ginseng.

Tigers, which are endemic to Asia, are one of the most endangered animals, trade in which is banned by CITES. High prices for tiger bones have induced poaching and reduced the animal's numbers, but over the last few years there has been a decline in the consumption of medicines using tiger bone owing to worldwide efforts for tiger protection, including a CITES resolution to both range states and consumer nations to institute protective measures.

The musk obtained from mature male musk deer is used in Chinese medicines and perfumes. This substance apparently sells for about ¥5,000 per gram. Demand for musk is high in East and Southeast Asia, but it also finds use in North America and Europe. Capture for musk extraction is the main factor behind the dwindling number of individuals, but habitat loss caused by forestry, fuelwood gathering, tourism, and other human activities also has serious impacts. CITES controls trade in all species of *Moschus* spp., and bans international trade in only the populations of Afghanistan, Bhutan, India, Myanmar, Nepal and Pakistan. CITES Permits are required for trade in all other populations.

Rhinoceroses, which are found in Africa and Asia, are highly endangered because their horns are used in traditional Chinese medicine. Currently domestic trade is prohibited in nearly all Southeast Asian countries, but not in Japan.

Some medicinal ingredients are also used in perfumes. Agarwood in particular is an essential scent ingredient. Some *Aquilaria* spp. trees of the family Thymelaeaceae produce the resinous wood that is used. Main producing regions include Assam and the Deccan Plateau in India, Myanmar, Vietnam, Thailand, the Malay Peninsula, Borneo, Sumatra, southern China, and Hainan Island. CITES controls only the trade in *Aquilaria malaccensis*. Formerly only the wood of dying or dead trees was used, but now many mature trees are being harvested, which is likely depleting their numbers.

*Food.* Wildlife is a valuable source of food to people around the world, who consume the flesh of many wild animals including insects, birds, turtles, primates, antelope, and hippopotamuses. Capture for food is thus one of the direct causes of wild animal population decline. Much freshwater turtle and tortoise meat is consumed especially in East Asia. Three-fourths of Asia's 90 freshwater turtles and tortoises are seriously threatened, and 18 of them are endangered.

Fish are the most familiar wild animals used as food. Most trade in these species is legal, but in recent years fishers have continued the illegal, unreported, and unregulated taking of tuna, Patagonian toothfish, sharks, and other fish. Detailed treatment of this follows.

*Decorations and ornaments.* Many kinds of animal products are traded around the world as decorative items. Some examples are ivory ornaments, coral gift items, marine turtle carapaces, shells, and specimens of insects such as butterflies and stag beetles.

Ivory has been used throughout Asia for a variety of products including carvings, ornaments, chopsticks, and personal seals because of its milky white color, luster, and ease of carving. In the past ivory from Asian elephants was used, but most ivory later came from African elephants. High East Asian demand especially in Japan, Hong Kong, and other places induced continued poaching of African elephants in the 1970s and 1980s, reducing the population by half. In 1989 the CITES conference of the parties banned the international trade in ivory from African elephants, while commercial trade involving the Asian elephant had been banned since 1975. But in 1999 some existing stocks of ivory were exported to Japan by Botswana, Namibia, and Zimbabwe in accordance with certain conditions under CITES. As this shows, some Asian countries are primary consumers of ivory and therefore the ivory trade in Asia is always monitored when considering how to protect African elephants.

Poaching is reducing the populations of the Tibetan antelope, a bovine that is hunted for its wool. Known as shatoosh, this wool is used in clothing that sells for high prices in Western countries, Hong Kong, Japan, and other consumer areas.

It is clear that Asian countries are both leading consumers and suppliers of wildlife products. If Asians learn the value of their natural resources and use wildlife sustainably, this will help maintain Asia's biodiversity.

## 2-2  International Trade in Marine Products: Focus on Tuna

### Tuna and the Overall Trend in the International Marine Products Trade

*The State of the Environment in Asia 1999/2000* describes the Asian trade in marine products.[1] The main points are as follows.

First: Asia's share of the world seafood trade is rising. Both imports and exports are increasing, but the rate of increase is higher for imports. Second: Japan continues to be one of the world's major seafood importers. Third: However, statistics on the value of exports from the South China Sea region (Indonesia,

Malaysia, the Philippines, Thailand, Taiwan, Hong Kong, and Singapore) in the 1980s and early 1990s show that increasing exports of shrimp and other products to the US have given exports to Japan a relatively lower share of the total. And fourth: The same statistics indicate rising exports to China and South Korea, suggesting that they are using imports to cover the increase in their marine product consumption.

Generally speaking, the emergence of new marine product consumer nations such as China, South Korea, and Western countries decreased Japan's relative share of seafood imports. And from the perspectives of natural resource management and food, the above trends point to a situation in which, worldwide, the fish catch and marine trade value are increasing, while, in Asia too, fish stocks are being overexploited and international food resources are being plundered.[2]

Fishing needs to be directly managed to provide for sustainable use of marine resources, but when that is difficult, limiting trade ought to be considered as one way to complementarily and indirectly conserve resources as the marine product trade expands around the world.

## Increasing Trade and the Impact on Tuna Stocks

This section examines the marine product trade by focusing on Japanese imports of tuna,[3] one of the representative products traded. UN Food and Agriculture Organization (FAO) statistics show that the world tuna catch (including that for stocking) rose from 240,000 tons in 1950 to 1,840,000 tons in 2000,[4] pointing to the need for management and conservation. For instance, Atlantic bluefin tuna have west and east (including the Mediterranean Sea) stocks, and in the 1960s the western population catch exceeded that of the east, while now it is only 1/20th that of the east. It is estimated that the number of western population parent fish decreased by 80 to 90% from 1975 to 1995.[5]

In the mid-1970s international trade (imports) in fresh and frozen tuna accounted for 30% the tuna catch, but increased to about 50% in the 1990s (the percentage of the catch traded internationally is likely even higher when calculated as live weight including canned tuna). Trade in tuna including canned and other tuna products is almost 10% of the world trade in seafood in terms of both value and quantity. Japan's imports account for 20% in volume and between 30 and 40% in value, but when limited to fresh and frozen tuna these figures rise to 40% and at least 60%, respectively. Japan is a major importer of high-priced tuna for sashimi, but one must also bear in mind that Western countries import much tuna as canned tuna or to make it.[6]

Tuna is distributed in fresh (i.e., refrigerated but not frozen) or frozen form. Fresh tuna is chiefly imported to Japan through Narita and Kansai international airports, or other ports of entry. In 2000 the sashimi tuna supply was 125,000 tons frozen and 61,000 tons fresh from domestic production, and 193,000 tons frozen and 71,000 tons fresh from imports, putting imported tuna above that domestically supplied in both fresh and frozen types (Table 2). In 1999 and 2000 domestic production of fresh and frozen tuna decreased or was unchanged, and imports

TABLE 2. Supply of Sashimi Tuna (Excluding Albacore) to Japanese Market

| Frozen tuna | 1990 | 1993 | 1994 | 1996 | 1998 | 2000 | Tons |
|---|---|---|---|---|---|---|---|
| Domestic | 197,220 | 180,348 | 182,878 | 143,514 | 151,289 | 125,321 | |
| Imported (Including other) | 169,269 | 197,244 | 172,339 | 190,069 | 214,883 | 193,282 | (100%) |
| Taiwan | 56,891 | 108,806 | 74,507 | 86,084 | 81,546 | 74,233 | (38.4%) |
| FOC nations | 34,213 | 30,393 | 36,644 | 38,032 | 47,665 | 46,369 | (24.0%) |
| S. Korea | 55,174 | 42,823 | 43,519 | 45,696 | 66,060 | 52,378 | (27.1%) |
| Total | 366,489 | 377,592 | 355,217 | 333,583 | 366,172 | 318,603 | |
| Fresh tuna | 1990 | 1993 | 1994 | 1996 | 1998 | 2000 | Tons |
| Domestic | 65,636 | 72,320 | 67,616 | 56,102 | 54,142 | 61,582 | |
| Imported | 43,092 | 63,390 | 70,917 | 71,759 | 71,369 | 71,152 | |
| Total | 108,728 | 135,710 | 138,533 | 127,861 | 125,511 | 132,734 | |

Note: According to statistics of Federation of Japan Tuna Fisheries Co-operative Associations. From 2000, FOC countries include China but not Indonesia and the Philippines. Domestic fresh salmon production figures from *Annual Statistical Report on Marine Product Distribution* (edited by Japan's Ministry of Agriculture, Forestry and Fisheries) were double under the assumption that its figures are 50% actual production. This table uses a vessel's flag state to determine whether tuna is domestically produced or imported. Hence tuna caught in the same fishery, either in international waters or within the EEZ of another country, is domestic if caught by a Japanese-flagged vessel and imported if caught by a foreign-flagged vessel.

Source: *Bonito and Tuna Yearbook* 1995 and 2001 (in Japanese).

increased. Especially noteworthy is the approximately 70% increase in fresh tuna imports.

## Factors Underlying Rising Frozen Tuna Imports

Imports of fresh and frozen tuna are both increasing, but for different reasons. This section deals with frozen tuna.[7] In the 1970s and thereafter Japan's pelagic fishing catches peaked due to factors including the advent of 200-nautical-mile zones, while domestic demand for marine products increased, inducing an increase in frozen tuna imports. Japanese trading companies and large seafood companies responded with attempts to monopolize or oligopolize import and sales routes.[8] By the early 1980s this situation had progressed so far it was said that most frozen tuna was purchased by about 10 major wholesalers that were almost all affiliated with trading companies or major seafood companies, or directly managed by the major seafood companies.[9] Subsequently major supermarkets and other businesses became major players,[10] and in recent years it is estimated that 70 to 80% of domestic and imported tuna and marlin is controlled by giant corporations.[11]

Frozen tuna are mainly produced by ultra low-temperature tuna longliners (ULTLLs). The main exporters to Japan are South Korea and Taiwan, and their ULTLL numbers grew for the following reasons.

In the mid-1970s South Korea began using ULTLLs to supply Japan's sashimi market, and their catch grew through the 1980s.[12] Studies performed at that time found that South Korean tuna exports were predominantly driven by Japanese trading companies, and that South Korean tuna operations became closely affiliated with Japanese trading companies through the provision of capital to buy fishing vessels and through loans of operating capital.[13] Major fleet expansion of independent tuna ULTLLs followed, mainly in the 300- to 500-ton range, and in the 1980s and following years operators gave most of their ships South Korean registry while operating mainly in the Pacific, but also in the Indian Ocean, Atlantic Ocean, and other areas, but in recent years South Korea is losing its advantage to Taiwan, Indonesia, and other competitors due to its high labor costs.[14]

Meanwhile, in the late 1980s and the 1990s Taiwan added ULTLLs (Table 3) and boosted its production of bigeye and yellowfin tuna for the Japanese sashimi market.[15] But oversupply caused the price of sashimi tuna in Japan to tumble, which led to a December 1993 agreement between Japanese and Taiwanese tuna fishing organizations that exports of frozen sashimi tuna to Japan would be held under 99,000 tons in 1994.[16] However, the Taiwanese are still very eager to expand their fleet, and there are still many used fishing boats in Japan because of the large number of new ships built during the bubble economy years. Meanwhile, in addition to the 99,000-ton limit on exports to Japan already in place, in 1997 the International Commission for the Conservation of Atlantic Tunas (ICCAT) capped Taiwan's Atlantic bigeye tuna catch at 16,500 tons. As a result, the used ULTLLs reflagged to countries including Honduras, Belize, and Panama, becoming flag-of-convenience (FOC) ships that operated outside Taiwan's quota.[17]

TABLE 3. Numbers of Ultra-Low Temperature Tuna Longliners

| Year | Japan | S. Korea | Taiwan | Others | Total |
|------|-------|----------|--------|--------|-------|
| 1980 | 943 | 219 | 72 | — | 1,234 |
| 1985 | 773 | 156 | 75 | — | 1,004 |
| 1990 | 758 | 203 | 196 | — | 1,157 |
| 1995 | 703 | 201 | 334 | 198 | 1,436 |
| 2000 | 529 | 197 | 370 | 266 | 1,362 |

Note: "Others" are Indonesia and FOC countries. The number of vessels is calculated from the amount of imported sashimi tuna assuming 200 tons from each vessel.

Source: *Bonito and Tuna Yearbook*, 1999 and 2001 (in Japanese).

Most large FOC tuna longliners operate from Taiwan. Although accurate information on the number of these ships is lacking, there are at least 240 including 130 used vessels exported from Japan, about 100 ULTLLs built in Taiwan, and vessels from South Korea and other countries.[18] Taiwanese FOC vessels operate in the Pacific, Atlantic, and Indian oceans. It is estimated that 24% of Japan's total frozen sashimi tuna imports are from FOC boats, a figure that runs to over 60% when Taiwanese FOC boats are added (2002, Table 2).

FOC vessels seek an edge by avoiding restrictions such as no-fishing seasons and zones, and they avoid international stock management regulations. Worldwatch Institute's *State of the World 2001* observes that fishing by FOC boats not only leads to unfair competition, but also undermines the validity of multilateral fisheries agreements.[19] If FOC boats are not members of regional fisheries management agencies that manage tuna resources, they operate free of regulations that other ships must follow, and even if a country is a member of such agencies, stock management measures will be ineffective unless institutions have been put in place to implement regulations. Progress in efforts to shut out imports from FOC ship catches is to be applauded, and indeed they are succeeding in reducing the number of these vessels (see Essay 1), but there are at least an estimated 100 FOC boats remaining, excluding fishing boats that are making efforts to abide by international stock management measures.[20]

## Tuna Stock Management and Marine Environment Conservation

At CITES COP 8 held in Kyoto in March 1992, Sweden proposed that to counter the decline in Atlantic bluefin tuna stocks, the western subpopulation should be listed on Annex I and the eastern subpopulation on Annex II because environmental NGOs believed that ICCAT management measures were not effective and were seeking to restrict trade under CITES. Not only were the quotas of Japan and other pelagic fishing countries higher than those recommended by ICCAT scientists, the US catch far exceeded its quota, inducing NGOs to get behind this proposal.[21]

Ultimately this proposal was retracted, and although limits on bluefin tuna catches by ICCAT members did not achieve the progress envisioned under the proposal,[22] some improvement was realized by means including import limits by non-ICCAT countries.[23] With agreement also from a shipowners' association of Taiwanese FOC boats, international private-sector efforts are currently underway to abolish FOC boats with part of the scrapping costs covered by the Japanese fishing industry (Essay 1).

Meanwhile, environmental NGOs are expressing doubts about the effectiveness of resource management under the ICCAT and other regional fisheries management organizations. If the purpose of marine resource management is conserving the marine environment, then governments should consider developing arrangements under which environmental NGOs indirectly influence resource management through trade controls, but also can directly participate in resource management.

## 2-3   Deforestation and the Timber Trade

### The Timber Trade as a Cause of Forest Loss

Tropical forest loss can be attributed to social factors such as institutions, population pressure, traditional practices, and the demands of economic development, and to natural factors such as forest fires and climate. Some institutional deficiencies are underdeveloped land systems and environmental protection systems, as well as insufficient administrative management capacity, while an especially significant population pressure factor is increasing rural population density. Economic development involves agricultural development, industrialization, and earning foreign currency, as well as non-traditional shifting cultivation (opening land by burning its vegetation), livestock grazing, and fuelwood gathering. But considering these from a different perspective as direct causes shows the problem in a new light: excessive export-oriented commercial logging, the creation of agricultural land for commercial plantations to grow oil palms and coffee, urbanization-driven land use changes, and non-traditional shifting cultivation by landless peasants, as well as large forest fires.

Few would disagree that the timber trade is closely linked to the decline of forests, but it is wrong to simplistically assume that deforestation is the fault of importing or exporting countries. It was not just international market demand that brought the timber boom to Southeast Asian countries; other factors were that governments and political forces of the involved countries badly wanted this trade, and that laws controlling logging and distribution were not properly enforced.[24] For instance, beginning in the 1980s Indonesia restricted roundwood exports and instituted measures to build a plywood industry and facilitate exports, which resulted in excess plywood production facilities. Raw log procurement induced widespread logging that violated operating standards, which led to illegal logging and rapid deforestation. Malaysia's Sabah state became overly dependent on timber royalties primarily from roundwood exports, which amounted to 70% of timber income in some years during the 1980s, and it sometimes tried to increase royalty income by granting many logging licenses with shorter-than-usual terms.[25] This clearly led to excessive logging.

### Recent Forest Statistics and Timber Trade

Deforestation continues in Southeast Asia, including the region's insular portion.[26] Over the decade from 1990 to 2000, Indonesia's forested area decreased from 118.1 million to 105 million ha (down 1.2%), Lao PDR from 13.1 million to 12.6 million ha (down 0.4%), Malaysia from 21.7 million to 19.3 million ha (down 1.2%), Myanmar from 39.6 million to 34.4 million ha (down 1.5%), the Philippines from 6.7 million to 5.8 million ha (down 1.5%), and Thailand from 15.9 million to 14.8 million ha (down 0.7%). All of these decreases are much greater than the average world annual deforestation rate of 0.2%, and Asia's average of 0.1%. And while the Philippines and Thailand were once timber exporters, they are now importers. Apparently even Malaysia's Sabah and Sarawak import roundwood from other countries.

World roundwood production in 1998 was over 3.26 billion cubic meters, of which more than 1.75 billion cubic meters were fuelwood and 1.51 billion cubic meters were industrial wood.[27] Exports of these amounted to 0.25% and 5.4%, respectively. Exports of industrial wood processed into sawnwood were 415 million cubic meters or 27.4%, those of wood-based panels were 150 million cubic meters or 32.0%, those of pulp for paper were 176 million cubic meters or 19.1%, and those of paper and paperboard were 293 million cubic meters or 30.6%. Most timber-exporting countries are industrializing timber processing to increase employment and add value, with the concomitant trend being less trade in round-wood and more in processed wood. Southeast Asian examples are Indonesia and the Malaysian states of Sabah and Sarawak, which are putting special effort into labor-intensive plywood production, and also building other industries such as plywood adhesive manufacturing.[28]

## Illegal Logging and Timber Trade

As the timber trade continues to expand, international society's most important task is to curb illegal logging and timber trading. The G8 Action Program on Forests agreed upon at the 1998 G8 foreign ministers' meeting included an initiative on illegal logging and often appears on the agendas at international conferences including meetings of the International Tropical Timber Organization.

NGOs provide a great deal of information on illegal logging and observe that there is much of it in countries such as Brazil, Indonesia, Malaysia, the Philippines, Papua New Guinea, Russia, and Cameroon, and that many developed countries import illegally logged timber as roundwood or processed wood. Depending on the source of information and the country or region, the proportion of illegally logged timber in total logging or import and exports differs, but some reports say that in certain countries 80% of the logging is illegal.

A report issued in summer 2001 by EU Forest Watch writes that in 1999, 62% of Britain's tropical timber imports, 27% of Germany's roundwood imports, 25% of Spain's tropical timber imports, at least 10% of France's roundwood imports and 48% of its sawnwood imports are illegally logged, and that their sources were Brazil, Indonesia, Malaysia, and Cameroon. In the Asia-Pacific region, Forest Trends points out the strong connection between China's rapidly growing timber imports and illegal logging in the Russian Far East.[29]

Although some illegal logging is minor covert logging by locals, much is performed in violation of regulations or permit periods, without permission, or with forged permits. Illegal trade and processing use timber that has been illegally logged, violate permit terms, lack permits, use forged permits, or involve smuggling. Some of the main entities involved are public employees (including military, police, and authorizing government agencies), politicians, forestry companies, and foreign and domestic investors.

Forest certification systems with labeling, described below, would be a good way to address illegal logging and promote sustainable forestry. It will also be vital to use timber tracking systems, such as by stationing NGOs, experts, or local people at inspection stations and other strategic timber distribution locations to monitor timber in accordance with the laws of each country.

## 2-4    Recycling and the Waste Trade

### Expanding International Trade in Wastes for Recycling

Industrialization in Asia has been inducing expanded imports of wastes for recycling. Although some wastes are traded within Asia, many are imported from Western nations. These wastes cover a broad spectrum including consumer appliances, personal computers, motor vehicle batteries, wastepaper, and plastic (Table 4). China's waste imports are expanding at a pace commensurate with its economic growth. Wastepaper imports skyrocketed from 20,000 tons in 1990 to 3.7 million tons in 2000, and waste copper from 20,000 to 2.5 million tons. Symbolic of major scrap exports was the 70,000 tons of steel from the World Trade Center towers to India.

### Two Sides of the Waste Trade

In some cases international recycling is seen to be reducing the environmental burden. For instance, in China straw was being used for pulp and paper production, but this polluted rivers. Replacing the straw with imported wastepaper turned out to be less-polluting.[30] Wastepaper imports are increasing as straw pulp mills are subjected to more stringent environmental regulations, and imports have jumped especially since 2000, when it appears that environmental regulations became more effective. In 2000, wastepaper imports grew 47% over the previous year, and in 2001 they increased 72%.

At the same time, some wastes are shipped abroad for disposal while claiming they are to be recycled, and sometimes recycling itself ends up polluting. In 1999 it was found that medical waste had been illegally exported from Japan to the Philippines under the guise of wastepaper. Japan's government paid to have the waste repatriated and incinerated, and the presidents of two Japanese companies received prison sentences in this incident.

Basel Action Network (based in the US), and the Silicon Valley Toxics Coalition (US) worked with several Asian NGOs to investigate exports of discarded personal computers from the US to China, India, and Pakistan for recycling. Their report revealed that only recyclable parts are recovered, while the remaining parts and their hazardous substances are carelessly dumped, that recycling facilities have no pollution control equipment, and that heavy metals are melted for recovery without heed for their hazards. Domestic reports by Chinese news media have led to enhanced restrictions on waste imports (Essay 2). Other studies have shown that discarded personal computers exported to China from South Korea are likewise improperly processed.[31]

TABLE 4. Waste Imports by Major Asian Countries/Regions for Recycling (2000)

Tons

| | S. Korea | China | Taiwan | Indonesia | India | Philippines | Thailand | Japan |
|---|---|---|---|---|---|---|---|---|
| Waste plastic | 18,052 | 2,007,164 | 68,555 | 2,888 | 72,224 | 5,241 | 734 | 14,717 |
| Recovered paper | 1,963,493 | 3,713,596 | 1,052,702 | 2,436,800 | 750,854 | 416,079 | 953,029 | 278,084 |
| Scrap iron | 6,851,902 | 5,099,104 | 1,957,884 | 1,263,505 | 1,754,942 | 3,570 | 741,331 | 321,379 |
| Scrap copper | 203,607 | 2,501,167 | 44,316 | 3,965 | 87,396 | 1,042 | 4,358 | 190,224 |
| Scrap nickel | 558 | 33 | 645 | 0 | 120 | 44,000 | 31 | 15,665 |
| Scrap aluminum | 101,948 | 804,629 | 138,602 | 2,837 | 58,323 | 437 | 11,485 | 7,767 |
| Scrap lead | 1,187 | 4 | 0 | 64 | 10,513 | 0 | 371 | 0 |
| Scrap zinc | 346 | 47,784 | 49,124 | 289 | 40,808 | 107 | 78 | 68 |
| Scrap tin | 4 | 19 | 308 | 73 | 182 | 62 | 0 | 195 |

Source: Trade statistics of each country/region.

## Basel Convention

The only Asian countries that have ratified the export ban amendment adopted by the 1995 Basel Convention Conference of Parties are Sri Lanka (1999), and China and Malaysia (2001). Even worldwide, not enough countries have ratified the amendment for it to enter into force.

However, banning the transboundary movement of hazardous wastes from developed to developing countries for recycling, as the amendment would do, would not necessarily prevent the smuggling of hazardous wastes or promote environmentally friendly recycling technologies in Asia. For example, the amendment could not have prevented the aforementioned shipment of medical waste from Japan to the Philippines because it was shipped as wastepaper, which is not controlled by the Basel Convention. Governments must beef up customs inspections and exporting countries need to create systems for proper management of hazardous wastes.

Caution is also needed regarding the relationship in developing countries between the formal sector, which has pollution control facilities, and the informal sector, which does not. India, for example, has not ratified the Basel Convention amendment, but in 2001 its Supreme Court responded to a lawsuit by NGOs by banning the import of hazardous wastes. This decision ended up forcing a few lead recycling plants into closing because they had used old vehicle batteries imported from abroad. Those plants had been comparatively large operations with some pollution abatement equipment. On the other hand the domestic informal lead recyclers do not pay taxes and have no pollution abatement equipment, and it is the informal sector that is more competitive in collecting hazardous wastes domestically. In 2001 India's government adopted a policy of restricting traffic in old vehicle batteries to help the formal sector, but it has not been very effective.[32]

## Creating an Asia-wide Cyclical Society

Even wastes not specified as hazardous might fail to be collected domestically when imports increase. Certain Western countries, Taiwan, South Korea, Japan, and other mostly high-income economies are passing laws and introducing policies that encourage recycling in an effort to build cyclical societies. Arrangements that provide for collecting recyclable wastes while having consumers and producers shoulder some of the economic burden are thought to be effective in lowering the market prices of collected wastes.

Although it is not clear if easy waste imports have ever slowed domestic collection of recyclables, even comparatively low-income countries have launched initiatives to improve resource recovery. For example, Thailand anticipates that rising incomes will cause shrinkage of the informal resource recovery sector, and therefore needs to make changes in the way resources are recovered. China too is considering arrangements for recovering end-of-life vehicles and waste consumer electronics, and in 1994 it began offering tax breaks to companies that collect, separate, and recycle wastes, but insufficient consideration has been given to the impact that limiting the importation of recyclable wastes would have on domestic waste collection.

Important criteria for Asian countries to consider, while keeping in mind the state of recycling in each geographical area, are how to build an Asia-wide cyclical society, and how to regulate the waste trade while building that society.

# 3.  Building and Modifying Institutions

## 3-1  Labeling

### Why Label?

Formerly both consumers and producers likely had complete information on the quality of goods traded on the market, but the distribution of goods became more complicated owing to the emergence of social, geographical, and temporal distance, and consumers began incurring huge costs in gathering information as transaction routes diversified. Consumers are therefore no longer able to obtain information as complete as that held by producers. Used cars are symbolic because in a market where ordinarily both good- and poor-quality goods would be supplied, good-quality goods will not appear if prices are determined by the presence of asymmetrical information. In other words, the market for those goods will disappear because that market no longer functions effectively. Currently distribution is so complicated that there are few markets without information asymmetry, and economies are challenged with how to make information sharing possible.

Some ways of avoiding or diminishing such information asymmetry are the certification or labeling of goods. Systems based on set criteria make it possible to trace distribution routes back to where goods were produced, which gives consumers reliable information. Examples in Japan are the Japanese Agricultural Standard and the Japanese Industrial Standard. Ecolabels can be placed in this group, and even brands can work the same way against information asymmetry because a track record built by a company or engineers over many years becomes a reputation that guarantees reliable information to consumers. Direct sales by producers to consumers is another way of avoiding information asymmetry.

Around 1990 many developed countries introduced environmental labeling. Japan's Eco Mark was created in 1989.[33] Ecolabeling has also been instituted by other Asian countries and regions including Singapore, Taiwan, South Korea, Thailand, Hong Kong, and China.[34] If labeling schemes are designed to include application to foreign-produced goods, it is possible that forms of production will become more eco-friendly in response to eco-conscious demand. The first initiatives on international ecolabeling were for building forest certification systems.

### Forest Certification Systems

Forest certification was catalyzed by tropical timber boycotts in the second half of the 1980s, primarily in Europe. This led to unintentional losses for the timber industry, which took the lead in developing forest certification systems to show that timber came from well-managed forests. At the same time, however, consumers had declining trust in the proliferation of marks and labels using "eco."

In addition to the Statement of Principles on Forests and the provisions for sustainable forests in Agenda 21, which were adopted at UNCED in 1992, there are a number of other initiatives that together cover about 80% of the world's forests, such as the International Tropical Timber Organization and the Helsinki Process.

While progress has been made throughout much of the world by the Forest Stewardship Council (FSC), Pan-European Forest Certification, and ISO 14001, forest certification efforts have just begun in Southeast Asian countries owing to problems such as the difficulties of observing legal systems and the inadequacies of legal systems covering forests and forestry.[35] In Indonesia the Indonesia Eco-labelling Institute has been working on forest certification since its founding in 1998, but it has yet to achieve its original goal of mutual approval with the FSC, and so far it has certified only one natural forest jointly with the FSC. As of July 2002 the FSC had three certified forests remaining after four lost certifications in 2001 because of illegal logging. In Malaysia the Malaysia Timber Certification Council has started its own certification system based on an experimental initiative in three states carried out jointly with Holland since 1996, but NGOs have noted deficiencies with regard to the rights of native peoples. As of January 2004 there was one FSC-certified forest in Sabah state, and there were two in Peninsular Malaysia.

Forest certification systems offer a number of benefits,[36] and there are hopes among forest owners, wood consumers, and the general public that ecolabeling will play a role in environmental conservation by helping bring about sustainable forest management. Attention will therefore focus on what kind of institutions that countries or regions will build to serve as better certification systems, and the cooperation of experts, citizens, and local people will be indispensable in building those institutions. Further, members of the public, in their role as consumers, must pay attention to the wood products they use so as to consume only wood produced in properly managed forests. Effective use of wood from sustainable forests will undoubtedly help conserve the global and living environments, and it is hoped that, in terms of both forest management and the timber trade, forest certification systems will be a leading means of effectively using forest resources while also conserving them.

## 3-2  Import and Export Controls

Labeling, international conventions, and protocols are some of the ways of controlling international trade to restrict the movement of banned items, and while they have some effect, the limited number of shipment inspections nets only a small number of violations and lets considerable contraband through (Table 5). Much imported contraband is discovered after entering domestic distribution systems.

TABLE 5. Recently Discovered Instances of Smuggling

| Date | Substance/amount | Exporter | Importer | Specifics |
|---|---|---|---|---|
| 01-Apr-'01 | 1,200 pangolins | Malaysia | Vietnam | Malaysian customs authorities discovered them hidden in fish containers. |
| 01-Jul-'01 | 500,000 cans (150 tons) of R12 refrigerant | China | Japan | 100,000 cans discovered by Yokohama customs; 400,000 had already been imported. |
| 01-Sep-'01 | 36,000 cans (12 tons) of R12 refrigerant | China | Japan | Discovered by Moji customs. |
| 01-Mar-'02 | Crabs | Russia | Japan | Sent away when authorities at Mombetsu and other ports discovered that Russian-issued shipping certificates had been forged. |
| 01-Aug-'02 | 160 kg of ivory | Guinea | Thailand | Discovered at Bangkok international airport. |
| 01-Sep-'02 | 450 tons of waste computer parts and color TVs | U.S. | China | Found by Wenzhou customs officials. To be returned. |
| 01-Oct-'02 | 2,700 kg of pangolins | Indonesia | China | Discovered by Hong Kong customs authorities. |
| 01-Oct-'02 | 3334.6 kg of ivory | Kenya | China | Discovered by Shanghai customs. Falsely reported as timber. |
| 01-Oct-'02 | Radioactive waste | Russia | China | Discovered by Kazakhstan customs authorities when about to cross border into China. |

Source: Prepared from *Asahi Shimbun*, *Bangkok Post*, *Jakarta Post*, and other press reports.

Customs and quarantine inspections should be more actively used help solve trading partners' environmental problems. In January 2002 the EU banned imports of shrimp and other products from China due to safety concerns. Some Chinese vegetables and other produce imported by Japan have been found to have residual pesticides exceeding standards. Despite visits by Japanese delegations to China to look into pesticide use, and reports[37] on this issue going back to the 1980s, Japanese and Chinese authorities finally began full-blown discussions on this issue in 2002. Asian countries should build a system for the close exchange of such information throughout the region and its use in furthering their environmental goals.

To counter the export to Japan of crabs and sea urchins poached by Russian fishing vessels, the Russian government has a system for informing Japan by email of the vessels issued shipment certificates, but problems including the entry of Russian vessels with North Korean forged certificates require that, in addition to trade regulations, the four countries of Russia, North and South Korea, and Japan create an arrangement to share information.

Although CITES restricts trade in some marine and forest products, international initiatives taking advantage of environmental trade measures are needed because CITES controls are not comprehensive enough. Especially since the prohibition on trade between parties and non-parties of multilateral environmental agreements encourages non-parties to join, this inducement should be part of an effort to create international systems to manage marine and forest resources. Asia should take the initiative to form an international consensus because Asian countries import and export many marine and forest products.

## 3-3 Mediation of Disputes over Environmental Trade Measures

Implementing environmental trade measures requires attention to agreements such as the WTO-administered General Agreement on Tariffs And Trade (GATT). By their very nature, such measures could violate basic GATT principles such as "most favored nation" and "national treatment." Environmental trade measures instituted by Asian countries have never been taken to the WTO for dispute arbitration.

An example in which Asian countries filed for WTO dispute arbitration against an environmental trade measure instituted by a non-Asian country is an April 1996 US measure limiting shrimp imports from countries that do nothing to prevent turtle bycatch. India, Malaysia, Pakistan, and Thailand responded in October of that year by asking the WTO to resolve the dispute. Although the 1998 decision held that an exception to GATT Article 20 could be applied even to measures to prevent environmental problems arising in other countries, it did not justify the US action for reasons including: it could not be said that the US was adequately implementing policy for transferring technologies to prevent turtle bycatch, and it was according discriminatory treatment to countries in South America and elsewhere.

In response, the US revised its guidelines and made some improvements, but continued its trade restrictions. Malaysia claimed that the US was not abiding by dispute resolution agency recommendations or the ruling, and filed again under the provisions of Article 21.5. The panel ruling came in June 2001 and that of the Appellate Body that October, judging that as a result of guideline revisions and other changes by the US, its trade restriction was justified under GATT Article 20.

These rulings show the conditions under which trade restrictions can be used for environmental purposes, but doubts remain as to the justifiability of this measure from an environmental perspective. As turtles in Southeast Asia do not eat shrimp, there are claims that trade restrictions against Southeast Asian countries are not helping conservation. Some observe that the WTO panels, comprising experts in economics and law, do not adequately address environmental concerns, which points to the need to reconsider the way disputes over environmental trade measures are resolved.

## 3-4  Summary

Trade can cause and aggravate environmental problems in conjunction with other factors. Trade restrictions are one way of protecting rare species and dealing with hazardous wastes. But while many cases of smuggling and the like are reported, there are few efforts to conduct studies with international cooperation or efforts to prevent smuggling. Further, environmental trade measures apply to only a few marine and forest products.

In view of this situation, encouraging still more trade liberalization could invite more environmental damage. When governments enter into free trade or economic cooperation agreements, lower their export taxes or import tariffs, or take other such actions, they must predict the consequences and take steps against anticipated environmental problems, or limit the extent of trade liberalization.

Initiatives by businesses and consumers are also important. Some companies in developed countries require raw material suppliers to take certain environmental precautions. Others demand that companies in developing countries have effluent standards as strict as those in developed countries, and some developing country companies are accommodating their trading partners' wishes by beefing up their environmental measures.

Increasing efforts are being made to supply consumers in developed countries with products whose modes of production are sensitive to environmental conservation and working conditions in developing countries. Such initiatives include what is called "fair trade" or "alternative trade," and the promotion and trade of organic agricultural produce. Sixty Asian organizations are members of the International Fair Trade Organization,[38] and there are also many consumers' organizations including 17 from India and 11 from Bangladesh. In Switzerland the market share of products with fair trade labels is estimated at 15% for bananas and 4% for tea. A representative Japanese fair trade organization, Alternative Trade Japan, had sales of well over ¥1.5 billion in 2000, an amount that rivals that of Britain's Oxfam Fair Trade. Even major corporations like Chiquita and Dole have begun pesticide-free and organic production of bananas with the cooperation of NGOs. The International Federation of Organic Agriculture Movements (IFOAM) too is promoting organic agriculture in developing countries, offering training courses and other help. When consumers buy environmentally friendly imported products, they help promote environmental conservation in the exporting countries.

As Asia has the biggest imports of many goods, governments should work toward conservation by reviewing their trade policies and using the information obtained through trade to solve environmental problems throughout the region.

International NGO networks have played a major role promoting fair trade and identifying problems in the international trade of rare species and hazardous wastes. Promoting trade which conserves the environment will require that NGOs augment their international exchange of information and perform more joint international studies, that intergovernmental arrangements for this purpose are created, and that enforcement systems are strengthened.

Kojima Michikazu, Kiyono Hisako, Yokemoto Masafumi, Izawa Arata, Yamashita Haruko, Tachibana Satoshi

## Essay 1:  Japan's Import Restrictions on Catches from Large Flag-of-convenience Tuna Longliners[39]

Flag-of-convenience (FOC) vessels could hurt the effectiveness of international management measures in the large tuna longlining industry supplying tuna for frozen sashimi, and for this reason measures are being put in place to deal with these vessels.

After withdrawal of the proposed regulation on Atlantic bluefin tuna at the 1992 CITES conference of the parties, progress was realized in restricting imports of bluefin tuna from FOC nations by the International Commission for the Conservation of Atlantic Tunas (ICCAT) and other organizations. Based on Japan's Special Law on Strengthening the Preservation and Management of Tuna Stocks (1996), the Japanese government has followed ICCAT recommendations by restricting imports from certain countries, such as the 1996 general ban on importing Atlantic bluefin tuna from Panama (rescinded on April 3, 2000).

Nevertheless, Taiwanese FOC vessels increased during these years. Because these vessels were after mainly bigeye tuna, they were not affected by restricting bluefin imports. ICCAT therefore created similar procedures to be applied to bigeye tuna, and in 2000 it adopted a recommendation to ban bigeye tuna imports from Belize and other countries.

One weakness of this import restriction is that it would be ineffective if an FOC vessel changed its flag state, which makes it necessary to identify individual FOC vessels and stop the importation of their catch. In this respect Japan took action in December 1999 when, based on a resolution adopted at the ICCAT regular meeting that year, the Fisheries Agency Director-General issued a notification to importers and other parties urging them to voluntarily refrain from buying catches from FOC vessels. The Fisheries Agency website publishes the names of businesses found to have purchased such catches. The ICCAT has a list to identify FOC vessels, but owing to the extreme difficulties involved in keeping this list, efforts are underway for what might be called switching the burden of proof so that, instead of making resource management authorities responsible for preparing the list, the export of catches would be allowed only when shipowners have declared their will to follow resource management measures and have been placed on a list of authorized vessels. A special ICCAT meeting held in October and November of 2002 decided to implement from July 1, 2003 a system that would exclude from international trade any bluefin tuna, bigeye tuna, and swordfish taken in the North Atlantic by vessels not on the list of authorized vessels.

In parallel with this, measures are being taken to more directly eliminate FOC vessels. In February 2001 a basic accord was reached between the Organization for the Promotion of Responsible Tuna Fisheries, whose members include the Federation of Japan Tuna Fisheries Cooperative Associations, and the Kaohsiung City Foreign-Flagged Pelagic Fishing Vessel Association, an organization of Taiwanese FOC vessel owners. They agreed that 62 of the FOC vessels imported as used vessels from Japan would be scrapped, at Japan's expense for the time being, by the end of 2003, and that 65 of the FOC vessels built in Taiwan would be reflagged to Taiwan by the end of 2005.

Yokemoto Masafumi

❖            ❖            ❖

## Essay 2:  Chinese Technotrash Imports

A report titled *Exporting Harm: The High-Tech Trashing of Asia*[40] details how exports especially from the US to Asian developing countries of discarded personal computers, consumer electronics, cellular telephones, and other such E-waste are creating environmental disasters in China, India, and Pakistan. The report ends with recommendations for the US to honor the Basel Convention, for extended producer responsibility, and for greater consumer awareness. Packed with photographs of the recyclers in action, this shocking report has no doubt had a considerable impact on governments and the industry.

China's media also gave this much attention. The February 27, 2002 *Liberation Daily* wrote, "Between 50 and 80% of the electronics recovered in the western US ultimately end up in Asian countries. Specifically, an estimated 80%, or 10.2 million of the more than 12.7 million personal computers discarded in the US in 2002 were taken to Asia over a one-year period," and "To recover the precious metals in E-waste, Asian workers are 'using 19th-century techniques to process 21st-century wastes.'" In March *Nanfang Daily* and China Central Television went newsgathering in the town of Guiyu, Guangdong Province, and at The Chinese People's Political Consultative Conference held at about the same time, one of the agenda items in the spotlight was the quick establishment a system for the domestic collecting and recycling of discarded consumer electronics. Such news coverage and events raised awareness of this issue overnight.

After the release of *Exporting Harm* the Chinese government took major action. Officials of the State Council, who took the matter very seriously, sent an investigative group comprising personnel from the State Environmental Protection Administration and other agencies. Their investigation concluded that nearly all the scrap was from domestically made computers, only a little had been imported, and *Exporting Harm* was exaggerated. Nevertheless, Guiyu Town carried out a major reorganization, closed factories, and took other actions for environmental recovery. In June 2002 the government toughened customs regulations, and on June 4, 2002 the government expanded its import ban list by adding "parts and shredded waste of used consumer appliances" (enforced beginning on July 1), a measure that followed the total import ban on discarded consumer appliances and other products that had been implemented on February 1, 2000.

Quickly creating a system to collect and process discarded E-waste was put high on the agenda, and China's government has begun tackling this problem with domestic measures. Because used consumer appliances are bought and sold, it is common for many such products to be in use past their anticipated lifetimes, and for people to practice "discretionary disposal," which involves removing usable parts and discarding the rest. These practices are unsafe, wasteful of energy, and environmentally unsound. Legally mandated collection, processing, and resource recovery of E-waste should be achieved quickly, but time and preparation will be needed to decide on payment of costs, raise consumers' consciousness, and overcome other obstacles.

The 2000 import ban on discarded consumer appliances suggests more of an effort to cope with these problems than to protect domestic manufacturers. It is not desirable for imports of consumer appliances and electronics to increase year by year without a domestic system for proper E-waste management. Although it is desirable to satisfy the resource demands of domestic manufacturers by first using recycled materials from abroad, these regulations and bans on imported waste likely show China's strong will to resolutely reject the imposition of wastes on developing countries.

Governments and businesses in waste-exporting countries must ascertain the recycling situations in developing countries and ensure that wastes are properly managed there. If exporting developed countries share information with developing countries that accept wastes for recycling, it will facilitate waste acceptance and the creation of management systems in accordance with each country's circumstances. Some possible measures in Asia might include using the Basel Convention regional centers in China and Indonesia, and that slated for India, as information resource centers.

Yoshida Aya, Kojima Michikazu

❖              ❖              ❖

## Essay 3:  Illegal Trafficking in Ozone-depleting Substances in Asia[41]

An international issue in recent years is the smuggling of CFCs and halons, which are restricted under the Montreal Protocol on Substances That Deplete the Ozone Layer. CFC production and consumption (production plus imports, minus exports) in developed countries were totally phased out in 1996 except for essential uses and the "basic domestic needs" of developing countries—the so-called "Article 5 countries"—which were given total-phaseout extensions up to 2010 because their per capita consumption of ozone-depleting substances (ODSs) is low. They were allowed to increase production until the start of restrictions in 1999.

ODSs that are newly produced or exported by developed countries to satisfy developing countries' basic domestic needs, then later illegally exported back to developed countries, are labeled as recycled ODSs that are exempted from the phaseout under the Montreal Protocol or labeled as CFC substitutes. Since these ODSs are traded at prices lower than those of the transaction partner's market price, they encourage the longer use of ODS-using equipment and thus prolong ozone layer depletion. Although accurate figures are hard to come by, it is estimated that smuggled substances accounted for about 15% of the world CFC trade in the mid-1990s. In the 1990s it was believed that the main source of contraband CFCs was Russia, which could not comply with the protocol and therefore continued production, but in the 2000s the main source might now be the developing countries, especially China (a CFC producer). Already in the years 1995 to 1997 the EU found 800 tons of contraband Chinese-made CFCs and halons through a network including a German company and brokers. During the smuggling peak from 1994 through 1996, Taiwanese authorities estimate that 2,000 tons of ODSs were being smuggled in from China each year.

Imports of contraband ODSs by developing countries have been growing as the effect of the 1999 freeze on their ODS production takes hold. India, which is a CFC producer, is also believed to have received vast illegal imports of ODSs mainly from nearby countries in 1999 and 2000. Illegal trafficking has also been discovered in Pakistan, Indonesia, and Malaysia. The Environmental Investigation Agency, a UK-based NGO, estimates that 80% of Vietnam's CFC-12 imports are illegal.

Why does ODS trafficking happen, and why is there still demand for CFCs and halons after their total phaseout? The Montreal Protocol phased out the production, import, and export of ODSs, which means that once ODSs have been produced or recovered and recycled, their use is permitted in countries other than those whose domestic laws establish a phaseout deadline on use. Most contraband CFCs are used as refrigerants. Because replacing or retrofitting ODS-using equipment is an expense, the availability of cheap ODSs encourages people to put off switching to equipment using ODS substitutes.

In 1997 the protocol was revised to address this situation, requiring all parties to institute licensing schemes for ODS imports and exports until January 1, 2000 or for the three months following ratification. Moreover, customs officers in developing countries are given training under a UNEP program.

The grace period for phasing out ODSs in developing countries is cited as the main reason for the illegal trade, but at the time of the protocol's adoption it was economically and politically unrealistic to expect developing countries to participate without this provision. Illegal trading in ODSs might be encouraged by the protocol's total lack of restrictions on ODS use, the fact that recycled ODSs are exempt from trade restrictions, and that ODSs contained in other products are not subject to trade controls between protocol parties. Or perhaps the grace period should have been accompanied by some other measures. Analyzing what went wrong could be a valuable lesson for future international environmental controls.

Matsumoto Yasuko

# Chapter 3
# Food, Farming and the Environment:
# The Development of Sustainable
# Agriculture

Farm workers loading rice for shipment after weighing and grading. October 2001, Punjab state, India
Photo: Sugimoto Daizo

# 1.    Introduction

Since agriculture is far more dependent on the environment than industry and other sectors, it must maintain a delicate balance to avoid damaging the environment while at the same time pushing the envelope to produce enough food to feed a growing population.

Yet in Asia major changes are working against maintaining this precarious balance: an unprecedented population increase, rapid economic growth in Southeast and East Asia, and the internationalization of economic relationships through trade and investment. This chapter explores the agriculture-related environmental problems arising out of these changes in Asian developing countries and examines their background.

Sections 2 and 3 look at the environmental problems arising from the production of grains and other basic foods. Specifically, Section 2 surveys the rise in grain consumption induced by population and economic growth in recent years and summarizes the impacts on agriculture and the environment, while section 3 uses examples from several countries in discussing the environmental impacts of agriculture on marginal land[1] and the increasing use of modern agricultural technologies. Section 4 discusses the production of commodities for export, with particular attention to agribusiness and how in recent years it is rapidly pushing to build international business networks. While traditional family farming emphasizes the sustainability of agricultural production, to agribusiness land and water are mere inputs because its purpose is profit maximization, and such a system readily lends itself to environmental plunder. Section 4 will explore this aspect through the examples of shrimp and eel farming, palm oil, and the production of vegetables for export. This should reveal the true face of the agro-food system fashioned by agribusiness. Finally, section 5 examines three areas: the future orientation of agricultural production and development, what kind of regulations should be imposed on agribusiness and the trade in agricultural produce and food, and creating the actors who will bring these about. Based on these areas the section then sets forth the challenges involved in rebuilding sustainable agriculture.

# 2.    Asia's Increased Food Demand and the Environment

### Diets Change, Grain Demand Rises

Over the three decades from 1965 to 1999 Asian grain demand increased 2.6 times from 340 million to 890 million tons.[2] This rate of increase in grain demand is especially large when compared to the two-fold increase for the world as a whole.

There are two main factors behind increased grain demand. One is population growth. During these years world population increased 1.8 times, while that of

Asia grew 1.9 times.[3] The other is dietary changes in some countries of Southeast and East Asia where people are eating more livestock products and instant noodles. Trends in livestock product consumption indicate that it has considerably increased in countries other than India, Indonesia, and Vietnam, that especially in China the consumption of meat and eggs is markedly higher, and that Malaysia's meat consumption has eclipsed that of Japan (Fig. 1).

Two reasons underlie these changes in diet. One is diet diversification resulting from the rising incomes begot by rapid economic growth, and the other is the vigorous efforts by agribusiness to promote new styles of food consumption, as seen in the advance of fast food. An example of the latter would be the number of McDonald's outlets in China, which skyrocketed from one to 252 between 1990 and 1999, and in Malaysia from 22 to 129 over that decade. The advance of fast food could have a major impact on the future of food consumption in Asia.[4] In particular one cannot overlook the effect of impressing dietary habits, tastes, and preferences on growing children.

FIG. 1. Per Capita Annual Consumption of Livestock Products in Asia

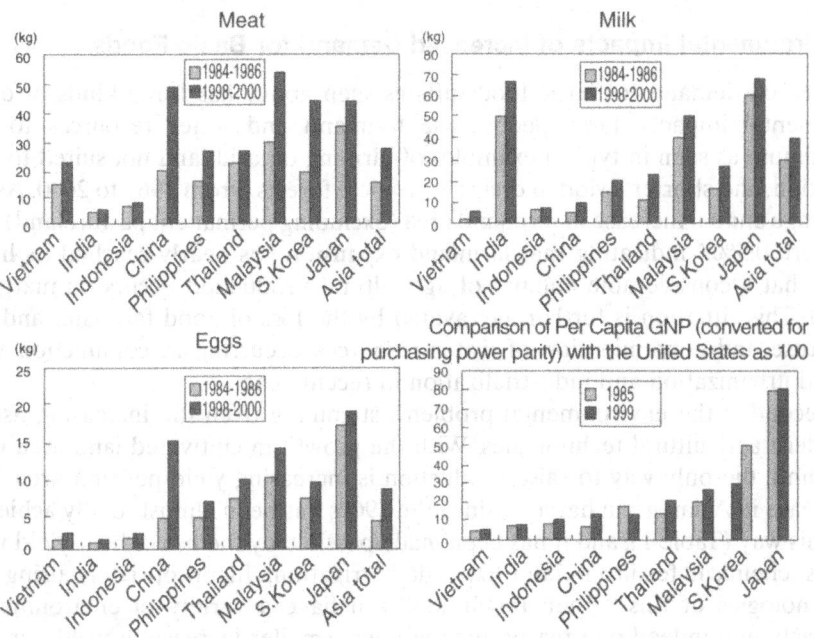

Notes: 1. Asia totals in the graphs showing livestock product consumption are averages for Asia as a whole. For the extent of Asia, see Part I Chapter 3, note 2.
    2. Livestock product consumption is given as the averages of two three-year periods: 1984 to 1986, and 1998 to 2000.
Sources: For livestock product consumption: FAO, *FAOSTAT*. For per capita GNP (converted for purchasing power parity): World Bank, *World Development Indicators, 2001*, CD-ROM, 2002.

The amount of corn required to produce 1 kg of beef is 11 kg, for pork 7 kg, for chicken 4 kg, and for eggs 3 kg. As such, the increase in livestock product consumption has a large impact on grain demand. Take the example of China, whose meat consumption has soared. Over the decade from 1989 to 1999 grain demand rose about 50 million tons, of which 33 million tons, or 67%, were for livestock feed. During the previous decade up to 1989 that proportion was 19%, indicating that livestock feed consumption is the force behind China's mounting grain demand in recent years.[5]

But this does not meant that all Asians now enjoy affluent diets. Fig. 1 shows that diets have changed little in India and other countries with relatively slow economic growth rates.[6] Even in countries with increasing livestock product consumption, these figures are averages; in many cases the class gap is still wide. One measure of poverty in Asia is afforded by the percentage of children under 5 suffering from malnutrition, which is 20% in Malaysia, 34% in Indonesia, 16% in China, and 53% in India.[7] A significant portion of the population, including the poor, and the weak members of society such as children, are not benefiting from improvements in diet.

## Environmental Impacts of Increased Demand for Basic Foods

Increased demand for basic foodstuffs as seen above has three kinds of environmental impacts. First, people use farmland and water resources to the maximum as seen in typical examples of farming on arid land not suited to cultivation, and shorter swidden cycles in tropical forests. From 1961 to 2000, Asia's average annual increase in farmland area (excluding permanent pastureland) was a mere 0.5%,[8] indicating that farmland expansion has nearly reached its limit, and that a considerable amount of agricultural production occurs on marginal land. This situation is further aggravated by the loss of good farmland and the overuse and contamination of water resources occurring in conjunction with rapid urbanization and industrialization in recent years.

Second is the environmental problems stemming from the increasing use of modern agricultural technologies. With the growth in cultivated land area near its limit, the only way to raise production is increasing yield per unit area. The increase in Asian grain harvests since the 1960s has been almost totally achieved in this way (Table 1), and it has been made possible by the use of high-yield varieties, chemical fertilizers, and expanded irrigation. Inappropriately using the technologies of this "green revolution" can have a variety of environmental impacts, and indeed one reason for the much smaller increase in yield per unit area in the 1990s is likely the declining effectiveness of the green revolution. Throughout the 1990s grain harvests increased at a considerably slower rate due to the combination of the slowed growth of yield per unit area and the smaller area of grain farmland being harvested. Judging by this situation, little optimism is warranted for Asia's future grain harvests.

Third is the internationalization of the material cycle through trade in agricultural products. As a reflection of the growing consumption of livestock prod-

TABLE 1.  Grain Harvested Area, per-ha Yield, and Total Harvest in Asia

| Year | Harvested area 10,000 ha | Per-ha yield tons | Total harvest 10,000 tons |
|---|---|---|---|
| 1961 | 27,185 | 0.97 | 26,335 |
| 1980 | 30,480 | 1.68 | 51,211 |
| 1990 | 31,061 | 2.30 | 71,359 |
| 2001 | 29,813 | 2.65 | 79,081 |
| Average annual increase rate (%) | | | |
| 1961–80 | 0.61 | 2.86 | 3.49 |
| 1981–90 | 0.18 | 2.54 | 2.73 |
| 1991–2001 | –0.26 | 1.30 | 1.04 |
| 1961–2001 | 0.25 | 2.69 | 2.95 |

Notes: 1. Rice is calculated by milled rice weight. The average annual rate of increase
was obtained by applying the regression equation $Yt = Yo (1 + X)t$ to the annual
data for each period and solving for X.
2. See note 2 of this chapter for the definition of Asia in this table.
Source:  Prepared using FAO, *FAOSTAT*.

ucts, Asian countries tend to import more grain and soy beans from the US and
Australia in recent years. Imports of corn by Southeast and East Asia (excluding
Japan and China) in the second half of the 1990s were about 30 million tons, and
those of soy beans about 5 million tons.[9] If agricultural products are consumed
where they are produced, wastes from humans and livestock are returned to the
soil as nutrients for more production, but movement of agricultural products
across borders stops this material cycle. Today's agricultural trade results in
massive one-way flows of nitrogen, phosphorus, and other nutrients from countries such as the US and Australia to Asia. Even if the loss of nutrients by exporting countries can be compensated with chemical fertilizers for the time being,
the nutrient inflow to importing countries will ultimately contaminate water
resources.

The foregoing observations on the environmental impacts of increased
demand for basic foods have provided a framework which is used below to
discuss agricultural production on marginal land and the growing use of modern
agricultural technologies.

# 3.   Environmental Impacts of Basic Food Production

## 3-1  Agricultural Production on Marginal Lands

Examples of agricultural production on marginal lands are inappropriate
swidden cultivation in tropical forests and farming on steep land in mountainous

regions. This section discusses production on cool arid land, using the example of Datong City in China's Shanxi Province, in the northeast part of the Loess Plateau.[10]

Datong has a long history and was already a large city of over 1 million from the fourth to the sixth centuries when it was the capital of Wei, but deforestation caused by the burgeoning population, creation of farmland, overgrazing, repeated wars, and other causes created the present severely eroded landscape of gullies. China's entire Loess Plateau is testament to how long years of human activity can degrade the environment.

Datong now has about 3 million people. It is China's largest coal-producing area, and quite industrialized, but most of the people make a living with self-sufficient agriculture and raising livestock. Cold winters limit farming to the summer, and the main crops are corn, sunflowers, kaoliang, and millet. Average annual rainfall is 400 mm, nearly all of which comes between June and September. Crops are planted before the rains begin in earnest, making the area susceptible to drought damage and leading to highly unstable harvests. The region's topography features many basins of various sizes, at the bottom of which farmers grow crops with groundwater irrigation. Farming in the hills depends mainly on precipitation, and topsoil loss is a concern in the terraced fields cut into steep hillsides. A ban on cultivating land with a slope of more than 25 degrees notwithstanding, some fields are nevertheless on slopes of up to 30 degrees. Farmers additionally practice small-scale livestock raising of mostly sheep and goats, but the difficulty of making a living on crops and livestock alone has in recent years induced many people to seek outside work.

One characteristic of Datong agriculture is its low productivity. It is evident from the 1993 agricultural production in Sunjiadian Borough, which is part of Datong City (Table 2), that although the drought was fairly serious, it was an ordinary year in consideration of the frequent droughts that quickly ensued. Villages A through C were in basins and could irrigate with groundwater, while G through K are high in the hills where irrigation is difficult, and D through F are in between. Generally the yields in villages A through C were over 300 kg per 10 ares because of irrigation, while G through K achieved only about 20 kg, a yield so low that farmers might not even be able to save seed for the next year. Long years of topsoil loss and insufficient water have combined to make agriculture in hill villages surprisingly unproductive.

A second characteristic is frequent drought. Over the 12 years between 1990 and 2001 only four years had no drought, and damage was especially severe in 1999 and 2001. In 1999 nothing was harvested on 57% of Datong's farmland, and the harvest was down 82% from the average year. It was even worse in 2001. While tropical monsoon rice farming is said to be stable even if production is low, field cropping in this arid region features both low productivity and high instability.

A third characteristic is the falling water table. Datong's water table was apparently about 5 m below ground level until industrial development began when the People's Republic of China was founded, but excessive groundwater use for

TABLE 2. Cultivated Land and Food Production in Sunjiadian Borough, Tianzhen County Near Datong City in Shanxi Province, China (1993)

| Village | Households | Population | Cultivated land | | | Food production | | |
|---|---|---|---|---|---|---|---|---|
| | | | Total (ha) | Irrigated area (ha) | Irrigation rate (%) | Planted area (ha) | Production (tons) | Yield per 10 ares (kg) |
| A | 423 | 1,749 | 367 | 367 | 100.0 | 267 | 855 | 320 |
| B | 545 | 2,186 | 396 | 361 | 91.2 | 293 | 820 | 280 |
| C | 122 | 460 | 57 | 57 | 100.0 | 47 | 243 | 517 |
| D | 101 | 369 | 126 | 58 | 46.0 | 100 | 108 | 108 |
| E | 255 | 953 | 260 | 58 | 22.3 | 227 | 127 | 56 |
| F | 202 | 729 | 195 | 195 | 100.0 | 153 | 137 | 90 |
| G | 146 | 512 | 173 | — | 0.0 | 157 | 37 | 24 |
| H | 43 | 58 | 77 | — | 0.0 | 63 | 11 | 17 |
| I | 40 | 140 | 69 | 7 | 10.1 | 61 | 11 | 18 |
| J | 146 | 624 | 240 | 7 | 2.9 | 213 | 41 | 19 |
| K | 63 | 257 | 71 | — | 0.0 | 60 | 10 | 17 |
| Total | 2,086 | 8,037 | 2,031 | 1,110 | 54.8 | 1,641 | 2,400 | 145 |

Note: In addition to grains, food production includes potatoes (5 kg counted as 1 kg).
Source: Green Earth Network, *Study on Clean Development Mechanism Projects to Address Global Warming: Report on Study of Potential for Afforestation on China's Loess Plateau*, 2001, p. 29, Table 1.

industry and the digging of many wells in rural areas for residential and agricultural use has caused the water table to drop 2 or 3 m annually. Rural wells currently range in depth from 100 to 180 m in between the lowest and highest parts of basins. Groundwater cannot be counted on to boost agricultural production in this drought-stricken region.

Further, environmental deterioration on the Loess Plateau, as seen in the case of Datong, is having a grave impact on other regions. Passing through the northeastern part of the Loess Plateau, where Datong is located, is the upstream part of the Yongding River, which flows through Beijing and Tianjin and into Bo Hai. Although the Yongding has been the water source for the North China Plain, China's breadbasket, and for cities along its course, in recent years it has been dry in the vicinity of Datong. Recently it is common for rivers in China's north to run dry, and over the summer of 2000 in Shanxi Province at least half the rivers ran dry and many dams were empty. This northern China water shortage has arrived at a critical juncture, as seen in Yellow River dry periods. While the causes of river dry periods are not always clear, there is little doubt of a connection with decreased aquifer recharge capacity due to deforestation and the excessive use of groundwater.

To address the worsening water shortage and topsoil loss, China's government has initiated a "return farmland to forests" policy that involves planting trees and grass on land unsuited to farming. China's large number of poor people will make implementation of this policy very difficult, but afforestation efforts in China mostly by NGOs are beginning to make sizable contributions, such as by developing ways to raise seedlings suited to local natural conditions.

## 3-2    Environmental Impacts of Modern Farming Technologies

Increases in Asian grain production over the past three decades have been realized by increasing yield per unit area with high-yield varieties. The International Rice Research Institute estimates that currently 74% of Asia's rice land uses such varieties.[11]

But high-yield cultivars alone aren't enough. Bigger harvests also require chemical fertilizers and irrigation facilities, and sometimes pesticides. From 1961 to 1995 the amount of chemical fertilizer used in Asia per hectare grew from 9.3 to 147.6 kg, showing that the introduction of high-yield cultivars helped steadily widen the use of modern agricultural technologies.[12] The following section spotlights these technologies, especially irrigation and pesticides, and their environmental impacts.

### Irrigation

Asia's irrigated farmland area has mushroomed since the end of WWII, currently standing at about 170 million ha, most of which uses surface water.

Surface water irrigation in Asia's humid zone has long employed flow irrigation and irrigation ponds. Of concern now because of environmental impacts is not this traditional method, but rather large-scale channel irrigation because of the soil salinization it causes.

Post-WWII increases in irrigated land area were primarily due to the development of large-scale channel irrigation in arid areas, which are prone to soil salinization because irrigation water itself has high salt concentrations, and because water rapidly evaporates.[13] Salinization considerably reduces agricultural yields, which was shown by a study conducted in India's Uttar Pradesh state in which salinized soil produced 41 to 56% lower yields per unit area of wheat and rice than good soil.[14] Higher salt concentrations effectively made agriculture impossible.

Irrigation-caused salinization is very widespread. An example is three large dams and a network of 43 channels (total primary canal length of 56,000 km) built in Pakistan. In the late 1960s this was one of the world's largest irrigation systems and served an area of 14 million ha, but now about 30% of that area is damaged by salinization.[15] In China and India the amount of salinized farmland is, respectively, 11 million ha (23.2% of total irrigated farmland) and 4.7 million ha (11%).[16] As irrigated farmland is far more productive than unirrigated land, loss of the former has a sizeable impact on agricultural production.

Maintaining food production requires the recovery of salinized farmland and preventing further salinization, which in turn necessitate good drainage. Yet, drainage has not been considered important in past irrigation projects. For example, from 1983 to 1991 the World Bank invested a total of $6.17 billion in irrigation projects in Asia, but only 7.4% of that investment was drainage-related.[17] One likely reason is that while water-supply facilities quickly achieve the readily perceived results of expanded farmland and increased yields, the effect of drainage facilities is not immediately apparent. Reviews of large-scale irrigation projects should include this consideration.

Dependence on groundwater irrigation is high in Asian countries including Pakistan, India, China, and North Korea[18] owing to the difficulty of securing surface water supplies. Although salinization is again a problem, a greater concern is falling water tables caused by excessive withdrawals.

In the case of India, about 38% of all farmland was irrigated as of 1995, with surface water and groundwater irrigation each covering about half.[19] Tube wells account for about 60% of groundwater irrigation,[20] and increased mainly in the northwestern states beginning in the second half of the 1960s. One reason for the growth of tube wells is that the successful introduction of high-yield varieties in this region made it into one of India's breadbaskets, serving especially to supply the cities with grain. Another reason is that tube wells are easy to create in this alluvial area.[21] Additionally, electric power rates for pumps are set very low as part of a strategy for agricultural development.

Yet the sustainability of groundwater irrigation is in doubt. A report by India's Ministry of Water Resources states that the proportion of blocks with higher groundwater withdrawals than recharge rates is very high in the northwest region.[22] In the 1990s 40 to 50% of the blocks in the northwestern states of Punjab and Haryana were net withdrawers. In Punjab over the 15 years from 1979 to 1994 the water table fell at least 5 m on 29% of the state's total land area, and especially in the state's central region the percentage was 60 to 70%.[23] Clearly, the spread of tube well irrigation is causing excessive groundwater withdrawals. Of course this is not generally true of India as a whole, as nationally only 4% of blocks are running a deficit.[24] Nevertheless, the fact that many deficit blocks are concentrated in grain-producing areas casts a dark shadow on India's future grain production, and poses a major turning point for the country's policy of increasing food production because of the heavy dependence on groundwater irritation.

## Environmental Impacts of Pesticides

Environments made suitable for crop cultivation through the use of chemical fertilizers and irrigation are also susceptible to insect pests and weeds. Formerly, developing Asian countries coped with agricultural pests by developing and introducing resistant cultivars, but recently increasing use is made of insecticides and fungicides. Similarly, weeding is increasingly reliant on herbicides. Still, the use of pesticides in most Asian countries is likely far below that in Japan. Taking India as an example, in 1998 insecticides were used on 47% of total planted area, and herbicides on 22%.[25]

But more serious problems in Asia than amount are that farmers use acutely toxic and highly persistent pesticides, and that insufficient safety precautions are observed during application. A study conducted in Bangladeshi farming villages in the late 1980s found that 80% of the pesticides used were chemicals that in Japan are designated highly toxic substances. Despite the use of such acutely toxic pesticides, hardly any farmers wore protective gear when applying them: 1% of farmers wore either gloves or goggles, and none wore raincoats. As a consequence, 87% of the farmers complained of toxic symptoms such as eye pain, itchy skin, or nausea.[26]

Environmental contamination by residual pesticides is another hazard. DDT, BHC, and other organochlorine insecticides are suspected of carcinogenicity and of diminishing reproductive ability, but they are also chemically stable and remain in the environment for a long time, and accumulate in fat when taken into the body. This led to their prohibition in developed countries beginning in the early 1970s.

Many Asian developing countries ban the use of such organochlorine insecticides in farming, but their use is often allowed against insects for purposes like fighting malaria vectors and other disease-carrying insects, and sometimes agents for use against such insects are diverted to agricultural use because of inadequate control. Such insecticides are actually used quite widely.

Owing to this situation people in developing countries have far higher concentrations of organochlorine insecticides in their bodies than do residents of developed countries. This problem is particularly grave in India, where the adult daily intake of DDT is 19.2 mg.[27] A comparison with Japan's 0.25-mg tolerable daily intake indicates the situation's seriousness.

But pesticide pollution is not limited to developing countries. With the growing quantities and types of agricultural produce being traded internationally, it is no overstatement to say that pesticide use in developing countries directly affects dinner tables in developed countries. A further impact is the long-distance trans-boundary movement of hazardous chemicals such as DDT and other organochlo-rine insecticides, which spread over the world on air currents and through the water cycle, and ultimately affect everyone.

Four major factors underlie the growing seriousness of pesticides in develop-ing countries.

First, farmers obtain too little information on the dangers of pesticides because of insufficient efforts by governments to provide guidance on proper use, and because of low educational levels, especially literacy, among farmers. As a con-sequence, farmers obtain knowledge about pesticide selection and use mainly from retailers, providing little expectation of safe and proper use.

Second, the use of organochlorine insecticides, which are meant for disease vector control, hampers a solution to pesticide problems in developing countries. While the eradication of diseases like malaria is still an urgent task, DDT and other antiquated chemicals are still being used. A solution will require the quick development of effective and inexpensive new insecticides to replace DDT and other organochlorines, and stronger controls to prevent the diversion of vector-control insecticides to agricultural use.

Third is the lack of pesticide management systems. Some Asian countries do not even have pesticide registration systems, and some systems are inadequate.[28] For example, India still allows the use, with some limitations, of dieldrin, lindane, and methyl parathion, which were banned in Japan in the 1970s.[29] In many coun-tries there is rampant smuggling of unregistered pesticides due to nonexistent or deficient controls.[30]

Fourth is exports of hazardous pesticides to the developing world by industri-alized countries, which take advantage of developing countries' inadequate pes-ticide regulations. An analysis of US customs statistics revealed that, of 1,453,000 tons of pesticides exported from US ports between 1997 and 2000, at least 30,000 tons were chemicals that are either banned or tightly restricted inside the US, and their main destinations were developing countries in Asia and Central and South America.[31] Other pesticides exported by the US include those which, although not banned domestically, are given the WHO rating 1A (extremely hazardous), those that are suspected carcinogens, and those considered to be endocrine disruptors.

An indication of how pesticides are an increasingly international issue is the signing of two international conventions: the 1998 Rotterdam Convention on Prior Informed Consent, and the 2001 Stockholm Convention on Persistent Organic Pollutants. The former stipulates that importers must grant consent before importing hazardous substances so as to prevent the unrestricted export to developing countries of pesticides and other chemicals that are banned or strictly controlled in developed countries, while the latter provides for the banning in principle of the manufacture and use of 12 persistent organic pollutants including organochlorine pesticides such as DDT and drin-family insecticides, and restrictions on their imports and exports.

There are considerable expectations that these two conventions will help stem the flow of hazardous pesticides, and indeed exports of such substances from the US have decreased since the conventions were signed. Yet, there has been no sign that trade in hazardous substances that are not subject to these conventions has abated. Addressing this situation will necessitate pushing for stricter regulations on hazardous pesticides and expanding these two conventions on the basis of the precautionary principle.

# 4.    Agribusiness and the Environment

## 4-1    Rapid Growth of Food Exports in Asia

The trend toward stepping up production of commodities for export is also in the spotlight for its connection with agriculture and food issues. From 1980 to 1999 food exports by Asian developing countries increased by a factor of 2.6 (Table 3). For the world as a whole the growth rate was 1.9 times during the same period, and it was 2.0 times for developing countries as a whole, which illustrates the substantial expansion of Asian food exports. Checking the destinations of these exports reveals the swift increase in Japan's imports, which eclipsed those of the EU in 1990 and now have a 22% share. Further, the portion of Japan's food imports from other Asian countries shot up from 24% to 35%, testimony to Japan's rising dependence on Asia for food supplies. In addition to increased Asian exports to the US in the 1990s, the growth of exports within Asia was also conspicuous.

Behind these globally expanding Asian food exports, aimed primarily at Japan, are the export promotion policies of Asian governments. Food exports have played an important role in earning the foreign currency needed for industrialization and economic development strategies, but especially in recent years these economies are working hard at exporting high-value foods such as livestock products, marine products, oil crops, and vegetables, in addition to traditional primary products including grains and tropical products. Similar to the label "newly industrializing countries" (NICs), representative food-exporting nations Thailand and China are sometimes called "new agricultural countries," or NACs.[32]

One other factor behind Asian food exports is the role of agribusiness, which holds the key to transformation into NACs. Agribusiness comprises groups of companies including farming supplies, food processing and distribution, and the food service industry. As such, it now exercises powerful influence on agriculture and food production in its capacity as the entity that forms an "agro-food system" which covers everything from production to consumption. Especially in Asia, a cross-border business network has formed in the process of direct

TABLE 3. Trends in Food Exports by Asian Developing Countries

| | Year | World | Developed market economy countries | | | | Developing countrie and economies | | Countries in Former USSR and Eastern Europe |
|---|---|---|---|---|---|---|---|---|---|
| | | | Total | EU | US | Japan | Total | Asia | |
| Exports ($1 million) | 1980 | 22,790 | 10,098 | 4,080 | 1,928 | 3,496 | 10,627 | 7,335 | 1,879 |
| | 1990 | 38,612 | 19,677 | 5,531 | 3,579 | 9,438 | 16,527 | 12,783 | 2,363 |
| | 1999 | 59,385 | 28,556 | 7,273 | 5,843 | 13,142 | 28,851 | 22,676 | 1,841 |
| Index (1980 = 100) | 1980 | 100 | 100 | 100 | 100 | 100 | 100 | 100 | 100 |
| | 1990 | 169 | 195 | 136 | 186 | 270 | 156 | 174 | 126 |
| | 1999 | 261 | 283 | 178 | 303 | 376 | 271 | 309 | 98 |
| Percentage of whole | 1980 | 100.0 | 44.3 | 17.9 | 8.5 | 15.3 | 46.6 | 32.2 | 8.2 |
| | 1990 | 100.0 | 51.0 | 14.3 | 9.3 | 24.4 | 42.8 | 33.1 | 6.1 |
| | 1999 | 100.0 | 48.1 | 12.2 | 9.8 | 22.1 | 48.6 | 38.2 | 3.1 |

Note: "Asia" includes South Korea but excludes Japan and countries of West and Central Asia (1995 and thereafter).
Source: Prepared using UNCTAD, Handbook of Statistics 2001, 2002.

investment by Japanese companies and the growth of local Asian capital, while at the same time the industrialization of agriculture, monoculture cropping, and the attendant environmental problems have created a new situation in production venues.

## 4-2    Shrimp and Eel Farming: The "Blue Revolution" and the Cycle of Ill-Conceived Development

Examining exported products can reveal the social and environmental impacts of agribusiness, and for that purpose the following sections spotlight certain exports.

Marine products are typical of Asian food exports,[33] and especially important are shrimp and eel. Shrimp currently has the second-highest aquaculture production in terms of value. Since the 1980s the ratio of aquaculture to shrimp capture has quickly risen to make up for the catch decline caused by trawling overexploitation, and as of 1999 aquaculture supplied 28%. Although eel production value does not approach that of shrimp, its aquaculture rate is very high at 95%. Especially in China eel production for export is quickly increasing, and is the biggest portion of food exports to Japan.[34]

Asia produces 80% of the world's farmed shrimp. Black tiger (*Penaeus monodon*) is the leading product and constitutes the majority of total farmed shrimp. Historically, production began in Taiwan and spread to Thailand, Indonesia, and the Philippines, which were the main producers, but lately production in these countries has stagnated, while South Asia and Vietnam are the rising stars. Also in Asia, China is the main producer of the Oriental shrimp (*Penaeus chinensis*), while recent years have seen a surge in the production of the whiteleg shrimp (*Penaeus vannamei*), whose main producers are Central and South American countries (Fig. 2). Meanwhile, Asia produces at least 90% of the world's farmed eel, mainly the Japanese eel (*Anguilla japonica*). The main producing area has shifted from Japan to Taiwan, and further to China (Fig. 3).

A characteristic of these producers and primary exporters is that instead of exporting only fresh shrimp and eel, they also add value through processing. For example, Thailand is a major shrimp processor using not only domestically produced shrimp but also those imported from nearby countries, and taking advantage of its cheap labor to perform shelling and other processing by hand, exporting frozen deep fried shrimp, sushi shrimp, and other shrimp products. Workers at Chinese eel processing plants skin the eels by hand and skewer them for broiling on assembly lines.

Both the shrimp and eel industries have grown large thanks to mass production through aquaculture, which has been made possible by the intensive aquaculture technologies that wrought the "blue revolution." Unlike extensive techniques, intensive aquaculture achieves high yields through high density and large quantities of feed and chemicals, and it has enjoyed the support of agribusiness investment, government support, and help with infrastructure from ODA and international aid agencies such as the World Bank. As a result of this support,

FIG. 2. Main Shrimp Farming Countries and Regions in Asia

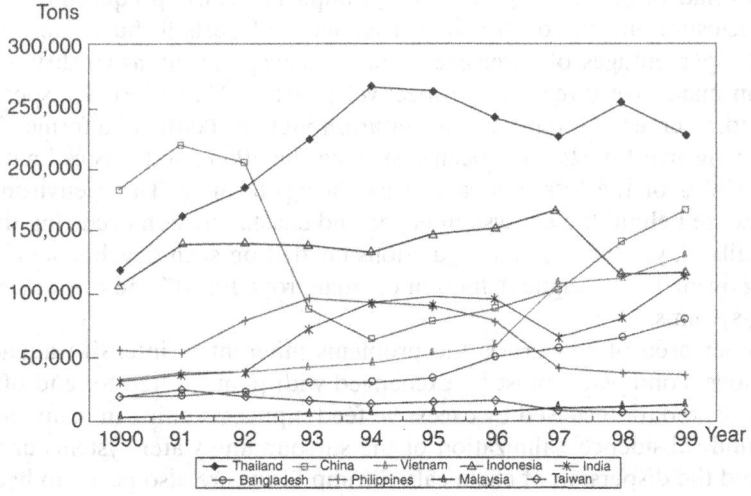

Note: Total of farmed shrimp, including black tiger and Korai prawns.
Source: FAO, *Yearbook of Fishery Statistics: Aquaculture Production 1999*, 2001.

FIG. 3. Main Eel Farming
Countries and Regions in Asia

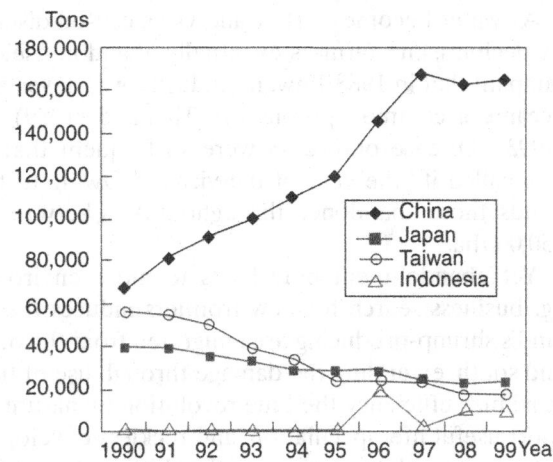

Note: These countries/regions
        are ranked the top four
        worldwide in aquaculture
        production.
Source: Same as Fig. 2.

at least 90% of Thai shrimp farming is intensive or semi-intensive, while in Malaysia and the Philippines it has attained 70%. Japanese and Taiwanese capital has financed "develop and import" schemes while harmonizing those efforts with the increasing adoption of intensive aquaculture. Meanwhile, local capital like Thailand's CP Group is expanding internationally by taking control of feed supply and processing.[35]

Yet the expansion and international commercialization of intensive aquaculture has had devastating environmental impacts.[36] Shrimp aquaculture entails land enclosure and the destruction of mangrove forests to build the ponds. In Asia the percentages of extensive, semi-intensive, and intensive shrimp ponds made in mangrove forests are, respectively, 43%, 45%, and 31%. Especially in Thailand, a center of intensive aquaculture, about 200,000 ha of a former 380,000 ha of mangrove forests have been lost since the 1960s, and it is estimated that at least 30% of the loss was caused by shrimp farming. These environmental changes are behind the ecosystem havoc and coastal erosion occurring throughout Thailand, yet the tougher regulations needed on such development are still lacking owing to the vague definition of "mangrove forest" and the lack of monitoring systems.

Another area of concern is the problems inherent in intensive aquaculture operations. Pond water must be exchanged with pumps often to fend off water quality degradation caused by excessive feed inputs, resulting in many instances of ground subsidence, salinization of the surrounding water systems and farmland, and the dispersion of chemicals. Shrimp farms are also prone to heavy use of antibiotics to prevent diseases, raising safety issues often as seen in the 2002 import ban and enhanced monitoring by the EU.[37] Although established procedures call for one-month solarization of ponds after shipping, farmers want to earn high revenues in a short time by increasing the number of rotations, and therefore in actuality often omit this solarization period.

As water becomes dirtier and virus-caused diseases take their toll, productivity declines, and farmers eventually abandon useless ponds. Disease became so rampant that in 1988 Taiwan's industry was devastated, and disease subsequently became a common problem in Thailand (1989), Indonesia (1992), and China (1993). Disease outbreaks were so frequent that in 1995 an industry publication called it "the year of the virus."[38] Owing to the short lifetime of intensive ponds, those abandoned throughout Asia between 1985 and 1995 totaled about 150,000 ha.

Yet abandonment only leads to more environmental damage because the agribusiness search for new frontiers induces the creation of new ponds. Thailand's shrimp-producing area migrated from the outskirts of Bangkok to the east and south, extending the damage through use of the same methods. By trying to maximize efficiency the blue revolution by nature creates a recurring pattern of short useful life, speculation, and reckless development.[39]

Some eel-related issues are the shift of production from Taiwan to China due in part to subsidence resulting from excessive groundwater withdrawals, and the clamor that arose in 2002 over the detection of synthetic antibacterial agents in Chinese-produced eel. But eel aquaculture has other problems peculiar to the industry. Raising eels starts with glass eels, but it is not yet possible to raise them artificially by reason of incomplete knowledge about their ecology. The dependence on natural glass eels and the attempts to boost eel production have in recent years induced a glass eel shortage and higher eel prices. Although the

Japanese eel was the main species raised in Asia, in 1993 China became a glass eel importer and began importing European eel elver (*Anguilla anguilla*) from Europe for raising. Glass eel demand raises concerns that stocks could become scarce worldwide.[40]

Governments are acting to regulate eel and shrimp farming. In 1995 India's Supreme Court banned the conversion of farmland in the eastern coastal states, while disputes and lawsuits have arisen between aquaculture farmers and local communities in South Sumatra, Indonesia and in Malaysia's Kedah state. In Thailand the government banned aquaculture in mangrove areas and set to work on an action plan with citizen participation, while even the industry has formulated a code of conduct to achieve sustainable aquaculture. Even the World Bank has initiated a coastal resource management project for Thailand, and is calling for mangrove forest protection. But a new aquaculture problem has arisen in Thailand: while mangrove forests might be spared further depredation, inland rice paddies are being converted to aquaculture ponds. In 1998 a ban was placed on aquaculture in eight central provinces due to concerns about salinization and deleterious effects on crops. A review of the ban started in 2001 in response to demands from the industry, but increasing pressure from farmers and NGOs led to a decision that October to keep the ban in place.[41]

## 4-3  Palm Oil: The Global Palm Connection and Another Asian Crisis

Oil palms are an oil crop ranking in importance with soy beans, and the palm oil extracted from fresh fruit bunches has experienced some of the biggest growth of agricultural exports from the developing countries. Palm oil growth is centered in Asia, mostly in Malaysia and Indonesia (Table 4).

Both countries started planting oil palms in the 1960s in newly created plantations and by replacing rubber trees, and they have surpassed Africa, the oil palm's place of origin, as the chief producing area. From 1970 through the end of the 1990s Malaysia's oil palm area enlarged from 290,000 ha to over 3.3 million ha, while Indonesia's expanded from 130,000 ha to over 2.9 million ha, and their production rocketed to 125 fold and 55 fold, respectively. Together, they account for at least 80% of world production and exports. India, China and other populous countries are importing more palm oil for the manufacture of cooking oil, while demand by industrialized countries is also rising to satisfy industrial uses such as cleaning agents because palm oil is said to be environmentally friendly. A global palm oil distribution network is growing with Southeast Asia at the hub.[42]

Requirements for palm oil production are that a plantation must have an adjoining plant to press bunches within one day after harvest to assure oil quality, and a plant must have a plantation of a certain minimum size in consideration of capacity utilization. In both countries these requirements have resulted in large-company oligopolies based primarily on plantations.[43] In the second half of

TABLE 4. Rankings of Top Five Producers and Exporters of Palm Oil

1,000 tons; figures in parentheses are percentages

| | 1961 | | 2001 | |
|---|---|---|---|---|
| Production | 1. Nigeria | 669.0 (45.2) | 1. Malaysia | 11,880.0 (49.1) |
| | 2. Democratic Republic of Congo | 224.0 (15.1) | 2. Indonesia | 7,950.0 (32.9) |
| | 3. Indonesia | 145.7 (9.9) | 3. Nigeria | 903.0 (3.7) |
| | 4. Malaysia | 94.8 (6.4) | 4. Thailand | 550.0 (2.3) |
| | 5. China | 40.0 (2.7) | 5. Colombia | 547.6 (2.3) |
| | World Total | 1,478.9 (100.0) | World Total | 24,172.8 (100.0) |
| Exports | 1. Nigeria | 167.2 (26.6) | 1. Malaysia | 10,002.5 (58.7) |
| | 2. Democratic Republic of Congo | 154.4 (24.5) | 2. Indonesia | 4,903.2 (28.8) |
| | 3. Indonesia | 117.3 (18.6) | 3. Netherlands | 379.1 (2.2) |
| | 4. Malaysia | 94.9 (15.1) | 4. Papua New Guinea | 327.6 (1.9) |
| | 5. Singapore | 29.5 (4.7) | 5. Thailand | 180.1 (1.1) |
| | World Total | 629.2 (100.0) | World Total | 17,026.8 (100.0) |
| Imports | 1. U.K. | 163.4 (25.3) | 1. India | 2,733.1 (18.0) |
| | 2. Netherlands | 89.0 (13.8) | 2. China | 1,606.3 (10.6) |
| | 3. Germany | 77.7 (12.0) | 3. Pakistan | 1,064.4 (7.0) |
| | 4. Belgium/Luxemburg | 42.5 (6.6) | 4. Netherlands | 989.6 (6.5) |
| | 5. India | 33.7 (5.2) | 5. U.K. | 619.5 (4.1) |
| | World Total | 645.6 (100.0) | World Total | 15,164.9 (100.0) |

Source:  Prepared using FAO, *FAOSTAT.*

the 1990s Malaysia's 10 largest companies had 45% of that country's total plantation area, while 16 companies accounted for 75% of palm oil refining by volume. Similarly in Indonesia, 10 companies own 30% of the plantations and five companies refine 60% of the oil. Each company tries to integrate everything from plantations to milling and refining, and recently they are forming joint ventures with multinational chemical companies that wish to secure supplies of industrial feedstocks, thereby expanding into "downstream" sectors.[44]

Palm oil-related environmental impacts must be seen in the context of this large-scale development mainly by companies. One significant impact is pesticide use on plantations. A study by a Malaysian NGO on 72 female workers reported that application of paraquat and other pesticides with portable pesticide sprayers and the effects of "pesticide cocktails" containing a variety of chemicals were responsible for toxic symptoms including fatigue, nausea, back pain, dizziness, breathing difficulty, skin diseases, headaches, and menstrual irregularity. Six of the women were hospitalized. Tasks are assigned by gender on plantations, with men performing harvesting, transport, and other heavy labor, and women applying pesticides, which tends to concentrate pesticide damage among women. Workers continue suffering harm from pesticides because they lack sufficient knowledge of the dangers, use inadequate protective gear, and rely on inferior first aid.[45]

Another impact is deforestation by plantation expansion. Oil palm plantations are vast monocultures that over the long term bring about an irreversible decline in teeming tropical forest ecosystems and biodiversity, but they also cause serious damage when being developed, as typified by the forest fires and haze in 1997 and 1998. This disaster covered Southeast Asia with smoke, visited calamity on 9.7 million ha and 75 million people in Indonesia, and caused damage estimated at between $4.5 billion and $10 billion. Its main causes were fires set to open land for plantations and to log for pulp and timber. Substantiation of the oil palm connection is the fact that 133 of the 176 companies charged with the fires are oil palm-related businesses.[46]

Further, 43 of those 133 companies were Malaysian, indicating that Malaysian capital had pushed into Indonesia to develop new plantations because they were running out of areas to develop in Peninsular Malaysia. Malaysian companies are also investing in Mindanao, Papua New Guinea, and other places. International investment for plantation development is having a substantial impact on forest resources.[47] Indonesia currently has 3 million ha of oil palms, but as plantation companies have been allotted 8.7 million ha, it is anticipated that development will be further expedited in combination with an export encouragement policy under the IMF regime, weakening central government control due to the introduction of local autonomy, and fiscal cuts.[48]

## 4-4 Fresh and Frozen Vegetables: The Vegetable War, "Poison Vegetables," and Food Insecurity

Vegetables have been a source of friction leading to a "vegetable war" among Japan, China, and South Korea. In 2000 Japan imported over 2.6 million tons of

TABLE 5. Vegetable Imports by Japan

| | | Imports | | Percentages of total | | Indexes (1990 = 100) |
|---|---|---|---|---|---|---|
| | | 1990 | 2000 | 1990 | 2000 | |
| | Total | 1,091.8 | 2,605.2 | 100.0 | 100.0 | 239 |
| Imports by | Fresh vegetables | 261.0 | 971.1 | 23.9 | 37.3 | 372 |
| category | Frozen vegetables | 345.1 | 772.8 | 31.6 | 29.7 | 224 |
| (1,000 tons) | Salted vegetables | 197.2 | 220.2 | 18.1 | 8.5 | 112 |
| | Total | 214.5 | 342.3 | 100.0 | 100.0 | 160 |
| Exports by | China | 46.1 | 150.5 | 21.5 | 44.0 | 326 |
| country | US | 51.6 | 83.1 | 24.1 | 24.3 | 161 |
| (1 billion | S. Korea | 12.8 | 20.5 | 5.9 | 6.0 | 160 |
| yen) | New Zealand | 12.0 | 15.1 | 5.6 | 4.4 | 126 |
| | Thailand | 9.3 | 13.0 | 4.3 | 3.8 | 140 |
| | Taiwan | 42.5 | 9.2 | 19.8 | 2.7 | 22 |

Source: Vegetable Supply Stabilization Fund, ed., *Trends in Vegetable Imports*, the Association of Agriculture-Forestry Statistics, 2000, 2001.

vegetables with a value of 340 billion yen from 75 countries (Table 5). Over the last decade the increase has been a considerable 2.4 times in quantity and 1.6 times in value. Especially big has been the 3.7-fold increase in fresh vegetables (37% of total imports), but frozen vegetables (30%) also increased 2.2 times. Owing to its fast growth in exports to Japan, China supplanted the US as the chief exporter, now claiming about half of Japan's total imports by value. Among Asian exporters, South Korea has emerged in third place and Thailand had a slight increase, while a substantial decrease brought Taiwan down from the third-ranked place it had enjoyed in 1990.

Soaring vegetable imports have had a heavy impact on vegetable farming and distribution in Japan. Especially Chinese vegetables, in conjunction with the ageing of Japanese producers, are a force behind the substantial decrease in Japanese farmers and the downscaling and reorganization of producing areas because Chinese vegetables have low unit costs and flow steadily into Japan regardless of bad harvests and off-crop seasons. Japan's vegetable self-sufficiency rate dropped from 95% in 1985 to 82% in 2000, and in 2001 the government invoked safeguards on Welsh onions, fresh shiitake, and tatami mat covers. Vegetable-related disputes have arisen between other East Asian countries as well.

It is important to see the role of agribusiness in building the vegetable trade routes involved in the increasingly stormy international trade disputes in East Asia. In fresh vegetable production, Japanese trading companies or volume retailers take Japanese seeds to producing areas and direct cultivation of vegetables which they then import. In the case of processed vegetables the restaurant industry will require that vegetables conform to their own specifications, while frozen food companies import through overseas subsidiaries and joint ventures. One large restaurant chain gets nearly all its garlic and shiitake from China, and it gets a year-round supply of Welsh onions through a Chinese producer from four provinces with different harvest times. Develop-and-import arrangements are behind the creation of a producing area for exports to Japan that is centered in Shandong Province. In particular, direct investment by Japanese and Taiwanese companies since the 1980s has spawned frozen vegetable factories in Shandong, concentrating processing companies into the province's coastal region and earning it the name "frozen belt."[49]

But new problems have emerged from this rapid development of areas producing for export, and one is soil degradation from continuous cropping. For example, intense competition between ginger-producing districts in Shandong and Henan provinces resulted in a 20% decrease in ginger cultivation area in Shandong's Laiwu City. Similarly, Jinxiang County began switching to onion production because it lost 15 to 20% of its garlic fields due to continuous cropping.[50]

Second is pesticides. Increasing attention is directed at the residual pesticides on vegetables from China in recent years for reasons such as the detection in 2002 of pesticides over the standard on frozen spinach bearing the JAS organic seal. In 1999 there were 11 cases in which Chinese vegetable imports were banned, rising to 87 cases in 2000 and 159 cases in 2001.[51] Practices reported from producing areas include the use of pesticide concentrate in undiluted form, the long-term use of DDT and other pesticides that degrade with difficulty, and pesticide application just before harvesting to improve appearance. Chinese local governments with exporting areas have at last begun controlling pesticide use by issuing regulations, which suggests that highly hazardous pesticides have been used in large quantities.[52] Post-harvest chemical applications are also implicated.

Third is the future of the food supply. More and more farmers in China's vegetable-producing areas are specializing in vegetables and buying basic foodstuffs instead of growing them. But due to the reaction from the "vegetable boom" and the effects of safeguards, farming operations and the production environment are heavily influenced by changes in international market conditions. For instance, in China the use of burdock root was originally limited to herbal remedies, but cultivation gained momentum in 1993 when seeds were imported from Japan. Yet only two years later an oversupply in the Japanese market resulted in a precipitous drop to one-third the cultivated area.[53] What is more, some vegetables are rejected because of strict Japanese requirements on size and other characteristics, especially in the case of Japanese cultivars grown in China. Rejects end up being discarded.[54] Having joined the World Trade Organization, China intends to encourage high-profit vegetable production, but this raises concerns about the profligate resource use that is a concomitant of export-oriented agriculture, and shrinkage of the basic food production base.

# 5.    Sustainable Agriculture in Asia

## New Trends in Agricultural Production Methods

Population growth and dietary changes in Asia are increasing food demand, and as symbolized by the Loess Plateau, there is a sequence in which population augments pressure on agriculture, which in turn affects the environment. Behind this sequence is poverty that drives people to use farmland and water resources beyond their limits, and the instability of productivity. To this are added new problems: the soil salinization, falling water tables, and pesticide contamination that are the consequences of modern agricultural technologies typical of the green revolution, and the appropriation and overexploitation of local resources by agribusiness against the backdrop of export-oriented agriculture. In short, this is the contemporary picture of Asia's environmental problems. Henceforth agriculture will be buffeted by a variety of forces including decreases in farmland and water resources by industrialization and urbanization, free trade negotiations in new rounds of WTO talks, more investment in Asia by Western agribusiness multinationals, and the commercialization of organic foods. As Asian agriculture and food production are exposed to further impacts from globalization and non-agricultural factors such as changes in the industrial structure, some new forces will come into play.

One is the development of biotechnology, and genetically modified organisms (GMOs) are especially in the spotlight. The biotechnology industry and international agencies believe that GMO crops are a powerful tool for saving people from poverty and famine, and for overcoming environmental constraints, and they are focusing on Asia as a potential region for GMO use. Bt cotton is being commercially grown in China, and experimental cultivation of other crops is already underway in a number of locations.[55]

But "technological breakthroughs" based on biotechnology involve a direct connection to market expansion by agribusiness, which is organized around the seed and pesticide companies that supply many small farmers, or can raise social concerns over ecosystem disruption or food safety through the release of GMOs into the environment. In fact, on Mindanao in the Philippines, 800 local farmers attacked a field in August 2001 to protest an experiment with Monsanto Bt corn, and in Indonesia there was a lawsuit and protest seeking a ban on planting Bt cotton in South Sulawesi by a Monsanto subsidiary. GMOs have become an issue, as seen in the creation of standards and the legislation of labeling requirements to address social concerns in Asian countries (see III-5).

Meanwhile, there are efforts to develop environmentally sound agriculture differing from modern agricultural technologies and how it has developed. Gaining prominence of late is integrated pest management (IPM), a field management method that minimizes pesticide amounts used (see essay). Behind the development and deployment of this practice are the health damage to farmers and laborers from excessive pesticide use, and worsening insect pest damage caused by nonselective pesticide use. To avoid the soil degradation due to commercial monocultures, and to get off the treadmill of high costs incurred in buying farming supplies, farmers are combining different crops as in agroforestry and wet-rice fish farming, and they are reassessing other types of farming such as low-input and organic farming.[56] Still another initiative is the fair trade movement (see I-2), which sets itself apart from agribusiness-led trade.[57]

## Challenges Facing the Development of Sustainable Agriculture

There are several challenges that must be overcome to achieve sustainable agriculture in Asia.

First, Asian countries must impose restrictions on environmentally damaging activities, review the ways they go about development, and create ways to support the development and deployment of eco-friendly technologies. For example, pesticide damage must be addressed by quickly teaching farmers how to properly use the chemicals, by carrying out comprehensive registration and management, and by banning or regulating hazardous pesticides. To address the problem of unsuitable irrigation schemes, it is imperative that small-scale environmentally compatible projects be implemented in place of standardized large-scale projects that ignore local environmental conditions. Promoting the development of IPM and other agroecological technologies is also essential. Prerequisites for these initiatives are support for the transition to eco-friendly farming, and the reevaluation of public research agencies, which should take the lead in developing basic technologies because these are not considered important in commercial development by businesses.

Second is enhanced monitoring of agribusiness, which is the primary entity behind the internationalization of agro-food issues, and international growth management for agriculture- and food-related trade and investment. As seen in section 4, the agribusiness-led development of export-oriented agriculture, and the internationalization of trade and investment have expanded rapidly, but the flip side is transborder environmental damage. As in the conventions on prior informed consent and persistent organic pollutants (see also I-2) and the resource management project for shrimp farming described in sections 3 and 4, there is a growing necessity to create an Asian system—with the same level of regulatory measures as those in other parts of the world—that can formulate business codes of conduct and environmental conventions in order to internationally control the chaotic development of agribusiness, and that can use capital investment returns to benefit local environmental conservation. An international framework like this and action based on it would make it possible to steer the growth of trade and investment in a sustainable direction.

Asian governments must also reevaluate their agricultural policy in connection with food imports. Some countries have become dependent on imports for basic foods because of their policy emphasis on industrialization or production for export, but since the Asian economic crisis some Southeast Asian countries have a renewed awareness about the importance of food security. Under the WTO system, domestic policies cannot be adequately implemented due to limitations imposed from above on protecting domestic agriculture, but the sustainable development of agriculture and food production is indispensable to attain food sovereignty.

Third is bringing together the actors who will achieve sustainable agriculture and food production. Fixing the current agro-food system, which is the cause of environmental damage and food uncertainty, requires that governments switch to eco-friendly policies that protect agriculture, receive the support of international agencies, and regulate and monitor agribusiness. But such policies will become reality only through collaboration among NGOs, farmers' organizations, labor unions, cooperatives, and other entities as they raise questions and exercise their influence toward creating that policy. Consumers have to rethink their lifestyles and how excessive food consumption and imports affect the environment. Producers must take advantage of both modern environmental science and traditional local knowledge while working toward eco-friendly farming and local resource management. It would be the first step toward achieving the development of sustainable agriculture in which both farmers and consumers take the initiative in cooperating globally and locally.

Sugimoto Daizo, Iwasa Kazuyuki

## Essay:  Integrated Pest Management in Indonesia

One way of preventing health and environmental damage by pesticides is to use them as little as possible and, when there is no choice but to use a pesticide, select the safest one. This is part of a practice that has come to be known as integrated pest management (IPM). Indonesia is ahead of other Asian countries in this area.

Rice production in Indonesia, where the green revolution has been in progress since the 1960s, has grown dramatically high-yield cultivars, chemical fertilizers, and pesticides, but due to heavy use of nonselective insecticides that killed even the insect pests' natural enemies, brown planthopper damage gradually became more frequent. In 1986 Indonesia's government responded by banning the use in rice paddies of 57 nonselective insecticides that kill natural enemies, and in 1989 it totally eliminated pesticide subsidies. In place of pesticides, IPM was given the central role in eliminating rice paddy insect pests.

Farmer Field Schools (FFSs) were established throughout Indonesia starting in the early 1990s with assistance from the FAO and other agencies, and they played a major role in implementing IPM in this country. A typical FFS is an outdoor school with an agricultural technician as advisor and about 25 participants, who actually raise crops in fields and learn skills and knowledge including how to observe crop growth, the types of insect pests and natural enemies and how to check their densities, and the types and properties of pesticides. In this way farmers become able to judge from their own observations how to eliminate insect pests.

FFSs also teach other skills such as managing fertilizer use, and better farm income through increased yields is also attractive to farmers. In an example from Lampung Province, farmers who had completed FFS training reduced their per-crop pesticide applications from an average of 5.3 times to zero, decreased their applications of nitrogen fertilizer and otherwise improved their fertilizer balance, and increased their per-hectare yields from 4.5 tons to 5.1 tons.

IPM initiatives are progressing in Indonesia in a wide variety of ways such as FFS-trained farmers passing on their knowledge to other farmers, the establishment of local farmer organizations to promote IPM, and the creation of FFSs with assistance from NGOs and international agencies other than the FAO. Farmers who have mastered IPM are attaining a significant place in their country's agricultural production.

Sugimoto Daizo

# Chapter 4
# Forest and Rice Paddy Biodiversity: Working Toward Comprehensive Management Based on Community Participation

An oil palm plantation. At left are oil palms, and at right trees have been cut to make room for replanting, exposing the soil (Malaysia, August 1996)
Photo: Iwasa Kazuyuki

# 1.    Introduction

"Conservation and Use of Biodiversity" (I-4) in *The State of the Environment in Asia 1999/2000* discussed the problems and challenges facing biodiversity, and described eco-farming in China and combined rice/fish farming in Malaysia. While international attention is focused on the loss and degradation of Asia's forests, peat bog development and rice paddy modernization are also involved in forest decline, and have a grave impact on wetland ecosystem biodiversity.

In view of this situation, this chapter discusses Asia's environment from the perspective of forests and rice paddies. The loss and degradation of forests and wetlands have been addressed as major international challenges at conferences including the 2002 Biodiversity Convention COP in the Hague, the Johannesburg Summit on Sustainable Development, and the Ramsar Convention COP in Valencia.

This chapter's first half analyzes deforestation and forest degradation in Asia, examines direct and indirect causes, and proposes remedial measures, while the second half explores the changes wrought by agricultural modernization in rice paddies, which are a traditional form of artificial wetland ecosystem. Finally, it proposes ways to reevaluate the traditional forms of rice paddies with attention to how the effects of modernization on paddies are linked to forest fires and to the decline in biodiversity.

# 2.    The Loss of Forests in Asia and Its Causes

## 2-1    1990s Initiatives

### Progress in Policy Dialog for Conserving Forests

Diverse initiatives on various levels have progressed over the 10-odd years since the 1992 Earth Summit. Globally, intergovernmental policy dialog on forests started with the Intergovernmental Panel on Forests and were carried over by the Intergovernmental Forum on Forests and the United Nations Forum on Forests, and major advances were achieved in negotiations on forest-related conventions such as the Biodiversity Convention and the Framework Convention on Climate Change (FCCC). On the regional and national levels the broad range of initiatives include policy dialog and activities, and the building of real cooperative relationships between international environmental NGOs and international aid agencies, such as the WWF-World Bank alliance for forest conservation. The UN Food and Agriculture Organization (FAO) has hailed these efforts as a decade of strengthened international cooperation,[1] and aside from whatever real progress has been made, such animated worldwide policy dialog on forest conservation is unprecedented.

## Forest Degradation

But will these efforts help stop the loss of Asia's forests? The FAO's *Global Forest Resources Assessment 2000* says that in the 1990s Asia lost 3.6 million ha of forests. This corresponds to an average annual loss of 0.1%, which is considerably less than the 0.6% loss[2] during the 1980s. According to the breakdown, rapid afforestation made up for the 20 million or more ha of natural forest that disappeared, but it is quite possible that this afforestation has far lower biodiversity than old-growth forest because it includes large-scale simultaneous planting of fast-growing species, resulting in lower forest age and stands with simplified structures. In sum, deforestation in Asia has slowed somewhat, but forest degradation is hastened.

## Attention to Underlying Causes

Deforestation in Asia continues and in fact degradation is accelerating despite the vigorous policy dialog on forest conservation. As the reason for this, some forest experts say that most attention is focused on the direct causes of deforestation while the underlying causes are being ignored.

The following section shall therefore explore the causes of deforestation in the tropical forest countries of the Philippines, Indonesia, Thailand, Lao PDR, Vietnam, and Cambodia, and in the southern Russian Far East (RFE) and its boreal forests. Instead of attempting a quantitative analysis, the authors had local experts identify direct causes which have had severe impacts over the last several decades, and worked through dialog with people responsible for forest policy in these regions to elucidate the anatomy of deforestation, including the main actors, cause interrelationships, and underlying causes.[3]

# 2-2   Direct Causes of Deforestation

The development actions constituting the main direct causes of deforestation in these regions (Table 1) were classified by purpose and placed within the series of processes leading from logging of old-growth forests to deforestation, showing that while the combination of actions leading to deforestation differs according to region, cause interrelationships are easy to understand (Fig. 1).

Deforestation starts with the commercial logging of old-growth forest, but this does not lead directly to loss. In tropical forest countries such logging is but the start of this process, which often continues with forest conversion projects, unsustainable swidden farming, or afforestation projects, then to degradation and loss.

Forest conversion projects that follow commercial extraction of useful timber from old-growth forests have development purposes such as creating rice paddies, other farmland, pastureland, plantations, residential areas, or dam sites, and they are common to all tropical forest countries. If a logged area is abandoned, it will become a secondary forest through succession and after several hundred years will become a forest similar to the original old-growth forest, but

TABLE 1. Primary Direct Causes of Deforestation in Six Southeast Asian Countries and the Southern Russian Far East

| Region | Primary causes |
|---|---|
| Philippines | Unsustainable commercial logging (including illegal logging) for export, the failure of industrial forestry, frequent forest fires set by local people, mine development, conversion of forest to expand farmland, highland farming, dams and other government projects, and opening of land for landless peasants |
| Indonesia | Unsustainable commercial logging (including illegal logging) for export, logging for the domestic plywood industry, large-scale migration and paddy development projects, non-traditional shifting cultivation, industrial plantation development, frequent large forest fires, oil palm plantation development (and illegal logging) |
| Thailand | Commercial logging, opening of forests for cash crop cultivation, shifing cultivation, rubber plantations, shirmp pond development, new land use categories and expansion of protected areas |
| Laos | Direct and indirect deforestation by Second Indochina Conflict, opening land for rice self-sufficiency, shifting cultivation, dam building, commercial logging (including illegal logging) |
| Vietnam | Direct and indirect deforestation by Second Indochina Conflict, opening land for rice self-sufficiency, migration within country, coffee plantations, migrating agriculture, and logging to fund the military |
| Cambodia | Opening land for cultivation, and logging by privileged classes and the military |
| Southern RFE | Unsustainable commercial logging (including illegal logging) for export, and large forest fires |

Source: Based on the first-phase strategic research report for "Forest Conservation" by the Institute for Global Environmental Strategies. <http://www.iges.or.jp/jp/pub/pdf/fc-j.pdf>

FIG. 1. Pattern Diagram of Deforestation Process in Asia

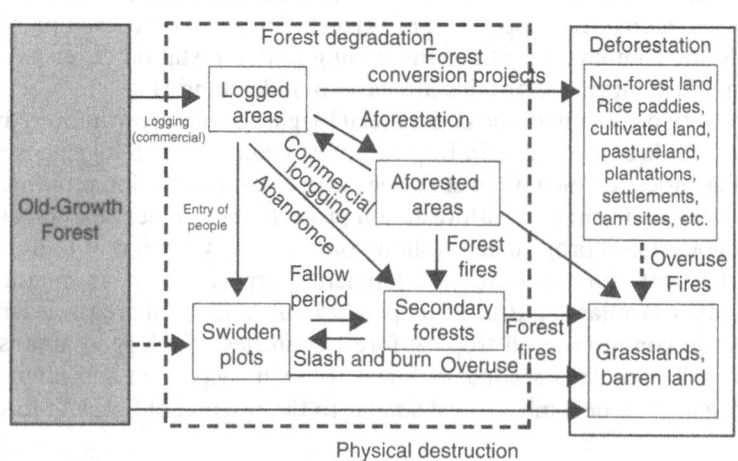

in tropical forests landless farmers will often enter these logged areas via logging roads and start swidden plots. Unsustainable practices, spreading fires, and other factors result in barren land where forest once stood. Large-scale afforestation with fast-growing species increasingly follows commercial logging of old-growth tropical forest, but in addition to creating large expanses of forest ecologically inferior to the original, such afforestation projects may sometimes likewise lead to barren land due to fires from pre-planting burnoffs or fires set intentionally over land conflicts. It is also not uncommon for old-growth forest to be lost directly or indirectly from physical destruction due to regional conflicts or other reasons.

In Southeast Asian tropical forest countries many activities combine in complex ways to cause deforestation. For example, the Philippines has lost vast tracts of forest due to nearly all the causes identified here. Indonesia too has all the identified causes except for physical destruction by war. In Laos, Vietnam, and Cambodia the process leading from logging to afforestation has not proceeded very far, but all other causes can be found. In the southern RFE deforestation is caused by a comparatively simple combination of unsustainable commercial logging and frequent fires.

## Unsustainable Commercial Logging

This is a major direct cause found in all these regions, and one can find strong mutual links among domestic actors including logging companies, central government authorities, the military, privileged classes, concession holders, and local forest management authorities, and outside actors such as companies in consuming and importing nations. Commercial logging has occurred in many cases when forests were nationalized and central governments had powerful authority over them, and natural resource policies have been implemented in a top-down manner for the purpose of exploiting useful timber. Awarding of logging concessions is sometimes a non-transparent procedure with ties between a few government officials and privileged figures, and with very little monitoring or control over logging practices or post-logging forest regeneration.

Hence many logging companies have adopted the "cut-and-run" strategy, which allows profit maximization. Examples of this are the Philippines, whose natural forests are nearly exhausted, Indonesia, which is losing its natural forests at a fast pace, and the southern RFE, which since the Soviet Union's collapse has been logging heavily for export to Japan, South Korea, and China. Since the days of the former Soviet Union, slipshod and wasteful logging to extract only good roundwood has been common in this region, but the situation worsened due to the weakening of the forest management system after the Soviet Union's collapse, and the Russian economic crisis of 1998. Further, illegal logging is feeding the rapid increase of log exports to China (see Essay 1).

Included here are instances in which a certain forest resource is exhausted owing to special timber demand from consumer countries. In Laos the logging of Lao cypress has proceeded without any particularly effective means of preservation to supply the large Japanese demand for old and large cypress trees, and as a consequence a number of cypress areas are nearly exhausted.[4] Lao cypress resources were developed in the early 1980s by Japanese timber importers to fill the void left after the Taiwan cypress resources were exhausted by exports to Japan.

In this region it is common for militaries to finance themselves with unsustainable commercial logging. In Vietnam, for instance, commercial logging has been heavy since 1986 mainly in the central highlands to generate funding for the military. Illegal and slipshod logging have been rampant in Cambodia since 1993 with the involvement of the privileged and the military, even under the United Nations Transitional Authority in Cambodia. In 1997 roundwood production attained 4.3 million cubic meters from an area of 7 million ha, the highest level ever, with illegal logging reportedly accounting for at least 92% of total production.

Logged areas are often planted on a large scale with fast-growing species for industrial applications, and serious land conflicts with local people are not uncommon. A typical example of this is the Philippines with its government-administered Contract Reforestation Program and the Industrial Forest Management Agreements. Large areas of land were arbitrarily planted without regard for forest use by local people, thereby fueling serious land conflict. Further, enormous project funds from international aid agencies brought about widespread corruption among involved government authorities, resulting in problems including no pay for some afforestation workers. Angry local people often resist by setting fire to planted areas.

## Forest Fires

Fires are a major cause of deforestation in the Philippines, Indonesia, and the southern RFE. Logged areas in the Philippines are often converted into commercial pastures, which entails repeated burnoffs that often result in the replacement of secondary forests with cogon grass (*Imperata cylindrica*). Additionally, fire from burnoffs often jumps to nearby secondary forests. Land conflicts over industrial afforestation with local people are also closely related to frequent forest fires.

In the southern RFE, large forest fires visited the Khabarovsk area and Sakhalin in 1998, affecting at least 2 million ha. Causes of these fires were the weakening of local firefighting systems by the political and financial chaos following the Soviet Union's collapse, and the increased citizen incursions into the forests.

Since the large 1983 forest fires in Borneo, Indonesia, there has been a trend toward a shorter forest fire cycle and more widespread damage, with the largest-ever fires occurring in 1997 and 1998. Over 2 million ha were affected, causing grave damage to human health, wildlife habitat, ecosystems, and the societies and cultures of forest dwellers. Two factors behind these immense forest fires are the widespread practice of forest burnoffs to prepare land for giant cash crop projects such as oil palm plantations, and institutional limitations such as inadequate firefighting systems.

## Plantations and Cash Crops

These types of forest development are now very common in Thailand, Indonesia, Laos, and Vietnam with the involvement of domestic and overseas actors such as governments, conglomerates, logging companies, foreign and domestic consumers, and international aid agencies.

Oil palm plantation area in Indonesia expanded quickly starting in 1978, attaining 2.5 million ha in 1997. Even now new plantations are being established. Not only do such projects cause the loss of forests, but the burnoffs for their creation are a prime source of forest fires. Reasons for the rapid increase in plantations include: demand for their products is growing; Malaysia, the world's largest producer, is running out of room for plantation expansion; the palm oil industry is promoted under government policy because natural conditions make most of Indonesia suitable for growing oil palms, and because assistance from international aid agencies is available because palm oil is a primary export to earn foreign currency in place of timber, and; domestic institutions help large-scale plantation development.

Major causes of deforestation in the Mekong River watershed are the industrial planting of rubber and eucalyptus trees, and the felling of mangrove forests to develop shrimp ponds. Shrimp farming in Thailand started mainly along the Gulf of Thailand coast in the second half of the 1980s after the industry's collapse in Taiwan, and recently is flourishing along the Andaman Sea coast, displacing mangrove forests. Causes of deforestation in Vietnam include clearing land for farming, and recently developing cash crop plantations such as for coffee in the central highlands, and developing shrimp ponds in the south (see I-3). A shared element in these forest conversion projects is strong consumer demand in the industrialized countries.

## Clearing Land for Relocation and Settlement

Deforestation for this reason is distinct in Vietnam and Laos recently, but it has been continuing in Indonesia for some time. The main actors are central governments, landless farmers, refugees, and returnees. One-fourth of the Laotian population was forced to relocate during the US bombings in the Indochina conflict, and along with the many domestic refugees from regional conflicts, they cleared much new land. The Laotian government has urged highland minorities to settle in lowlands to grow rice, while the government of neighboring Vietnam has encouraged lowland inhabitants to relocate in the highlands and open land.

These two countries' relocation policies aggravated conflicts over use of highland forests along their common border and resulted in unsustainable forest management practices over broad areas. Since 1975 Vietnam has maintained a policy of relocating 6 million people from crowded cities and densely populated rural districts to New Economic Zones in forested regions, and from the mid-1970s through the 1980s the central highlands were the destination of several million lowland Vietnamese. Deforestation has been widespread as a consequence of these actions. It is evident that deforestation through land clearing for relocation and settlement do not occur simply because of population growth and poverty, but also have strong links to other factors such as government policy and regional conflicts.

## 2-3    Underlying Causes of Deforestation

Diverse causes account for deforestation in Asia, but they can be classified into the four underlying cause groups of "forest governance," "trade and consumption," "international economic relations and capital transfers," and "valuating forest products and services."

### Forest Governance

This group includes non-transparent and unfair decisions on land use that ignore or discount the traditional rights of indigenous peoples and the reality of local people's forest use, and economic and forest development policies slanted toward industrial priorities. Other factors are decentralization of government decision-making, insufficient participation and transparency, and the legal and administrative foundations of sustainable forest management and community participation. The problems are found throughout Asia, and are likely major underlying causes of the incapacitation of forest governance, violation of regulations, and corruption. Still other conceivable factors are that earnings from forest development are not properly distributed, and that there is no mechanism to assure reinvestment in forest conservation.

### Trade and Consumption

Included in this group are the industrialized countries' lifestyles, which are based on the unsustainable mass consumption of timber and products from converted forest lands, and the down side of the system of free trade in forest products that supports that lifestyle. Although it is well known that Japan's soaring postwar timber demand accelerated Asian deforestation, China appears poised to travel the same path. The FAO says that conversion of forest to farmland is a major cause of tropical forest loss. It is important to note that production on such farmland is mainly for earning foreign currency, and that much of this has happened under economic and forest development policies slanted toward industrial priorities.

### International Economic Relations and Capital Transfers

This group includes current economic development models and strategies, structural adjustment programs, and their infringement on national governance. It is historical fact that many of the countries achieving speedy economic growth have lost or degraded their forests in a short time at the expense of environmental conservation and social equity. Structural adjustment programs are a major cause of deforestation when quickly phasing in market economies and instituting economic policy reforms without adequate forest governance. Economic and social chaos in Russia and Indonesia intensified the weakening of the forest sector and the incapacitation of forest governance, and accelerated forest development whose main aim was immediate foreign currency earnings.

### Valuating Forest Products and Services

Included in this group are insufficient understanding and a missing awareness concerning the diverse values of forests and the ecosystems that give them form, attitudes that see forests as places to produce timber and other products, or as places that are to be developed for other uses, and insufficient or lacking awareness that forests supply many kinds of products and services, and that local people are highly dependent on them. Improper or skewed valuation of forest products and services is the most basic of the underlying causes, and compels a reconsideration of basic attitudes about nontraditional forest uses.

## 2-4   Efforts Toward Regional Forest Conservation

### Changing Initiatives

As more and more people share an awareness of how these factors are causing deforestation, and as they demand more realistic forest conservation measures, forest conservation initiatives are gradually changing.

First, there is a rising consensus on seeing the importance of diverse forest values and local stakeholders as a prerequisite of forest conservation. For example, many forest development projects are beginning to adopt participatory approaches that provide for sustainable use of forests based on local realities and with the participation of various stakeholders. Such approaches are now at the foundation of national forest policy in the Philippines and Thailand, where rates of forest loss are high.

Second, the Biodiversity Convention COP5 in 2000 adopted the Ecosystem Approach guiding principles, which constitute a strategy for using fair and equitable methods for encouraging the integrated management and sustainable use of land, water, biota, and other national resources while providing for a balance with social acceptability, economic feasibility, and ecological soundness. Based on the premise that nature is intrinsically complex and dynamic, the conference elected to use adaptive management that can cope with uncertainty, and agreed on the need for preventive initiatives. This increased the possibility that even international society will transcend the traditional antagonism between development and nature protection, and accept ideas such as environmental cost allocation, harmonization among stakeholders, and others based on respect for the diverse values of natural resources.

Third, many more people are seeing how domestic and international distribution and trade under free-trade regimes are directly linked to deforestation, and efforts are underway to obtain World Trade Organization (WTO) approval for forest product trade management based on principles for maintaining forest sustainability. There is also rapid progress in addressing illegal logging and the illegal international trade in forest products, which environmental NGOs have long identified as causes of deforestation.[5]

## The Role of Environmental NGOs

Environmental NGOs have played a significant role in the process leading to these substantial and positive changes. For instance, they have consistently called attention to the problems of free-trade regimes and have influenced changes in stance at the WTO and other organizations by sometimes combining the use of radical protest actions. Further, NGOs around the world have collaborated strategically on everything from determining local situations to getting issues on the international political agenda. In relation to elucidating the causes of forest degradation and loss, forest NGOs successfully backed an initiative called "Addressing the Underlying Causes of Deforestation and Forest Degradation" at the Intergovernmental Forum on Forests.[6]

Asian environmental NGOs must help see that rapidly expanding afforestation is properly done. It is anticipated that afforestation in Asia will proceed at a faster pace because it is allowed for carbon sink projects under the clean development mechanism, but so far large-scale, top-down afforestation has ignored local differences because of project uniformity, deprived people of their land, and brought social unrest. A further consequence is deforestation because displaced people burn forest to create farmland in new locations.

Additionally, corruption and conflicts of various kinds are likely to occur when forest governance is inadequate. Urgent tasks to counter this are the development of methods for the assessment, coordination, and proper implementation of afforestation plans that are also acceptable internationally. These would include measures needed to resolve disputes, means for the fair and equitable distribution of earnings from afforestation, and ways to conserve biodiversity.

It is hoped that NGOs will make a contribution to creating mechanisms conforming to real-world situations. For example, in projects that claim to be transparent and have the participation of local communities and stakeholders, it will be important to monitor the actual extent of participation because the extent of real participation is quite different depending on whether people have the authority to determine forest management plans themselves, or are just are carrying out government decisions as laborers. There are also expectations for additional NGO efforts such as participation in studies to determine local conditions and facilitating inter-regional information exchange through networks.

# 3.  Wet Rice Agriculture and Wetland Ecosystem Conservation

## 3-1  Rice Paddy Biodiversity and Its Role

Rice paddies are artificial wetlands that supply people with food and natural resources, and provide wildlife with habitat, breeding areas, refuge, feeding grounds, and other services. Paddies also help mitigate flooding caused by heavy rain,[7] and have other diverse functions including soil conservation,[8] recharging aquifers, maintaining water quality, and preserving traditional landscapes. The position of rice as a prime staple food in Asia is behind the development of cultures based on rice cultivation.

Traditional paddy cultivation using small-scale organic methods are especially important from the perspective of biodiversity because paddies create more habitat for fresh-water organisms,[9] and even now primitive ecosystems can be found.

Paddies for wet rice agriculture in East and Southeast Asia vary in form depending on topography. Small terraced paddies are found in steep regions, while in flat areas they are larger. In all countries generally paddies are becoming larger and farming is mechanized. In Cambodia, Vietnam, and the Mekong Delta paddies become swampland during the rainy season, constituting an extension of swamp ecosystems.

## 3-2  Current State of Asian Rice Paddies

Comparing rice paddy area and production between 1990 and 1999 (Table 2) provides a general idea of the current state of rice paddies in Asia, whose major rice-baskets in order of size are China in first place, followed by India, Indonesia, Vietnam, Bangladesh, and Thailand. In comparison with 1990, there was about 5% more paddy area in 1999, and production was up about 13%, showing that the increase in production outpaced that in area.

Vietnam had the largest area increase at about 27%, followed by an approximate 20% in the Philippines. Decreases occurred in the industrialized countries of Japan and South Korea, as well as in North Korea. The largest production increases were in Vietnam with about 63%, Cambodia with 52%, and Pakistan with 41%. In North Korea, which apparently suffers grave food shortages, production fell even more than paddy area.

The following sections discuss a selection of information from presentations made at conferences including the 6th Wetland Symposium of the International Association of Ecology (INTECOL VI), held in Quebec in 2000, and the Asian Wetland Symposium 2001.

### Indonesia: Reckless Paddy Development

In January 1996 President Suharto launched a project to convert 1 million ha of tropical peat swamps to rice paddies, but in July 1998 President Habibi suspended

TABLE 2.  Area of Rice Paddies and Rice Production in Asia

| Country/region | Area (100 ha) | | Production (100 t) | |
|---|---|---|---|---|
| | 1990 | 1999 | 1990 | 1999 |
| Asia | 132,328 | 138,503 | 479,480 | 540,621 |
| Bangladesh | 10,435 | 10,470 | 26,788 | 29,857 |
| Bhutan | 26 | 30 | 43 | 50 |
| Cambodia | 1,740 | 1,961 | 2,500 | 3,800 |
| China (incl. Taiwan) | 33,519 | 31,270 | 191,615 | 200,499 |
| India | 42,687 | 44,800 | 111,517 | 131,200 |
| Indonesia | 10,502 | 11,624 | 45,179 | 49,534 |
| Japan | 2,074 | 1,788 | 13,124 | 11,469 |
| DPR of Korea | 650 | 580 | 3,570 | 2,343 |
| South Korea | 1,244 | 1,059 | 7,722 | 7,271 |
| Laos | 650 | 718 | 1,491 | 2,103 |
| Malaysia | 681 | 674 | 1,960 | 1,934 |
| Myanmar | 4,760 | 5,458 | 13,972 | 17,075 |
| Nepal | 1,455 | 1,514 | 3,502 | 3,710 |
| Pakistan | 2,113 | 2,400 | 4,891 | 6,900 |
| Philippines | 3,319 | 3,978 | 9,885 | 11,388 |
| Sri Lanka | 828 | 829 | 2,538 | 2,692 |
| Thailand | 8,792 | 10,000 | 17,193 | 23,272 |
| Vietnam | 6,028 | 7,648 | 19,225 | 31,394 |

Source:  Based on International Rice Research Institute materials.

the project, and in July 1999 it was cancelled. Yet, already at least 500,000 ha of these primitive peat swamps have been developed, 4,600 km of drainage and irrigation canals have been dug, and 10,000 migratory farming households have been settled. The consequences are the loss of highly productive forests that were rich in biodiversity and natural resources, and could be sustainably used, as well as the destruction of landscapes. What is more, because logged forest land was burned off to make way for paddies, the peat caught fire and smoldered for long periods, spreading the fire to other forests and causing smoke damage.

Indonesia's government has plans to use 2.8 million ha of peat swamps in central Kalimantan, including the paddy development project described above, as unsustainable multi-purpose farmland.[10]

## Thailand: Regional Differences in Rice Paddies

Thailand is the world's biggest rice exporter. About half of its paddy area is in the northeast, 23% in the north, 19% in central Thailand, and 6% in the south.

Central Thailand has much natural and semi-natural marshland. It was here that large-scale irrigation started in the early 20th century. Although northeastern and northern Thailand have the most paddy area, these areas have low yields per unit area because of scarce water sources and a low irrigation rate. Thailand's central region is its ricebasket, and rice for export is mainly produced here because of its good water access. About 40 years have passed since modern agriculture came to the central region, which is important not only as a rice-producing area, but also for its role in conserving biodiversity (see Essay 2).[11]

## Japan: New System for Preserving Traditional Rice Paddies

Japan's paddy area was largest in the 1970s, but in the following decade the area declined under the government's paddy reduction policy that had begun in the 1970s, and abandoned paddies are now a matter of public concern. The Ishikari Plain on the northern island of Hokkaido is the northernmost extent of rice agriculture in the world, and Japan's largest rice-growing area. This plain comprises a gently sloped peatland and a flood plain that has been used for rice paddies, but as paddy land has been filled and used for other purposes the plain has had increased flooding.[12] Southeast Asian areas with similar topography may find this example to be useful.

Small-scale terraced rice paddies in intermediate mountainous areas constitute traditional paddy landscapes in Japan, but many of these paddies have been abandoned. Many organisms depend on paddies. For instance, the dragonfly life cycle is geared to the agricultural cycle.[13] Geese and ducks depend on paddies by feeding on shack and roosting in paddies.[14] Both the species and numbers of organisms have declined with the reduction in traditional terraced paddies.

In view of this situation, the Japanese government in 1999 passed the nation's first basic law on food, agriculture, and rural areas that assesses the diverse functions of agriculture, thereby allowing the payment of subsidies to farmers in intermediate mountainous areas. Some local governments pass ordinances tailored to their local situations. Examples are an ordinance paying farmers compensation for crop damage in the Miyagi Prefecture towns of Ichihasama, Tsukidate, and Wakayanagi, and ordinances for preserving agricultural reservoirs with a view to nature conservation and wise use, in places including Kasuga City, Fukuoka Prefecture.[15]

In other initiatives to preserve paddies, local government administrative departments, farmers with terraced paddies, NGOs, NPOs, and other parties sponsor events that involve elements of environmental education for city dwellers, especially children, to experience farm work such as traditional rice planting and harvesting.

## Bangladesh: Government and NGO Initiatives

Rice production is one of Bangladesh's major industries, accounting for about 30% of GDP. To cope with its rising population, Bangladesh introduced and expanded modern agricultural practices in the mid-1960s, and since that time the use of chemical fertilizers and pesticides has quickly increased. Their undisciplined use has had a deleterious effect on the ecosystem and human health, including a 3.2% decrease in rice yield from 1985 to 1995 due to soil degradation.[16] The widening poverty gap maintains a situation in which 60% of farmers have no work.

The government, research institutions, and NGOs are promoting a rice fish farming system that does not burden paddy ecosystems. Although it has the disadvantage of lower yields, costs are lower. There are also proposals for integrated agriculture that not only grows rice, but combines fruit growing, livestock raising, fishing, forestry, and other such pursuits.[17]

## Malaysia: The Importance of Modern Rice Growing

New paddy land is created in response to increasing food demand, and in that process traditional rice farming is being replaced by irrigated paddies. Nearly all large paddies are irrigated, but irrigation and drainage channels have become new habitat, serving as refuge for fish in the dry season and spawning sites for fish in the rainy season. Rice paddies, rivers, irrigation ponds, and channels now form a rice agro-ecosystem.[18]

A study of dragonflies in paddies in Bertam, northern Malay Peninsula found that even in the presence of herbicides and other harmful pressures on dragonflies, paddies provide an environment making it possible to maintain dragonfly numbers and species diversity.[19]

# 3-3  International Conventions for Conserving Rice Paddies as Wetland Ecosystems

It is useful to examine multilateral environmental agreements that help preserve rice paddies, in particular the Ramsar Convention and the World Heritage Convention.

## Ramsar Convention

The broad definition of wetlands under this convention makes it applicable also to rice paddies, irrigation ponds, and other artificial wetland environments. Some registered Asian Ramsar sites that include paddies are China's Honghe National Nature Reserve and Yancheng National Nature Reserve; Japan's Izunuma/Uchinuma and Katano Kamoike; the Philippines' Agusan Marsh Wildlife Sanctuary; Sri Lanka's Annaiwilundawa Tanks Sanctuary; Thailand's Bung Khong Long Non-Hunting Area, Kuan Ki Sian of the Thale Noi Non-Hunting Area, and Pru To Daeng Wildlife Sanctuary; and Vietnam's Xuan Thuy Natural Wetland Reserve. Other sites, such as India's Ropar and Pakistan's Drigh Lake, have rice paddies on their peripheries.

As yet only a small percentage of registered sites include paddies, but more should be registered as wetlands considering their importance in Asia.

## World Heritage Convention

Under this convention the rice terraces of the Philippines Cordilleras were registered as a cultural property in 1995. Two thousand years ago people created terraced rice fields on these steep mountain slopes, producing magnificent harmony between the artificial paddies and the natural environment, and a fabulously beautiful landscape. Yet 25 to 30% of the paddies have been abandoned because there are too few young farmers to take over their cultivation. In 2001 the area was put on the List of World Heritage in Danger for reasons including the risk of collapse.

### Networks and Other Conventions

The Biodiversity Convention, the Bonn Convention, and others also have some connection to rice paddy conservation. And although not legally binding, the Asia-Pacific Migratory Waterbird Conservation Strategy is also important in view of the fact that waterfowl migrate across national borders while taking advantage of rice paddies.

## 3-4   Current Initiatives and Needed Conservation Measures

In some countries traditional rice paddies are being lost due to modernization or to the decline of wet rice agriculture itself. But even paddies farmed by modern agricultural methods are still important to organisms that use them. Hence it is necessary to promote agricultural systems that do not burden paddy ecosystems, and especially to educate farmers how to properly use potentially damaging pesticides and chemical fertilizers. To assess and preserve the diverse functions of traditional rice paddies, Asian countries must create appropriate legal systems and provide subsidies for farmers on the national and local government levels.

Over the past decade more Asian countries became party to the Ramsar Convention, which can help paddy conservation, but more countries will have to join Ramsar and other such agreements. Japan must take the initiative on this matter.

But some countries block the involvement of farmers and local people when designating protected areas. Because the environment has been maintained and conserved through a deep involvement with people, efforts for wetland (including paddy) and wildlife conservation must be carried out with the cooperation of farmers, local people, governments, research institutions, NGOs, and other parties, and governments must ensure the participation of all actors in the decision-making process on conservation plans. Yet, many Asian countries still lack information disclosure and community participation. Japan and other countries that have institutions for information disclosure and community participation could offer material and personnel assistance to countries still lacking such institutions to create them. NGOs and other organizations have to educate local people and governments on the conservation of wetlands including paddies, serve as an intermediary by bringing together the wishes and suggestions of local people and presenting them to governments, and monitor the actions of governments.

# 4.  Sustainable Management of Ecosystems

Although forest- and rice paddy-related issues tend to be conceived as separate, the foregoing discussion has demonstrated that they are interrelated. In particular, desiccation of peatlands for conversion to rice paddies is said to be one cause of forest fires in recent years.

Forests often grow on peatlands in low-latitude areas. While such forests have been opened for farmland in the past, in recent years modern rice agriculture is encouraged to raise food production, and peatlands are often desiccated as part of national projects. While setting fires to burn off trees is a direct cause of forest fires, the drying of peatlands and dropping water levels facilitate the spread of fires to underground peat deposits, and putting out such fires is difficult. Burnoff fires spread to adjacent forests, and they are a prime cause of widespread air pollution.

At the same time, both forests and wetlands have high numbers of species and individuals, and high levels of biological activity per unit area, making them crucial ecosystems from the perspective of biodiversity. It is well known that tropical forests are good wildlife habitat, but the same holds true for temperate and boreal forests. Forest flora and fauna are valuable economic resources especially to local communities and indigenous societies, and have traditionally been used at sustainable levels. People have to examine traditional methods of using forests by reassessing the value of forest products other than timber, and they have to create management systems that take into consideration the conservation of forest ecosystem biodiversity.

Traditional forms of wet rice agriculture, which have long been stably maintained, are also important secondary ecosystems. Although traditional forms of rice growing remain, in some places farmers practice modern methods dependent on chemical fertilizers and insecticides, just as in Japan. These modern practices result in the same ecosystem damage and impoverished biodiversity as the conversion of peatland to paddies. Having thought better, people are now reassessing traditional forms of rice growing that combine other kinds of agriculture. In particular, academics, NGOs, and some governments are reassessing and encouraging combined rice/fish farming that uses networks of channels and ponds because it is both environmentally and economically beneficial.

Taking the actions outlined above requires, for both forest and wetlands, assuring the sustainable use of their resources, as well as management practices consistent with biodiversity conservation and based on the Biodiversity Convention, Ramsar Convention, and relevant Guidelines under the International Tropical Timber Agreement (ITTA) and the Criteria and Indicators for the Conservation and Sustainable Management of Temperate and Boreal Forests (the Montreal Process). Fundamental to this is the need to assure the participation of local people and all other stakeholders in decision-making and management processes, and to build collaborative relationships with NGOs that assist local people.

Isozaki Hiroji, Yamane Masanobu, Nakamura Yuriko, Inoue Makoto

# Essay 1:  China: Soaring Log Imports and Impacts on Russian Forest Degradation

Per capita wood consumption in the People's Republic of China is about 0.1 cubic meter, still far below that of Japan and the US, but China's nearly 1.3 billion people consume over 140 million cubic meters annually, making China one of the world's largest timber-consuming countries. Until recently most industrial logs were obtained from domestic natural forests, but dependence on imports is gradually rising. This increase has been rapid especially since 1996, with imports breaking all previous records at 16,860,000 cubic meters in 2001, and attaining 24,330,000 cubic meters in 2002 (Fig. 2). China's imports now have a heavy impact on the world timber market. Most of its imports are from Russia and the tropical countries of Malaysia, Gabon, Guinea, and Myanmar. Imports of softwood timber and hardwood species such as oak and ash, which are now increasing rapidly, are mostly from Russia. Recent growth in log imports owe much to strong demand because of economic development, and to the decline in domestic production. Natural forests along the mid and upper reaches of the Chang Jiang (Yangtze River) and Yellow River, and in the mountains of the Northeast district, serve as the main sources of timber, but they are shrinking due to overlogging, forest fires, conversion of forests on slopes to farmland, and other factors. Degraded forests have considerably diminished capacity to conserve water and soil, and the consequence is extensive natural disasters. Since 1990 flooding in the Chang Jiang and Yellow River watersheds has grown in severity and frequency, while at times there is no water at all, resulting in huge economic losses.

China's government has turned its attention to forest conservation, and in 1998 it launched a natural forest protection project that, in addition to preexisting state afforestation projects, aims to rejuvenate natural forests, conserve biodiversity, and achieve the sustainable development of regional economies and societies. This new policy affects 17 provinces and areas including the upper Chang Jiang, the mid and upper reaches of the Yellow River, and the state-owned forests in the Northeast, and it employs intensive investment through top-down management to administer prohibitions or restrictions on

FIG. 2. Log Imports by China (1992–2002)

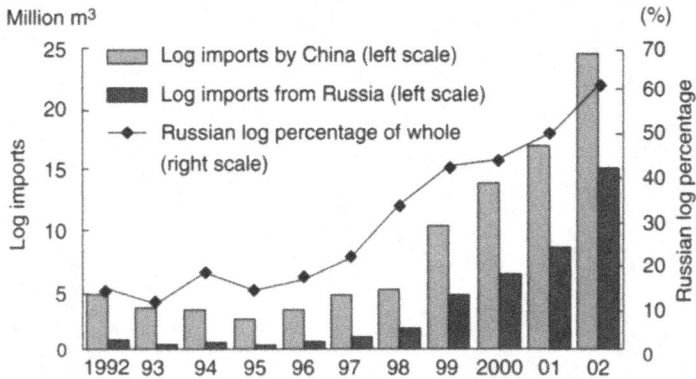

Source: Prepared from *Chinawood*, May 2001 and February 2003.

logging in natural forests, reassigning lumberjacks to forest conservation projects, and the planting and maintenance of forests. As a result, timber production in the affected regions dropped about 20 million cubic meters from 1998 to 2001, and the slack is being taken up by timber from eastern Siberia and the Russian Far East. This Russian timber is produced in an extremely careless fashion, with illegal logging and shipping not uncommon, meaning that substantially greater exports to China are accelerating the destruction of Russia's forests.

Yamane Masanobu

❖               ❖               ❖

## Essay 2:   Fish Farming in Ponds Adjoining Rice Paddies in Northeast Thailand[20]

Nearly everywhere in Thailand farmers raise fish in ponds adjacent to their rice paddies. In northeastern Thailand the countless streams and water channels connected to the Mekong River flow into irrigation ponds whose water is used in paddies. This river water brings various kinds of fish into the paddies and ponds. Farmers often eat the fish themselves, and meticulously tend to their ponds by removing things that capture the fish, feeding the fish, and catching those that have attained full size. When the rainy season comes, fish are released into paddies that have some water in them because during the irrigation season fish farming will be expanded by letting the fish eat whatever is available in the paddies.

Nakamura Yuriko

# Part II

# Asia by Country and Region

# Chapter 1
# Northeast Asia: Region Building Based on Environmental Cooperation

# 1.  Introduction

Northeast Asia retains the imprint of international confrontations even now after the Cold War, but for that very reason there is hope for the possibility of potential international cooperation. In this region the United States, the world's biggest military power, uses its military bases in Japan and South Korea to directly face off with Russia and China, which have respectively the largest land area and the biggest population, and with North Korea, which maintains its own distinctive socialist system. Through the Cold War era the Sea of Japan region was long a sphere of antagonism, which had a profound effect on Japan's economic development by making the Pacific side into the nation's nucleus and relegating the Sea of Japan side to incidental status. From this backdrop came a post-Cold War proposal for "Northeast Asia" as a regional concept upon which to base efforts for transforming antagonism into cooperation to regain social governance through regional collaboration that does not see lands across the sea in terms of "countries."

Characteristics of this region from an environmental perspective include: (1) the Sea of Japan is semi-enclosed and therefore readily accumulates pollutants, (2) to the surrounding nations, those portions of their countries facing the Sea of Japan are regarded as peripheral and generally have comparatively well-endowed natural environments, but are subject to development strategies entailing huge environmental burdens to compensate for low degrees of development, or are saddled with nuisance facilities, and (3) heavy environmental impacts are anticipated if development progresses in the future.

Two major conceptions of "Northeast Asia" are: (1) the part of Japan along the Sea of Japan, the Russian maritime region (Primorsky Krai), the eastern part of China's Jilin Province, and the part of the Korean Peninsula along the Sea of Japan, and (2) all of Japan, the Russian Far East (RFE), China's Northeast district (Heilongjiang, Jilin, and Liaoning provinces), and the Korean Peninsula.[1]

The first is the region whose water systems flow into the Sea of Japan. If the Amur River (which empties into the Tartar Strait) is included in the Sea of Japan water system, Northeast Asia includes southeast Siberia, eastern Mongolia, the northeastern part of China's Inner Mongolia Autonomous Region, and most of that country's Jilin and Heilongjiang provinces. River pollution in this region readily pollutes the Sea of Japan. And except for the parts of Jilin and Heilongjiang provinces in the Amur River watershed, economic development is relatively behind in all the countries and regions included in Northeast Asia (Japan, Russia, Mongolia, China's Northeast, and the Korean Peninsula). Unfortunately, efforts to help Northeast Asia catch up often lack environmental considerations, are overly ambitious, or have distortions. In fact, two of Japan's four worst pollution diseases, itai-itai disease and Niigata Minamata disease, occurred on its Sea of Japan side, and most of the large-scale development projects emphasizing the Sea of Japan side either ended in failure or had disastrous environmental consequences. In South Korea, polluting companies finding it hard to operate in heavily populated areas were located in the Ulsan-Onsan Industrial Complex in the southeast facing the Sea of Japan, which helped economic development but also caused Onsan disease and other environmental problems. Further, Northeast Asia is often saddled with so-called nuisance facilities because of its relative political and economic weakness. Many of Japan's and South Korea's nuclear power plants are on their Sea of Japan sides, and Russia has dumped nuclear waste in the Sea of Japan. As a whole this region still has a flourishing ecosystem, but much of it is fragile, like the cool temperate forests that recover only with difficulty once damaged. As these characteristics are held in common by all countries in the region, exchanges within this definition of Northeast Asia are led by local governments and citizens.

Definition (2) includes the entire countries facing the Sea of Japan in Northeast Asia (but only parts of Russia and China), but Mongolia is included under Tumen-NET, a program that is part of the Tumen River Area Development Programme.

Socialist systems remained in China and North Korea after the Cold War ended, making for large differences in political systems and economic development among countries. In all countries there is a strong tendency for central governments to control international exchanges, thereby frequently creating significant difficulties for local governments and even for contacts among environmental NGOs. Over the last few years, however, there are more citizen exchanges that cannot be entirely controlled by central governments.

After World War 2 there were efforts to break through these national-level policy frameworks. For example, in the 1960s during Japan's First National Land Development Plan, hub development projects such as a local industrial city development plan in Toyama Prefecture envisioned focal points for socioeconomic development in countries across the Sea of Japan. It was not until the 1990s that talk of the "Sea of Japan era" emerged, along with a vision for cross-border relationships in the search for symbiotic ties among countries with different pasts and social systems. This included discussion on the appropriateness of the name "Sea of Japan."[2] The 1994 founding of the Association for the Japan Sea Rim Studies anticipated its intellectual hegemony, and there are now efforts to attract anticipatory public investment and private capital into this region to create an arrangement for socioeconomic exchange among local governments.

Below, this chapter discusses Tumen River development as an example of how difficult government-private international cooperation is in this region, then examines the state of the Sea of Japan, which is itself a microcosm of Northeast Asia's pollution problems. Our window on marine pollution will be responses to oil spill accidents mainly from tankers. The chapter then moves on to an overview of Russia's forests and nuclear waste dumping, and finally it will review international cooperation arrangements on marine and atmospheric issues.

# 2.  Tumen River Area Development

## 2-1  Trends in Each Country

Most of the Tumen River flows along the eastern portion of the China-North Korea border, and its last approximately 15 km before emptying into the Sea of Japan are part of the North Korea-Russia border.

The Tumen River Area Development Programme is a project relying on multilateral cooperation for the large-scale development of the lower reaches of the Tumen River, making it into an entrepot with the same position in Northeast Asia as Hong Kong and Singapore have in Southeast Asia. Yet this area is a rich and ecologically valuable natural environment, and is like a "frontier" to China, Russia, and North Korea. Accordingly, large-scale development plans are over-ambitious both socioeconomically and in terms of the potential environmental burden. What is more, the involved countries and actors take quite different positions regarding development and the environment in this area, thereby creating an intricate mix of intents. This situation could be seen as a typical example or a microcosm of environment-development efforts in Northeast Asia as a whole, and also of international cooperation on such efforts. This section surveys the Tumen Programme as it affects involved countries and regions.

In China's Northeast nearly all foreign trade passes through Dalian and Yingkou ports in Liaoning Province. In response to this situation, in 1990 the Jilin provincial government announced a development scheme to make the Tumen River delta area into an international hub for distribution, trade, and industry. Subsequently a concept was formulated with the support of the United Nations Development Programme (UNDP) to facilitate development of the area where the borders of China, Russia, and North Korea meet on the lower reaches of the Tumen River by building an "international city" there.

China's government gave "open-border city" designations to Hunchun in the Tumen River area, Heihe and Suifun in Heilongjiang Province, and Manzhouli

Photo 1.  International border shared by China, North Korea, and Russia
(August 2001 photo by Aikawa Yasushi)

in the Inner Mongolia Autonomous Region. It further established the "Hunchun Border Economic Collaboration District" as an "economic collaboration district" (an area where international investment should be encouraged), and has been proceeding with the intention of leasing Russian and North Korean ports for access to the Sea of Japan. But as Hunchun has a population of only 200,000 including suburban areas, China's plan to make it into an "international city" and "Hong Kong of the north" (the same slogan as Dalian) with an urban population of 500,000 by 1999 was unrealistic. Many of the buildings started were therefore never finished, or even if finished remained vacant. Nevertheless, there are far fewer see-through buildings than along China's southern coast because construction never proceeded very far.

In a December 1991 experiment meant to introduce a market economy, North Korea's government designated the Rajin/Sonbong area on the Tumen River as a free economic and trade zone. In 1995 it opened a container ship route between Rajin and Pusan, and in 1997 some progress was made toward transporting Chinese wood chips to Ulsan in South Korea. Although North Korea's government is trying to attract foreign capital, its trade links with China are still weak.

In March 2002 the North Korean government established the Republic Land Planning Law because it had thought the 1977 Land Law was inconsistent with the with the 1986 Environmental Protection Law and the 1993 Underground Resources Law. The new law requires that agencies and organizations developing land or resources attach site survey reports, environmental impact assessments, or other documents when submitting applications to environmental protection agencies, and that those agencies can consider whether to grant or refuse requests (Article 26), thereby on the face of it expecting caution on large-scale development. Yet under North Korea's dictatorial and repressive political system it is hard to believe that development conducted as part of state policy would be halted due to the environmental rights of people not among the privileged, or even owing to general environmental concerns. Still, there is far too little reliable information to allow discussion of not only the Tumen River area, but also of the environment and development in North Korea. The only thing that can be said for sure is that, judging from examples of other socialist countries, one cannot be optimistic on the sole basis of press-agentry from their governments.

At first Russia's Primorye government supported the "Greater Vladivostok Free Economic Zone" concept, for which Vladivostok had asked the United Nations Industrial Development Organization to perform a study, but this concept fell by the wayside due to financing difficulties, and was replaced by a realistic plan to develop existing ports in the Nakhodka Free Economic Zone.

To accomplish the Tumen River Area Development Programme, which is promoted by China's Jilin Province and by South Korea, the UNDP in 1995 decided to create the Tumen River Trust Fund after confirming that the project was ready for implementation, and the 1996 meeting resolved that Japan would be invited to formally become a member. But achieving Tumen River development as initially conceived, which entailed building an international city at the three-country junction (Hunchun) and networking it with surrounding cities, would require a considerable amount of long-term investment that assumes stable international relations in the region, including North Korea.

However, North Korea's relations with international society are in fact deteriorating, and harsh economic situations rule in other involved and nearby countries, including South Korea and Japan. Hence it is not surprising that development has made little progress other than that in access from Hunchun to the Russian port of Zarubino (nevertheless, Russia has responded coolly to China's request to lease the port). Further, if development pays too little heed to the environment, no progress at all would be better from an environmental viewpoint.

## 2-2    The Environment and NGO Activities in the Tumen River Basin

The failure of these big development plans to proceed smoothly has barely spared the area's environment from catastrophic damage. In fact, to China this area represents one of its few well-forested places, but the countries that share this thriving but fragile ecosystem stretches do not have congruent policies on how to use it. In connection with the Tumen River Area Development Programme, from the outset in 1992 Russia had designated the wetland at the river's estuary as a nature reserve because it is a nesting area and habitat for migratory birds, and North Korea has not decided what to do. These are influences behind the cautious tone of debate on opening and developing this area.[3]

This region comprises several zones of diffent kinds: the southern part of a boreal forest called the taiga, a moist mixed forest of softwood and hardwood species stretching from the Russian coast to China, an area of brown soil, and China's largest wetland zone, 26,000 ha encompassing Russian territory and the Tumen River, and extending into North Korea. The visually beautiful marshes are visited by birds including the Manchurian crane, and girded by valuable natural landscapes including sand dunes. China created Xianghai Nature Reserve in its own territory, but enacted no special conservation measures. Development of pastureland in the western portion of the wetlands, and freshwater fish farming ponds, farmland, and residential areas in the eastern forest and wetland zone are causing topsoil runoff. By contrast, on the Russian side local people and NGOs are leading nature conservation efforts.

This area is also the habitat of the rare Siberian tiger, whose range extends across the three countries' borders. Widening and tunnel work is proceeding on the road to Hunchun from the tip of China's territory where the borders meet, which has elicited concerns from conservationists about blocking the tiger's movement.

There is strong interest in this region from environmental NGOs not only in Russia but also South Korea, while China's government has a strong predilection toward development, a situation that has long persisted. In the summer of 2000 an environmental NGO called Green Yanbian (Yianbian Green Federation) was organized in China mainly by ethnic Koreans returning from South Korea, and it quickly grew to 400 members in about a year. A primary reason behind this rapid growth is ethnic Korean identification with the Changbai Mountains and Tumen River in this region, and the will to protect them. Although ethnic Koreans account for half of this area's population, they are about 80% of the organization's members. Further, it has had much interchange with South Korean environmental NGOs from the outset in part because it is modeled on them instead of on Chinese organizations.

At first Green Yanbian made little effort toward contact with environmental NGOs in other parts of China, but later began actively working with Chinese NGOs and environmental experts primarily in Beijing and the Northeast. Owing to its strategic location, Green Yanbian could play an even bigger role if there is more international contact in this region, especially interaction between Chinese and South Korean environmental NGOs.

Yet one cannot be too optimistic about the future of Green Yanbian because environmental initiatives in Northeast Asia with its mix of geographical and historical interests often run up against diplomatic considerations. For example, in May 2001 Green Yanbian spent about a month testing water quality and studying the riparian environment along the Tumen River from its upper to lower reaches, creating a map showing river pollution and watershed damage, but Chinese authorities did not allow the organization to release their data and photographs, as it would have been impolitic to reveal the extent of North Korean pollution. Another example is the health damage to sheep farmers because of chemical weapons abandoned by the former Japanese Army near Dunhua in the northwest part of the Yanbian Korean Autonomous Prefecture. As the disposal of these weapons by Japan is underway pursuant to an agreement between the two governments, China is wary of raking up this issue again.[4]

In addition to the many restrictions on NGO activities already existing in China, these "diplomatic" concerns impose still more limitations. Geographically and ethnically this region has advantages in terms of international influences and opportunities for interchange, but it remains to be seen whether the advantages or the restrictions will win out.

# 3.  Marine Pollution in the Sea of Japan

## 3-1  Lessons of the Nakhodka Accident

Many people still recall the 1997 *Nakhodka* oil spill,[5] but pollution from tanker spills is not restricted to the North Pacific, and there are many more such accidents than people in general realize (Table 1).

In 1995 three major oil spill accidents occurred in South Korean waters: the *Sea Prince* near Yosu, Chon-nan in July, the *No. 1 Yuilu* near Pusan in September, and the *Honam Sapphire* near Yosu, Chon-nan in November.

Japan on the other hand was fortunate to have no spills on the order of several thousand kiloliters between the 1974 heavy oil spill at the Mizushima industrial complex and the *Nakhodka* accident. On the other hand, this lack of serious spills ironically delayed Japan's building of a system for responding to such accidents.

TABLE 1. Major Oil Spills in the Sea of Japan (East Sea) and the Yellow Sea after 1990

| Date (YMD) | Country | Place | Vessel | Amount | Type of oil |
|---|---|---|---|---|---|
| 1992.5.1 | Japan | Kushiro, Hokkaido | Shell Oil base | 246 kl | Unknown |
| 1993.9.27 | S. Korea | Jeonnam Yeocheonsi, east coast of Myo Island | Gumdong No.5 | 1,228 kl | Heavy B–C |
| 1994.10.17 | China | Qinhuangdao, Hebei | Fwa Hai No.5 | Unknown | Unknown |
| 1995.7.23 | S. Korea | Jeonnam Yeocheonsi Sori Island | Sea Prince | 5,035 kl | Crude/bunker |
| 1995.9.20 | S. Korea | Busan South Hoyongie Island | No.1 Yuilu | Unknown | Unknown |
| 1995.11.17 | S. Korea | Jeonnamyeosu Honam Oil Refinery berth | Honam Sapphire | 1,402 kl | Crude |
| 1996.9.19 | S. Korea | Nine miles from Jeonnam Yoso Island | Ocean Joedo | 207 kl | Heavy B–C |
| 1997.1.2 | Japan | Near Oki Island, Shimane Pref. | Nakhodka | 6,240 kl | Heavy C |
| 2001.1.17 | Taiwan | Kenting National Park | Amorgos | 1,150 kl | Bunker |
| 2001.3.30 | China | Mouth of Yangtze River | Deiyong | 700 kl | Styrene |

Sources: Based on reports including: Sao, Kunihisa, "Repeated Accidents"; Lee, Bon-Gil, "Changes in the Oil Spill Response System of Korea after the Sea Prince Accident," Oil Spill Intelligence Report.

South Korea learned a lesson from the three devastating spills of 1995 by over-hauling its laws on marine pollution prevention and completely changing its response systems, which are very similar to the system prescribed by the Oil Pollution Act of 1990 passed by the US after the 1989 Exxon Valdez accident. South Korea's system has the following major features.

(1) Centralizes response authority in the hands of on-scene coordinators.
(2) Just after an oil spill, it is classified according to anticipated spill volume into large (tier-3), medium (tier-2), or small (tier-1), according to which on-scene coordinators are assigned at the national and local levels.
(3) A progressional group of "scientific support coordinators" (SSCs) is orga-nized to support on-scene coordinators during spills.

One can readily see the significance of systems such as on-scene coordinators and control technology support groups when coming face to face with an oil spill. Japan's oil spill response system does not have the concept of "on-scene coordi-nators" who take charge of everything including final disposal of recovered oil, on land or at sea. In Mikuni Town, Fukui Prefecture where the *Nakhodka*'s prow drifted to shore, there were in fact five different headquarters: the Maritime Dis-aster Prevention Center, Fukui Prefecture, Mikuni Town, the fishing cooperative, and volunteer organizations. Recovery of oil washed ashore on a beach at Kaga City in Ishikawa Prefecture should have involved the maximum effort to avoid mixing oil and sand, but the use of heavy construction machinery contrary to advice mixed oil and coastal sand, producing a colossal volume of the mixture. Ultimately more than 10,000 cubic meters of this untreatable material were returned to the shoreline. Near the tip of the Noto Peninsula local people hurried their recovery operation by digging large holes and ditches on the coast and burying the recovered oil in them. Still another local government built a tempo-rary road along the coast for recovery work, then declared the job done after hardly performing any actual oil recovery. Inappropriate recovery efforts like these prolonged and worsened the environmental impacts more than if proper procedures had been followed.

One fundamental flaw of Japan's oil spill response system is that it is not clear where responsibility lies when oil washes ashore instead of being recovered at sea. Government agencies that directly respond to oil spills are the Japan Coast Guard and its auxiliary organization, the Maritime Disaster Prevention Center (MPDC). As their names indicate, their authority is clear at sea but ill-defined on land.

Japan's oil spill responses are governed primarily by the Law Relating to the Prevention of Marine Pollution and Maritime Disaster, which was partially amended in May 1998 because of the *Nakhodka* accident, and later in May 2000. These revisions allowed the Maritime Safety Agency to have the MPDC take whatever actions are necessary when a foreign-flagged ship has an accident outside of Japanese territorial waters, and opened the way for the government to pay response costs temporarily on behalf of other entities.

In December 1997 Japan's Cabinet adopted a national emergency response plan that includes express provisions for coordination with the relevant laws when a major oil spill occurs. As part of its efforts to make major oil spills subject to the national crisis management system the government also took other actions, such as creating a crisis management officer post in the Cabinet.

But unlike the contingency plans of the US and South Korea, Japan's hastily prepared plan is "spiritualistic," and certainly not anything like an action plan that anticipates the occurrence of specific accidents. Basically nothing has changed since the *Nakhodka* accident because there is still the question of who exercises the "unified command" that on-scene coordinators should have on land over national and local government agencies, local organizations such as fishing cooperatives, volunteer groups, and local citizens when accidents occur.

## 3-2    Sakhalin Petroleum and Natural Gas Development Projects

The RFE and eastern Siberia have a wealth of oil and natural gas. Most promising is Sakhalin's offshore area. In recent years promising offshore oil fields have been discovered in the Sakhalin area, and already several development plans are being fleshed out. Estimated recoverable reserves in the Sakhalin II Project, which has investment from Mitsui & Co., Ltd. and Mitsubishi Corporation, are 140 million tons of oil and 550 billion cubic meters of natural gas, thereby holding great promise for a stable supply of energy to Japan. But these oil fields are all in the Okhotsk Sea, which is also a bountiful fishery that yields world-class catches and underpins the economy of Japan's northern island of Hokkaido.

Oil extracted in Sakhalin II is first taken to a floating storage and offloading unit, and then transferred to tankers for delivery. Transferring oil between two vessels at sea makes it more likely that spills will occur, and in fact a minor spill in 1998 caused a massive dieoff of herring, according to reports by Friends of the Earth Russia, Sakhalin Environment Watch, and other environmental NGOs.

The necessity of preparing for major oil spills calls for specific steps such as: (1) set up a mechanism to encourage information disclosure on all resource development projects offshore of Sakhalin, and to ensure full implementation of environmental protection measures that include oil spill response actions and pre-boring studies; (2) all coastal local governments that might be affected by oil spills should develop emergency accident response plans under the international cooperation scheme of the North-West Pacific Action Plan that is discussed later; and (3) train personnel for organizing scientific support coordinator groups, and initiate official contacts with experts in countries that already have such systems, especially South Korea.

Recently studies have been performed on the profitability of building pipelines directly joining production and consumption sites. It is assumed that pipelines would be laid at least 1 km away from the coast by Hokkaido, and along the Sea of Japan and Pacific sides of Honshu, but the cost of compensation for fishery damage has not been taken into consideration. Yet even building pipelines at least 1 km from coastlines could have unavoidable impacts in case of an accident. For instance, fishermen stock seed scallops between 500 meters and 1 km off the Okhotsk Sea coast, a zone that would be directly affected.

Yet owing to the international situation of late, the heavy dependence of Japan and most ASEAN countries on Middle East oil is encouraging the rapid development of energy resources around Sakhalin and in the RFE. In February and March of 2003 Tokyo Gas, Osaka Gas, Nippon Oil Corporation, and other Japanese energy companies announced that they would be purchasing natural gas and oil produced in the Sakhalin offshore area. However, some of the environment-related documents released by the Sakhalin Energy Investment Company Ltd., the operating company for Sakhalin II, concerning the construction of pipelines and other phase II construction raise questions about the soundness of methods used in response to oil spills. For instance, high priority is placed on applying chemical dispersants for every kind of spill response.

# 4. Russian Far East

## 4-1 Maritime Province Forests (Taiga)

Russia has vast forest resources in the form of the taiga, one of the world's largest expanses of softwood forest. With the tundra, it corrals a large amount of the greenhouse gas methane underground. The taiga covers 650 million ha, of which the RFE and eastern Siberia alone have 210 million ha representing a timber volume of 26.1 billion cubic meters. Most of the timber produced there is for export, and exports to Japan are rapidly increasing. In the southern Far East tree species include Japanese ash and oak, which are ecologically important, as well as creeping pine and alder, which account for 85% of the taiga, while in the north is a vast tundra reaching the Arctic circle and populated by reindeer and many minorities who make their living by hunting and fishing.

Logging in recent years has cleared 150,000 ha annually, losing especially the economically accessible mixed forests of Korean pine and hardwoods, and forest of Yezo spruce and Saghalein fir, which have turned into secondary forests. This logging not only diminishes forest regenerative capacity, but also induces forest fires that spread over an average of 200,000 ha every year. Although determining the $CO_2$ emitted by these fires is difficult, it is estimated at between 100 and 200 million tons annually. Russia's forests can absorb about 100 million tons of $CO_2$ a year, but fires have triggered the thawing of the permafrost, which extends as far down as 800 meters in Siberia and down to 50 meters in parts of China and Mongolia.

Currently there is much young forest, and the taiga is degenerating both qualitatively and quantitatively. And although total RFE forest size remains unchanged, half of the Far East forest industry is concentrated in Khabarovsk, while forestry in Sakhalin and Amur is declining, and that in Kamchatka and Magadan is no longer viable. This situation threatens not only the timber supply, but also the maintenance of river water quality needed to regenerate fish populations (especially salmon), and efforts to arrest global warming. Far East forestry faces the challenge achieving sustainable forest management that abandons careless logging practices which engender forest degradation. To ensure the survival of minorities, the Northern Minorities Association (RAPON) was established during perestroika under the former Soviet Union, and it sought rectification of the economic intrusion into native people's areas by government ministries, distortions in policies on minorities, retarded social development, and rapid environmental deterioration. Logging was later banned in certain areas and for the Korean pine, but reports of illegally logged timber by importers have raised concerns that management systems are inadequate to the task.

Absolute restraints on ecosystem regeneration include (1) high soil acidity in the Siberian and Far East tundra, (2) the low capacity of permafrost bacteria to degrade pollutants, and (3) sparse vegetation. In the southern tundra it takes 400 years to grow to a thickness of 10 cm due to the low temperature, and these forests play a role in the global regulation of $CO_2$ absorption. Far East forests are said to have a sustainable yield of 100 million cubic meters a year, but this region needs an arrangement under which timber production will maintain a balance with the ecosystem and the livelihoods of native minorities.

Although China now consumes almost 280 million cubic meters of timber a year, its domestic production is only 142 million cubic meters, and the shortfall is covered by imports and illegal logging. The International Tropical Timber Organization forecasts that within a few years China will eclipse the US and Japan as the world's biggest timber importer. Fifty-seven percent of China's imported timber comes from Russia, where the decline of local sawmills allows logs to be exported without processing. Environmental NGOs point out that at least one-fifth of Russian timber is logged in contravention of laws, and attention is focused on exports of illegal timber to Japan as well as to China.

Forest fires broke out in eastern Siberia beginning in early April 2003, and by May 23 at least 100,000 ha had burned in Irkutsk Oblast, Buryat Republic, Chita Oblast, Amur Oblast, and Khabarovsk Krai. NASA satellite photos suggest that the fires spread to northern Mongolia and the northern part of China's Northeast. These fires are suspected of contributing to the haze over Japan's northern island of Hokkaido and the Tohoku region.

## 4-2  Nuclear Waste Dumping

Parts of the RFE suffer a continuing electricity shortage, prompting district officials to demand nuclear power plant construction. China and South Korea have aggressive nuclear power policies. China has six operating reactors, and although the Northeast has no reactors operating, under construction, or planned, China intends to have 40,000 MW of nuclear capacity by 2020. South Korea already gets 34.3% of its electricity from nuclear, and it has formulated an electricity supply and demand plan calling for developing 23,000 MW of nuclear capacity by 2010. Add trends in North Korea and Japan's reactors along its Sea of Japan side, and the general public's risk awareness is bound to rise. In view of the contamination that might spread in any direction depending on the wind, many people are saying it is necessary to quickly set up a system for international cooperation having joint monitoring of radioactive substances as one of its main elements.

A deep-sea area in the Sea of Japan offshore from Vladivostok is the final resting place of decommissioned nuclear reactors from submarines and ice-breakers dating back to the former Soviet navy. According to the February 1993 Yablokov Report, the former Soviet navy dumped large amounts of nuclear waste beginning in the mid-1960s, and the Russian Pacific Fleet dumped radioactive wastes in the Sea of Japan six times in 1992 alone. It was also found that in August 1985 at a Russian Pacific Fleet base near Vladivostok, an explosion happened because of an operational error when changing fuel in a nuclear submarine. In October 1993 the Russian Navy dumped 900 tons of low-level liquid radioactive waste generated by nuclear submarines into the Sea of Japan. Although dumping stopped after that, the Pacific Fleet commander said that the lack of places to store low-level waste would force Russia to resume sea dumping in the future.

In 1994 Russia and Japan sealed an agreement under which they would build a plant at a naval arsenal in Peter the Great Bay to treat liquid radioactive waste, and that Japan would finance it. But being a non-nuclear country places limitations on what Japan can do to help denuclearization. The expansion of Japan-US cooperation on nuclear wastes and the treatment of liquid radioactive waste from submarines comprise just one area related to nuclear power, leaving many other challenges to address, such as the treatment of spent nuclear fuel and other solid wastes, decontamination and disposal of ships and treatment facilities, and the construction of high-level waste storage facilities.

Cooperation projects already off the ground have also developed problems. For example, private companies with much experience in nuclear vessel ship-breaking, nuclear waste treatment, and environmental protection are in the US. Thus although the Japanese-financed construction of a nuclear waste treatment facility for superannuated nuclear submarines was begun under a commission to a British consulting company in 1995 by the Japan-Russia Committee on Cooperation for the Elimination of Nuclear Weapons, which is under the control of Japan's Ministry of Foreign Affairs, the project was delayed more than three years due to circumstances in Russia, and it has yet to gain momentum (Japan has already provided 3.6 billion yen in grant aid, and is to continue with the total 20 billion yen in aid).

Construction of the nuclear waste treatment facility was sparked by the Russian Navy's multiple dumpings of nuclear waste in the Sea of Japan in 1993. In November of that year the London Convention added an amendment prohibiting the dumping of nuclear waste at sea, but Russia refused to accept it, and there is an increasing possibility that Russia will resume dumping.

Because the Sea of Japan is a semi-enclosed sea in which pollutants are caught in deep-sea circulation taking between 100 and 300 years, the safe management of pollutants and their impact on fishery stocks are of great concern. The quickly growing number of nuclear power facilities in Northeast Asia and the marine dumping of radioactive waste by the former Soviet Union and Russia pose new risks of radioactive contamination in this region's marine environment. Among the artificial nuclides introduced into the sea are plutonium 239, which is most closely watched because of its half-life of 24,000 years and high toxicity due to the radiation it emits. The Japan Nuclear Cycle Development Institute is involved in the engineering work and financial assistance for recovering the highly enriched uranium and plutonium yielded when dismantling the former Soviet Union's nuclear weapons.

# 5.    Building Systems for International Cooperation

This section briefly describes the main frameworks for international environmental cooperation and their actors in Northeast Asia, especially with regard to marine and air pollution.

## 5-1    Main International Cooperation Efforts

The United Nations Environment Programme is running Regional Seas Programmes under which the countries around certain marine areas work together to monitor and conserve enclosed seas and coastal and border zones. One of these programs is the North-West Pacific Action Plan (NOWPAP), a framework set up jointly by China, South Korea, Russia, and Japan to conserve the marine environment in the Sea of Japan and the Yellow Sea. North Korea participated in preparation, but did not join in adopting the plan. However, it did participate in the seventh intergovernmental meeting in 2002 as an observer.

This action plan's purposes include (1) preparing databases of monitoring information and other data, and an information management system, (2) monitoring of coastal and marine environments, (3) systematic management of coastal and marine environments and resources, and (4) preparations for mutual assistance in emergencies. Since the initial intergovernmental meeting in South Korea in 1994, meetings have been held at slightly greater than one-year intervals.

Under NOWPAP, Regional Activity Center (RACs) were established in the four participating countries to develop immediate priorities and assign specific tasks. In Japan the Northwest Pacific Region Environmental Cooperation Center (NPEC) in Toyama City was designated as the Special Monitoring and Coastal Environmental Assessment Regional Activity Center (CEARAC), and there are expectations that it will benefit NOWPAP activities with expertise, technologies, and other information on local-level conservation measures and international environmental cooperation. NPEC has additionally been commissioned to run the satellite receiving station built by the Environment Ministry in 2002, and it is building a system to receive and analyze marine environmental remote sensing data (including sea surface temperature and color) from satellites, then provide it to entities in Japan and other countries, including NOWPAP participants.

The Sea of Japan has a high potential risk of accidents because of its comparatively heavy tanker traffic, which is why the *Nakhodka* accident served as a powerful catalyst for current efforts to quickly prepare for international cooperation on oil spill accidents, and these have been the most effective of NOWPAP activities. International society also invests considerable expectations in NOWPAP because it is no overstatement to call it the only political framework in Northeast Asia for international cooperation on oil spills. Under South Korea's leadership NOWPAP has already prepared the NOWPAP Regional Oil Spill Contingency Plan, which could be considered a manual for the purpose. At the Seventh NOWPAP Focal Points Meeting held 18–21 May 2004 at MERRAC (Daejeon, Korea), participants from NOWPAP members discussed the progress of tasks related to regional cooperation on marine pollution preparedness in the NOWPAP region and the Sea of Okhotsk. Navigation monitoring, crew education, enhanced inspections and maintenance, studies of submarine topography, and other basic efforts are also being made to prevent tanker accidents or their worsening when they do occur.

Although some activities are conducted only on the government level, local governments with high awareness or NGOs initiate activities within the NOWPAP framework without central government involvement. For example, objects that are buried in or drift to shore are a serious problem along not only the coast of Japan (especially the Sea of Japan side), but also those of South Korea and China, but governments in this region are not taking positive action. Owing to the seriousness of the situation, local governments and people in mainly the four NOWPAP countries took it upon themselves to survey the types and amounts of wastes that drift ashore.

Transboundary acid rain also has direct impacts on the general public in Northeast Asian countries. A framework for international cooperation to address acid rain is the Acid Deposition Monitoring Network in East Asia (EANET), which officially began operating in 2001, and has the participation of 10 Asian countries. This network's purpose is to create a framework for joint regional action on acid rain based on scientific knowledge. Under the framework, countries employ uniform methods to collect and analyze monitoring results on the dynamics of SOx and other substances, and quantitatively elucidating the state of acid rain in East Asia, how it arises, and what its impacts are.

Behind EANET is the Acid Deposition and Oxidant Research Center, which was established in 1998 in Niigata City, Japan, and serves as the network's de facto secretariat. The center actively accepts researchers and government administrative personnel from other countries as trainees in an effort to have all countries use the same methods to monitor air pollutants, thereby serving as a total information provider on acid rain. Recently it is not only studying SOx, but also beginning work on the generation and movement of other air pollutants such as NOx, particulates, ozone, and mineral dust.

Still, although no one denies the long-distance transport of air pollutants from China to Japan, a more constructive discourse on transboundary acid rain unquestionably requires accurately determining the quantitative relationship between the sources and recipients of cross-border air pollutants. Nevertheless, there is no long-range air pollutant transport model for this region that can serve as the basis for a shared understanding on quantitative relationships. To address this shortcoming, the center is leading an initiative that includes researchers from Japan, China, South Korea, the US, Taiwan, and other places to compare their transport models, and it is hoped that in a few years this will yield

TABLE 2. Municipalities Participating in the Association of North East Asia Regional Governments

| Country | Regional Government |
| --- | --- |
| China | Heilongjiang Province, Shandong Province, Ningxia Hui Autonomous Region |
| Japan | Prefectures of Aomori, Yamagata, Niigata, Toyama, Ishikawa, Fukui, Kyoto, Hyogo, Tottori, Shimane, Yamagata, and Akita (observer) |
| Mongolia | Central Province |
| S. Korea | Gangwon-do, Chungcheongbuk-do, Chungcheongnam-do, Jeollabuk-do, Gyongsangbuk-do, Gyongsangnam-do, and Busan-Gwangyoksi |
| Russia | Primorsky territory, Khabarovsk, Kamchatka, Sakhalin, and the Interregional Association Far East & Zabaikalye (observer) |

Source: The Association of North East Asia Regional Governments.

a common understanding on the relationship between pollutant origins and destinations.

In 2000 The Association of North East Asia Regional Governments (NEAR), which was created to enhance collaboration among local governments in Northeast Asia, held the Conference of the Association of North East Asia Regional Governments on Awaji Island in Japan with delegates from 26 local governments of China, Japan, Mongolia, South Korea, and Russia (Table 2). This conference reaffirmed the value of NEAR, which is to work toward the advancement of Northeast Asia as a whole by enhancing mutual understanding through the encouragement of interregional exchanges, and by sharing technology and experience. Participants agreed that future participation by North Korean local governments in NEAR is welcome because it would contribute to the common advancement of Northeast Asia and to world peace.

Further, the Economic Research Institute for Northeast Asia (ERINA),[6] a think tank that specializes in this region, annually hosts the Northeast Asia Economic Conference with the participation of delegations from throughout the region. ERINA's primary areas of concern are Tumen River development and finding solutions to environmental problems. On the NGO level, at the First East Asian Citizens Environmental Conference held in November 2002 as part of as part of a trilingual information-sharing and exchange project among Japan, China, and South Korea,[7] delegates sought participation from Russia's coastal region, Mongolia, and North Korea. These and other ideas and initiatives hold promise that bilateral and Japan-China-South Korea cooperation projects will have a spillover effect on all of Northeast Asia.

## 5-2  Challenges for International Cooperation

Northeast Asia as a whole faces at least five challenges in relation to international cooperation.

First is the difficulty of creating organizations. Especially when governments are involved, each government's role and order of precedence are points of contention in international cooperation. In not only Northeast Asia but Asia as a whole, there are considerable dissimilarities in socioeconomic systems, levels of economic development, cultural and sociological context, political motivation, a sense of human rights including environmental rights, and perceptions of environmental problems that require urgent responses, which are all impediments to creating organizations. In both EANET and NOWPAP, in fact, Japan and South Korea have fought over who will be in the driver's seat, and in the case of NOWPAP, there was disagreement to the very last on which country would host the Regional Coordinating Unit. Ultimately the problem was solved by a compromise in which centers would be located in both countries. The unfortunate reality of environmental cooperation in Northeast Asia is that much time and energy are wasted in political negotiations and in creating a shared understanding, and the region is finally starting regular monitoring that covers a sizable area, that uses the same methods to measure the main pollutants, and that can serve as the basis for all discussions.

Second is insufficient funding. While this is a difficulty for all governments, it is more pressing for local governments and NGOs as they must often depend on central government funding. For example, there is a plan for joint monitoring of marine water quality within the NOWPAP framework in the near future, but hardly any of the funding has been secured. Further, owing to Japan's economic prominence in this region, the Japanese government or the prefectural governments of Toyama and Niigata pay large shares of the operating costs for EANET and NOWPAP. Such being the case, these are not equitably funded sustainable institutions for international cooperation.

Third is the lack of effectiveness. Both NOWPAP and EANET are "plans" depending on volunteer participation, not conventions for reducing pollutant emissions as in the Mediterranean region and Europe. Hence there are always questions about the effectiveness of these plans because they have no numerical targets for emission reductions, and they are not legally binding. While the European experience cannot be directly applied to Northeast Asia, at some point it will be necessary to build more effective frameworks.

Fourth is that because pollution-caused damage in Northeast Asia is not as apparent as in other regions, there is a low awareness of the importance of international cooperation for effecting improvements. More than being a challenge itself, it is a circumstance underlying the previous three challenges. Specifically, water quality deterioration in the Sea of Japan is not as serious as that in other semi-enclosed seas like the Mediterranean, the Baltic, and the Aral seas, and at least in the rivers and lakes of Japan there are no fish die-offs caused by transboundary acid rain as in Europe. Of course it is possible that in the future red tides like those in the Yellow Sea could appear in the Sea of Japan, and there are concerns about the impacts on fisheries of heavy metals such as lead and cadmium, and of persistent chemical substances like PCBs and pesticides. If China's coal consumption continues to grow, there is a definite possibility that the impact of transboundary acid rain on Japan's ecosystem will become apparent in the future. At present, however, the relationship between source and recipient countries is not quantitatively clear.

Fifth is territorial issues. At least now, these do not affect environmental cooperation very much, but in the future, territorial issues such as Takejima (Dok-do Island), claimed by both South Korea and Japan, could be a serious intrusion on international cooperation in the environmental arena. Territorial issues could, for instance, precipitate serious bickering over what country should be in charge of joint water quality monitoring in affected ocean areas, making it possible that such areas would be intentionally avoided. In fact, for political reasons NOWPAP does not use the name "Sea of Japan"; as it would be a waste of time to hold discussions in NOWPAP venues on what the official name should be, participants have provisionally settled on "North-West Pacific Sea."

# 6.   Summation: The Need for Engagement

Following is what we see as a comprehensive concept that should underpin environmental cooperation in Northeast Asia.

In the 21st century national and local governments, NGOs, and citizens should, either through mutual cooperation or on their own, propose new frameworks as arrangements for regional cooperation, and carry out comprehensive, tangible environmental cooperation covering the gamut from assistance for the sustainable development of China's coastal zone, to low-impact development of central Asian and eastern Siberian resources, reform and opening of the northern Korean Peninsula with sufficient care for the environment, RFE development, and nuclear waste management including arms control. Especially with regard to environmental problems that require collaborative action because they are shared by all of Northeast Asia, countries in the region must provide for solutions on the global citizen level and pursue regional development that has close relevance to livelihood and a low environmental burden. In doing so it is important to build partnerships that include local governments, NGOs, businesses, and other parties.

To make the Sea of Japan into the "Sea of Cooperation," it stands to reason that relations among rim countries must be improved, and the most important matter is building (or rebuilding) relations with North Korea. South Korea has adopted a "sunshine policy" since the government of Kim Dae-jung and has brought the two heads of state together, but subsequent improvement in relations has not gone well. North Korean relations have also soured with Japan and the US. Combined with these factors, other contentious issues such as North Korea's nuclear status have conspired to further isolate the nation, making it difficult to shepherd an improvement in relations as far as social development and environmental conservation.

Improvement of relations will necessitate finding many different channels for candid dialog. In particular, there is a great need to urge North Korea to participate in academically oriented activities, in NGO/NPO networks, and in exchanges among local governments at NEAR meetings, and to support North Korea's involvement in Northeast Asian cooperation through environmental conservation. Although North Korea does not formally participate in NOWPAP, it is an observer, and the importance of carefully tending such channels cannot be overstated.

Sawano Nobuhiro, Long Shi-Xiang, Jin Jian, Katsuragi Kenji,<br>
Aikawa Yasushi, Asuka Jusen

## Essay:  Songhua River Minamata Disease[8]

China's Songhua River watershed is the scene of organomercury contamination and resulting illness like that of Minamata disease. The polluter was Jilin Petrochemical Company in Jilin City, Jilin Province, which dumped waste into the Second Songhua River, a tributary of the Songhua River that flows north across Jilin Province from near the border with North Korea, joins the Nen River, and becomes the Songhua River. Organomercury pollution affected mainly the border area between Jilin and Heilongjiang provinces, about 300 km downstream from the source. In fact, two cases on which postmortem examinations were performed were fishermen living in this area. In two other cases of fishermen living in the same area, declining values for total mercury in hair (90.00 mg/kg to 15.00 mg/kg, 84.00 mg/kg to 3.71 mg/kg) and total methylmercury in hair (69.00 mg/kg to 8.30 mg/kg, 64.00 mg/kg to 1.80 mg/kg) found by intermittent hospital examinations from 1976 to 1982 were reported, along with long-term reports on whether patients had symptoms such as constricted visual field, loss of peripheral sensation, and reduced hearing (nearly all symptoms were observed).

But 1986 data from measurements of methylmercury in the hair of fishers in the Songhua River watershed in Heilongjiang Province where the river runs west from the confluence had values of 26.94 mg/kg maximum and 10.52 mg/kg average (116 people) at five sites on the river's upper reaches near the provincial border, which overlaps the area described above, while at four sites in the midstream area near Harbin (about 500 km downstream from the pollution source) the values were 16.94 mg/kg maximum and 6.58 mg/kg average (61 people), and at three sites near the Russian border (about 1,200 km downstream from the pollution source) they were 34.91 mg/kg maximum and 11.27 mg/kg average (47 people). Data from 1990–91 at five sites on the upper reaches had values of 8.88 mg/kg maximum and 5.31 mg/kg average (116 people), while at four midstream sites the values were 5.11 mg/kg maximum and 3.52 mg/kg average (46 people), and at three sites on the lower reaches they were 7.12 mg/kg maximum and 5.37 mg/kg average (28 people). Hence, instead of decreasing as the distance downstream increases as one would expect, the values increase with the distance from Harbin both in the upstream and downstream directions. These results lead to the inference that differences in the hair mercury values of fishers in this watershed arise from socioeconomic factors such as the dependence on the Songhua river for food and the underlying economic structure of Heilongjiang Province with Harbin at its center. They also suggest that harm is more widespread.

At the border with Russia the Songhua River joins the Amur River, which further joins the Ussuri River at the northeastern extremity of Chinese territory, flows through Russia's Maritime Province, and then empties into the Sea of Japan at the Tartar Strait. The high mercury values for hair of people living along the lowest reaches of the Songhua River (where it joins the Amur River) give rise to concerns that there may be undiscovered mercury contamination and cases of Minamata disease along the lower reaches of the Amur River as well. So far there have been few such wide-area studies, and especially no international studies including Russia, but as part of the initiatives toward Northeast Asia cooperation, Japanese researchers are beginning to propose international studies and other undertakings on Songhua River Minamata disease in view of Japan's heavy historical involvement in this region and its experience with Minamata disease.

Aikawa Yasushi, Tani Yoichi

# Chapter 2
# The Mekong Region: Incorporating the Views of Regional Civil Society

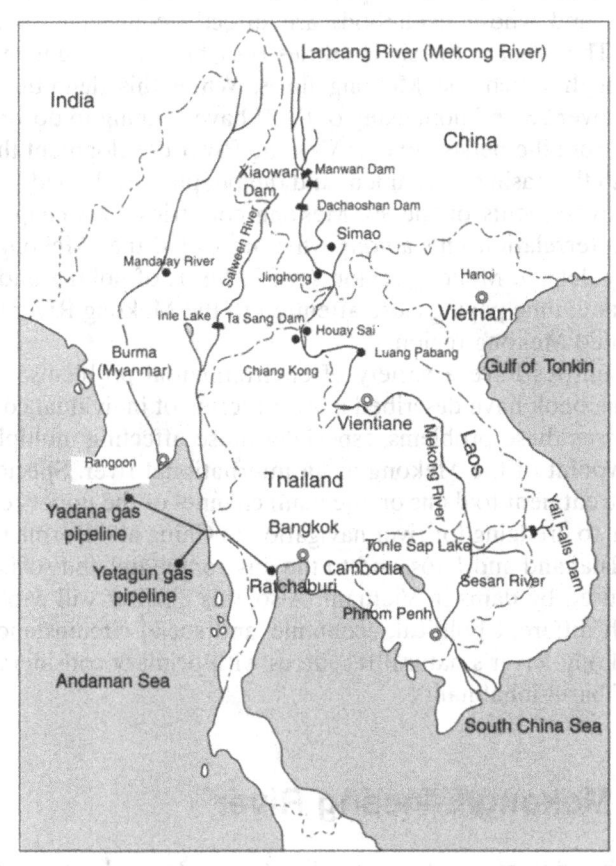

# 1. Introduction: The Environment and Development Across Boundaries

The length of the Mekong River could be 4,400 or 4,800 km, depending on where its source is placed. Either way, in length and flow rate it is one of the world's largest rivers, and the largest in Southeast Asia, with a basin covering 795,000 square km, or more than twice the land area of Japan. Arising on the Tibetan Plateau, which is also the source of the Yangtze and Salween rivers, the Mekong flows through China's Yunnan Province, Burma (Myanmar), Lao PDR (hereafter, Laos), Thailand, Cambodia, and Vietnam on its way to the South China Sea.

"Mekong Region" has two definitions. One is the natural geographic region bound by the river's basin. It has a population of 60 million, many of whom live in rural areas and whose livelihoods are directly connected to the riverine environment. The other is the six countries (in China, only Yunnan Province is counted) through which the Mekong flows. While this definition includes a population of over 230 million, many of them have nothing to do with the river. Nevertheless, from the perspective of Mekong River development that is closely associated with the basin environment and the people's livelihoods, we must also include the governments of the six Mekong countries. This chapter therefore explores the interrelationships among what are called the "Mekong countries," or the broadly defined Mekong region seen in terms of politics and economics, while at the same time paying close attention to the Mekong River basin, or the narrowly defined Mekong region.

Mekong countries have a variety of environmental problems, and previous volumes of this book have described a few in terms of individual countries. This chapter discusses these problems, especially those affecting multiple countries, from the viewpoint of the Mekong as an international river. Specifically, it will give detailed treatment to dams on the main channel of the upper reaches of the river in China, to dredging for river navigation in China and Burma with the participation of Thailand and Laos, and to the environmental and social impacts in Cambodia caused by dams in Vietnam. Also, this chapter will explore how six countries with different political, economic, and social circumstances can conserve the Mekong River's natural resources with primary consideration for the livelihoods of basin inhabitants.

# 2. The Mekong/Lancang River

The Mekong has different names according to country and region. In China it is the Lancang Jiang, and in Thailand and Laos it is Mae Nam Khong (Mother of Waters), while downstream it is called Tonle Thom (Great Water) in Cambodia and Cuu Long (Nine Dragons) in Vietnam. This section deals with the environmental and social impacts of main-channel development in China (Lancang) and the upper-Mekong countries.

## 2-1    Mekong Upper-Reaches Dredging

### Freer Commercial Navigation in the Four Upper-Mekong Countries

The four upper-reaches countries of China (Yunnan Province), Thailand, Laos, and Burma signed a commercial navigation agreement in April 2000 to energize international river traffic. Based on this agreement, in March 2002 they commenced work to remove rapids, shoals, and reefs that impede river traffic. Blasting is in progress along a 331-km section of the river from Number 243 Boundary Stone on the China-Burma border to Ban Houay Sai in Laos. This section has over 100 rapids, shoals, and reefs, and the plan calls for blasting the 11 major rapids and 10 reefs that pose the greatest encumbrance to river navigation. China provided $5 million to finance the work. China has been dredging its part of the river since the 1960s, and has made a 293-km stretch of the river navigable for 100- to 300-ton vessels. China hopes to see the navigable length of the river extend southward.

According to a four-country joint survey,[1] by 1993 there had been 50 ship accidents on the river's upper reaches. Nine of them were caused by collisions with reefs, resulting in sinking or loss of cargo. In addition to preventing such accidents, the plan aims to make the 886-km length of the river from Simao in Yunnan Province to Luang Pabang, Laos navigable to vessels of 300 to 500 tons, and to invigorate economic interaction in the upper-Mekong region. Presently the four upper-Mekong countries have 110 vessels that transport between 150,000 and 200,000 tons of cargo annually. It is predicted this will grow by a factor of 10, to 1.5 million tons of cargo and 400,000 people annually by 2010.

### Environmental Impacts

The greatest concern over the impacts of dredging is that destroying rapids and shoals to facilitate the safe passage of shipping traffic will seriously damage the ecosystem that fish have depended upon for spawning sites and habitat. In an EIA[2] performed before the dredging, the four participating countries set forth these study results and precautionary measures:

- Explosions are set off in advance to warn fish away.
- No blasting is done during fish migrations or spawning.
- Blasting rapids slows the river's flow, which aids fish migration upstream.
- Blasting shoals and reefs affects only very small areas, nearly always no farther than 300 meters from the blast.
- Changes in river flow velocity are limited to at most slightly over 1 meter per second.
- Because the river banks are rock, they will not collapse from soil erosion.
- The flood peak will pass Ban Houei Sai, the most-downstream location of the dredged length, only 58 seconds sooner than now.

On this basis the four-country study team concluded that there would be no serious damage to the fishing industry or downstream environmental impacts.

But the Mekong River Commission (MRC), a regional international agency that coordinates water use in the Mekong watershed, commissioned experts to perform a review of the EIA, and released three reports,[3] which made the following harsh observations on the EIA report's problems.

- There is hardly any biological analysis.
- No scientific basis at all is offered to substantiate the conclusion of no impacts.
- The EIA lacks even a list of fish living in the river.
- There are no references at all to aquatic organisms other than fish.
- The five-month study (two days field work) is too short. At least two years are needed.
- The study violates the assessment rules of Laos' Science, Technology and Environment Agency (no alternative analyses, and no examination of the connection with domestic Laotian law).
- The study analyzes the direct impacts of blasting, but does not at all analyze the long-term impacts.
- There is no mention of secondary impacts that are the concomitant of increased river navigation and tourism.
- The report does not reveal methods used to study social impacts.
- There were no consultations at all with river people in Laos and Burma.
- The EIA is speculation and therefore entirely unacceptable.

These reports were written at the behest of the Laotian government, a member of the MRC. Although they were submitted to Laos through the MRC, this has not led to a redo of the EIA.

## Concerns of Regional People

In response to reports of the first rapids blasting in March 2002, opposition rapidly spread among local people and NGOs mainly in Chiang Khong District of Chiang Rai Province in northern Thailand along the upper Mekong. Rapids blasting was temporarily interrupted in late April because of the rainy season. During the stoppage in June the chairman of Thailand's Senate Environment Committee and other officials visited the site and demanded that the Thai government stop the blasting. Twelve thousand people in Chiang Khong District held a protest demonstration against the project.

Citizens' groups in Chiang Khong District and northern Thailand NGOs carried out their own study of the project's environmental and social impacts, and issued a report[4] spelling out their grave concerns:

- Aquatic plants that feed fish, and the freshwater algae (order Schizogoniales) that locals eat will be lost.
- Rapids blasting will destroy fish habitat, which will decrease the numbers of the endangered giant Mekong catfish and other fish, thereby affecting 100,000 people who depend on small-scale fisheries.
- Passage of large vessels might pollute the river and present hazards for small fishing boats.

On July 31, 2002, 76 NGOs and citizens' groups from 25 countries, mainly Thailand, sent a letter of protest to Thailand's Deputy Prime Minister and Defense Minister General Chavalit Yongchaiyudh, urging him to immediately stop the dredging project and to perform a comprehensive environmental impact study including the effects on Vietnam, Cambodia, and other downstream areas.

### Concerns of Governments and International Agencies

Mr. Joern Kristensen, then-CEO of the MRC, also has voiced strong concerns about upper Mekong dredging, and made critical observations that dredging the river bed affects fish reproduction and could change the downstream flow of water.[5] But as China is not an official member of the MRC, such criticism has hardly any effect.

Although Laos agreed to upper-Mekong dredging, it said that Vietnam and Cambodia had not, making it necessary to obtain their agreement.[6] For this reason the blasting of rapids and reefs was temporarily stopped. From the beginning it was said that China and Thailand would benefit the most from this project, while Laos was not very enthusiastic.

Even Thailand's government started raising objections for reasons of national defense, not the project's environmental and social impacts. On July 18, 2002 the Cabinet Screening Committee, headed by General Chavalit Yongchaiyudh, said it was possible that dredging would change the Mekong River boundary between Thailand and Laos, and asked the government to review the dredging plan. "The needs of all countries sharing the river, including downstream countries Cambodia and Vietnam, should be considered," said the general.[7] Additionally, the Thailand-Laos Joint Committee on Border was supposed to have defined the 976-km Mekong border between the countries by the end of 2003. Until that is done, it is possible that at least one rapid in Thai territory will not be blasted.[8] This was welcomed by citizens' groups and NGOs concerned about environmental and social issues, which suggests the reason why the recipient of the July 31 letter to the government was General Chavalit—who also served as defense minister—instead of the prime minister.

Yet, Thailand's government nixed only the blasting of the rapid near Chiang Khong because of the border issue, and has said nothing about the blasting of reefs and other rapids involving Laos and Burma. Criticism has emerged from within the Thai government of the early 2002 agreement to the blasting based on the four-country commercial navigation agreement.

## 2-2  China: Lancang Mainstream Dam Development

### Xiaowan Dam

In January 2002 China started construction of the Xiaowan Dam on the Lancang River main channel in Yunnan Province. This large dam is China's second-largest after the Three Gorges Dam, having a height of 292 meters, projected generating capacity of 4,200 MW, and water impoundment of 15 billion cubic meters. All

PHOTO 1. Dachaoshan Dam, the second dam on the Mekong River's mainstream (February 2003 photo courtesy of Mekong Watch)

of its approximately $2.68-billion price tag is to be raised domestically. The dam will inundate 370 square km and, according to official Chinese government figures, force the relocation of 38,000 people.

This is the third dam on the Lancang main channel after the Manwan Dam (completed in 1993) and the Dachaoshan Dam (nearly completed). All three are meant to generate electricity. China's government is considering another five dams, and if they are all built, Yunnan Province alone will have a generating capacity of 15,600 MW. Hydropower is 70% of Yunnan Province's potential generating capacity of 90,000 MW. Power is supplied to Guangdong Province and other places in high-demand eastern China as part of "Western Electricity Sent East," one of four large projects under China's current Tenth Five-Year Plan.

In 1995 the Mekong River Commission was created to reconcile the interests of Mekong countries in the development of water resources on the Mekong River's main channel, and for that reason the MRC members of Vietnam, Cambodia, Laos, and Thailand are not building dams on the main channel. But as China is not an MRC member, it is pushing ahead with dam construction on the Lancang main channel in its own territory without discussing these projects at all with downstream countries.

## Claimed Benefits for Downstream Countries

Little outside information was available on the colossal Xiaowan Dam until May 2000 when the Asian Development Bank (ADB) released a report on research conducted as part of its technical assistance for the project, titled "Policies and Strategies for the Sustainable Development of the Lancang River basin," whose fourth chapter dealt with water resources and hydropower.

This report claims that this large-capacity dam will benefit downstream countries because it will increase river flow during the dry season and alleviate flooding at the rainy season peak. He Daming, one of the ADB report's authors and director of the Yunnan University Asian International Rivers Center, wrote in a newspaper article[9] that this dam will decrease water in the downstream Mekong 17% during the rainy season and increase it 40% during the dry season. The ADB report concludes that main-channel dams will have little effect on flood mitigation in the Mekong delta and other areas owing to the large contribution by Mekong branches in Laos, but that they will make a big contribution to dry-season irrigation.

## Problems with Lancang Main-Channel Dams

Of the many concerns voiced about such dams' environmental and social impacts, a serious one is siltation. The ADB report writes that in the first three years of its operation the Manwan Dam, the first mainstream dam to be completed, lost the same effective impoundment volume originally forecast to be lost in 15 years. Some experts are even saying that the Xiaowan Dam is needed just upstream to stave off the crisis of the Manwan Dam filling up with sediment.

Stopping siltation with the Xiaowan Dam will release sediment-hungry water downstream, and the ADB report observes that this relatively sediment-free water tends to erode downstream riverbeds, raising concerns about physical damage to bridges and other structures.

Of even greater importance is the role that sediment from China has played in the Mekong River downstream. Although China's contribution to total Mekong River flow is just 16%, half of the 150 to 170 million tons of annual sediment is estimated to originate in China. The ADB report therefore expresses concerns that the Xiaowan Dam will have a damaging impact on downstream rice-producing areas, especially the Mekong delta.

During the dry season people living along the Mekong grow crops on the banks left fertile by the river when it floods. While there are no statistics on production amount, many field workers attest to the importance of this practice. Even though increased river flow in the dry season could make this river-bank agriculture impossible, this has not been considered at all.

## Concerns of the Mekong River Commission and Downstream People

In January 2002 just after China announced that it would start work on the Xiaowan Dam, MRC CEO Kristensen expressed the following concerns:

"China needs to realize that the Mekong River system is one ecological system that should be cared for cooperatively." The four-member commission was concerned most about any reduction of downstream water flow and a drop in water quality. "If the flow is going to be altered then that could have a severe effect on the lower Mekong region," "And if the water's quality is altered, then that could impact downstream fisheries which provide the single most important source of protein for millions of Cambodians."[10]

Reduced flow occurred soon after these remarks in March 2002 when the Chiang Saen hydrographics center in northern Thailand near China observed a water level drop of more than 1 meter in just 10 days, and a temporary sudden rise even though it was the dry season.[11] One cause was a water release from dams. Fishermen at Chiang Khong in northern Thailand have observed two or three changes in water level daily since about 1998, and claim that amounts of fish and other aquatic life have decreased as a consequence.[12]

### Chinese Government Data

On April 1, 2002 China's Ministry of Water Resources agreed to supply the MRC with Mekong River hydrological data once each day via a computer network. Specifically, this information is the water levels and precipitation amounts at two observation stations in Yunnan Province, Yunjinghong and Man'an, to be provided during the rainy season at 8:30 a.m. from June 15 to October 15. The MRC agreed to help make improvements at these two observation stations. However, this agreement between China and the MRC did not sufficiently take the needs of downstream people into consideration because changes in dry-season water volume affect downstream fishermen just as seriously as floods do.

Although the MRC applauds China's cooperative attitude, many people express doubts about the effectiveness of this arrangement. A September 6, 2002 signed article in the Bangkok Post described the grave flooding and erosion during the 2002 rainy season and cited water discharges from main-channel dams as one of the causes. While main-channel dams have severe environmental and social impacts, of even greater seriousness is that water discharges and impoundment are decided solely at the discretion of the Chinese government with no consideration whatsoever for the concerns of downstream-country governments and the people who are affected.

# 3.    Bilateral Issues: Cross-Border Flooding

Cross-border environmental problems in the Mekong River basin are complex and diverse. For example, the ecosystem of Cambodia's Tonle Sap Lake is likely impacted by a combination of factors including the logging of forests in the catchment, and development on the main channel of the Mekong River and its tributaries. The Thai government's ban on logging in the late 1980s is said to have encouraged deforestation in other Mekong basin countries. Further, dam opposition campaigns and rising electricity demand in Thailand accelerated dam development in neighboring Laos, causing environmental and social problems there.

An example of bilateral cross-border environmental problems, discussed below, is the environmental and social impacts arising in Cambodia because of a hydropower dam built in upstream Vietnam.

## 3-1  Yali Falls Dam

A wire service story about drownings and inundated farmland in Cambodia[13] alerted the world to this instance of cross-border artificial flooding.

One of the largest Mekong tributaries, the Sesan River, flows out of Vietnam's central highlands into eastern Cambodia, and empties into the Mekong. Behind this human-caused disaster was the Yali Falls Dam on the Vietnam part of the Sesan River, built with assistance from Russia and the Ukraine. It is Vietnam's second-largest dam and has a generating capacity of 720 MW. Although it was completed in April 2002, the flooding disaster occurred two years prior to that.

The governor of northeastern Cambodia's Ratanakiri Province reportedly said that three districts along the Sesan were inundated, breaking levees, claiming three lives, and seriously damaging farmland, domestic animals, and fishing. The Cambodia Daily further reported that Cambodia's official in charge of agriculture asked the MRC, whose members include both Vietnam and Cambodia, to determine why the dam's floodgates had been opened.

## 3-2  Study by Ratanakiri Province and an NGO

Beginning on April 15, 2000 the Ratanakiri Province Fisheries Office conducted a month-long questionnaire study[14] in 59 villages in 15 communes and four districts with the cooperation of an NGO called the Non-Timber Forest Products (NTFP) Project, which had been active in grassroots efforts for managing local natural resources. Owing to the urgency of the study, the Rapid Rural Assessment (RRA) method was chosen, and 12 study team members were given advance training in how to conduct the study. Because members divided into groups when working, they came together a number of times during the study to coordinate their effort and ensure the study's consistency and soundness. Linguistic and gender differences were taken into consideration by having three women and people who could speak minority languages on the team.

The study determined that 20,000 people in 3,500 Ratanakiri Province families had been seriously affected by the Yali Falls Dam. Briefly, the study's findings included these points:

- Unnatural changes in water level since 1996
- At least 32 human lives lost, many drowned livestock, and inundated farmland due to unnatural flooding
- Serious impacts on fishing, gold panning, and food gathering
- Health damage and unnatural wildlife deaths due to worsened water quality
- Impacts on fishing gear and boats

At the same time, the Ratanakiri study was not able to definitely attribute all such problems to the deleterious impact of Yali Falls Dam due to the lack of information on water releases from the dam. The following section explores specific problems raised in the report.

## 3-3  Unnatural Fluctuations in Water Level

Cambodians living in the Sesan River basin noted unusual water levels about October 1996 because despite heavy rains in September on the river's upper reaches, flooding was a month late but happened quickly when the water level rose suddenly over just a few hours.

Sometimes water level changed over just a few days, and sometimes over many. In one instance the river's water level rose 7 meters in one day. People using the river found that the inability to foresee water level changes had a heavy impact on their livelihoods.

Surges caused by unnatural water level increases claimed at least 32 human lives, according to the Ratanakiri report. In a number of instances, people using the river were carried off by suddenly rising waters, and drowned domestic animals number in the hundreds and thousands.

## 3-4    Impacts on Agriculture, Fishing, and Gold Panning

It was reported that in the 1999 rainy season alone, unusual flooding inundated 1,830 ha of rice paddies and 629 ha of swidden land, although these figures are likely not accurate. At least 50 tons of stored rice were washed away by flood waters.

Local people grow tobacco, corn, and other crops for consumption and sale on the land exposed along the banks of the Sesan River during the dry season, but this fertile land is inundated by rising water events. The Ratanakiri report says that about 1,800 families, or about half the basin's population, practice this river-bank agriculture in the dry season.

Fish catches decreased by 10 to 30%. A deep-water pool that used to be 7 or 8 meters deep is now about one-half meter deep and unable to accommodate catfish and other freshwater species requiring deeper water. Boats and fishing gear have also been washed away by nighttime surges. Thousands of gill nets, basket traps, and other items have been lost to these surges, at great expense to indigent villagers with little wherewithal for such purchases. Many dugouts and even a few engine boats have been lost or damaged.

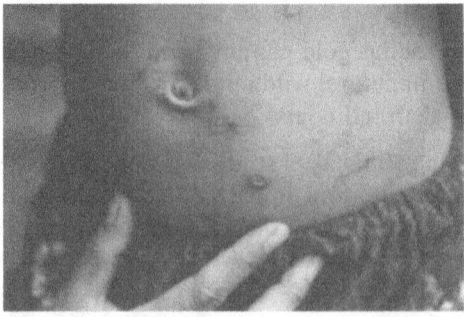

PHOTO 2. Skin lesions like these are increasing due to deteriorating water quality in eastern Cambodia caused by the Yali Falls Dam upstream on the Sesan River (Photo courtesy of Mekong Watch)

Dry-season gold panning is an important industry in two of the four surveyed districts, but higher water levels and surges during the dry season have completely prevented this occupation. As people in 47 of the 59 villages engaged in this activity, its cessation is a serious economic blow to them.

## 3-5  Health Damage from Water Quality Deterioration

Villagers insist that since observing unusual changes in water level in the 1996 dry season their health has been under siege. Sesan River water is used for laundry, bathing, and sometimes for drinking. Symptoms that villagers most complain of include diarrhea, abdominal pains, pain and itching of the throat and nose, dizziness, nausea, coughing, and rashes. While not actually proving that deteriorating water quality is the cause, the report points out that these symptoms are unusual in comparison with the afflictions generally found in this region, such as cholera, measles, chicken pox, malaria, and diphtheria. Dr. Lena Vought, a stream limnologist at Lund University in Sweden, suggested that the villagers' symptoms may be related to toxic blue green algae in the Yali reservoir that have entered the Sesan River, and indeed, November 1999 photographs of the dam suggest that the spillway has released surface water, where algae usually grow.

Villagers also claim that poor water quality has caused unusual deaths among livestock, killing thousands of animals such as water buffalo, cattle, and pigs in the two dry seasons since the 1996 rainy season alone. It is not certain, however, that all of those deaths are attributable to water contamination by the Yali Falls Dam.

Additionally, villagers maintain that wildlife usually found killed by wild dogs or other animals is now often found dead of disease near the Sesan River. This wildlife includes valuable species such as wild boar and the barking deer.

## 3-6  Government Response

From March 16 to 19, 2000 prior to the Ratanakiri Province study the MRC, which is responsible for coordinating water management in the Mekong River basin, sent a fact-finding mission to Ratanakiri and received a report on cross-border flooding from both Vietnam and Cambodia. The MRC mission found unusual fluctuations in water level and a report by the MRC said that sudden ups and downs of 4 to 5 meters per hour in water level occurred a number of times from January through early March of 2000. Six people drowned over this period of time.

In April 2000 Vietnam's government recognized and apologized for the deaths of at least five people in Cambodia due to flooding and surges from the Yali Falls Dam. Local government officials from Vietnam visited Ratanakiri Province, the hardest-hit area of Cambodia, and said that they would never again release water downstream without sufficient warning.

At about the same time a delegation consisting of the Cambodia National Mekong Committee secretary-general and officials from Cambodian government agencies and Ratanakiri Province visited the Yali Falls Dam and met with Vietnamese government officials. Agreement was reached on five items regarding the sharing of dam information and management.[15]

(1) Vietnam will issue appropriate warnings and information about water releases from the dam.
(2) Water will be released gradually.
(3) A 15-day advance water-release warning will be issued under normal circumstances.
(4) Water will be immediately released in emergencies.
(5) An environmental mitigation study must be completed.

## 3-7  Protests Against New Dams

NGOs active in Ratanakiri Province have continued studying environmental impacts with provincial cooperation. A new study in July 2001 found that although damage had lessened, unusual flooding was still occurring. In addition to continued monitoring, the NGOs formed a local people's network that wants to keep any more dams from being built on the Sesan River.[16]

In December 2001 and January 2002 NGOs interviewed 1,913 people in Stung Treng Province, downstream from Ratanakiri Province, and found that although not to the extent of Ratanakiri, the lives of local people had been affected by unusual flooding and poor water quality since 1996. In June 2002 Vietnam announced that it would build a new 260-MW dam, Sesan 3,[17] followed by approval in August 2002 for construction of the 110-MW Plei Krong hydropower dam. Other dam projects on the Sesan are also in the offing.[18]

There is still much to be learned about the harmful cross-border environmental and social impacts along the Sesan River, and the provision of the information pursuant to the agreement between the Vietnamese and Cambodian governments is, to remote villages, not exactly a realistic way to address the problem. While Vietnam continues to build dams on the Sesan with its own funds, what processes will the two governments and the MRC use to formulate solutions that benefit the affected Cambodians? This could be a test case for bilateral cross-border environmental and social problems along the Mekong River.

# 4.    Cross-Border Environmental Problems and Local Agencies[19]

International watercourses like the Mekong number 214 worldwide. Their watersheds make up 47% of the Earth's land area, and hold 60% of the world's population. Yet, there are no international conventions with binding authority over the allocation and use of water resources in international watersheds. It wasn't until May 1997 that the UN General Assembly adopted the Convention on the Law of the Non-navigational Uses of International Watercourses. It provides expressly for fair and rational water use and the participation of watercourse states, which are not supposed to cause "significant harm" to other watercourse states. Considering that China was one of the three countries opposing the convention's adoption, it will have only limited effectiveness in solving upstream-downstream problems among countries in the Mekong River basin.

On a regional level, there are three frameworks capable of tackling the cross-border environmental problems among Mekong-basin countries discussed in this chapter: the MRC, the ADB's Greater Mekong Subregion (GMS) economic cooperation, and ASEAN. This section discusses, in connection with these three frameworks, the possibilities and challenges involved in dealing with the cross-border environmental problems in Mekong-basin countries.

## 4-1   Mekong River Commission

The commission's predecessor was the Committee of Coordination and Investigation of the Lower Mekong River Basin (Mekong Committee), formed in 1957 with UN cooperation. Hardly any of its development plans came to fruition because of the Vietnam War and Cambodia's civil war. In 1995 the MRC grew out of the Mekong Committee and came to coordinate water use among member states rather than to take over the previous role of promoting water resource development. Recently the MRC established a Working Group on Transboundary Issues.

The biggest impediment to coordinating water use is that only the four lower-Mekong states (Thailand, Laos, Cambodia, and Vietnam) are among the MRC member countries, while the upper-Mekong countries of Burma and China are only dialog partners. At this time China shows no interest at all in joining, and the MRC can only ask for cooperation in individual situations, such as having China's government share hydrological data.

Moreover, the MRC's role is also limited in dealing with environmental problems arising among member countries. Although the MRC sent a fact-finding mission to investigate the Yali Falls Dam situation, the commission has not necessarily been able to take steps for preventing another such occurrence. Pursuant to the agreement on commercial navigation by the four upper-Mekong countries, Thailand developed a plan to expand the harbor at Chiang Khong on its side of the river, but Laos expressed concerns that this would accelerate river bank erosion on its side. Although the MRC asked for a temporary suspension of the project, the actual solution was left to the two governments.

## 4-2   Greater Mekong Subregion Economic Cooperation Program

Just after peace returned to Cambodia in 1992 the ADB initiated the Greater Mekong Subregion Economic Cooperation (GMS) program, which is a framework in which the six countries sharing the Mekong River cooperate and further cross-border development. Major features of this program are that it involves all six of the Mekong-basin countries, and that it focuses exclusively on economic cooperation without getting involved in politics. Over the past decade it has created projects in the seven sectors of transport, tourism, the environment, personnel training, communications, energy, and trade and investment, provided $770 million in financing for infrastructure construction, and given $56 million in assistance for technical cooperation.

The 10-Year Strategic Framework, endorsed by the 10th Ministerial Conference in November 2001, has five strategic thrusts, one of which is "protect the environment and promote the sustainable use of shared natural resources." Among the steps for stopping environmental damage in the Mekong region is "collective action to resolve cross-border environmental problems," and it observes that "cooperation with neighboring countries is critical to resolving unintended negative outcomes of development activities that go beyond national borders." One specific action is that Mekong-basin countries should help monitor cumulative environmental impacts.

## 4-3   ASEAN

Founded in 1967, the Association of Southeast Asian Nations (ASEAN) is shifting the focus of its activities from political cooperation to economic development based on a regional approach. In 1985 ASEAN members concluded the Agreement on the Conservation of Nature and Natural Resources, creating a framework in which members would in some cases have to offer funds for environmental conservation in other member countries such as when causing cross-border environmental problems, but it has yet to be enforced. Decision-making based on non-intervention and the agreement of all members has created circumstances in which it is difficult to make innovative political judgments on such delicate issues. And although the increase in members has brought five Mekong-basin countries into ASEAN, all-important China is not a member.

One possibility for dealing with cross-border environmental problems in this region is ASEAN + 3, which is ASEAN with Japan, China, and South Korea. ASEAN's 30th anniversary summit in 1997 was the first to invite the heads of state from those three nations, and since that time ASEAN has held summit meetings once a year and various cabinet-level meetings that include discourse on a wide variety of East Asian topics including economic and social issues, politics, and the environment. Thanks to the participation of the six Mekong-basin countries, and the ability of Japan to state its opinions as an East Asian country and as the region's biggest aid donor, these meetings can serve as venues to discuss regional environmental issues.

# 5.   Civil Society Problem-Solving Initiatives

While section 4 described the possibilities and limitations of regional international frameworks for solving cross-border environmental problems in the Mekong River basin, this section surveys the initiatives and approaches used by civil society, especially NGOs, to overcome these country-level limitations.

China's Yunnan Province has a number of environmental NGOs working for nature protection and participatory natural resource management, but it is not easy to expect provincial authorities to address environmental and social harm from upstream development. According to staff members at an international environmental NGO in Hong Kong, a representative of a Yunnan environmental NGO had been searched by the authorities after he had made critical remarks about upstream development in lectures and newspaper interviews abroad.[20] On the other hand, the international NGO Oxfam Hong Kong studied how local communities were negatively impacted by the Manwan Dam on the Lancang River's main channel and asked the Chinese government for remedial measures, but instead of making its report public, Oxfam approached the government directly. This made it possible to raise the problem without being criticized by the authorities.

Burma is under the control of an authoritarian military government, and people find it hard to openly criticize government development. But pro-democracy groups working just across the border in Thailand are gathering information from inside Burma and appealing to international society about the impacts of upstream development on the lives of minorities in northern Burma.

In Thailand people's organizations, NGOs, and researchers collaborate in addressing problems. Small-scale fisherfolk in Chiang Khong District affected by dredging in the Mekong formed their own group and cooperated with environmental NGOs and researchers in the north-central city of Chiang Mai in studying the impacts of dredging on the fishing industry, then approached the government through their parliamentary representatives to ask that dredging be halted in Thai waters. While their efforts have a measure of influence over the Thai government, they have been unsuccessful in swaying the Chinese government on upstream development.

There are no local NGOs in socialist Laos, and international NGOs with offices there find it hard to deal with politically sensitive issues such as the environmental and social harm caused by development, and they cannot even work with local people to perform their own impact studies, as is done in Thailand. Instead, Thai NGOs lobby Thai parliamentary and government officials who share their concerns to pursue dialog with Laos.

Although Cambodian NGOs have vague concerns about the impacts of upstream development on the ecosystem of Tonle Sap Lake, their immediate focus is problems with Vietnam over the Sesan River as described in section 3. Ratanakiri Province, people's organizations carrying out community development in the area, and NGOs are joining hands in asking the Cambodian government to take action, and in November 2002 they held the first "Sesan Workshop" in Phnom Penh with the participation of Vietnamese officials. While this will not directly lead to a solution, there are significant expectations that NGO-initiated networks involving the concerned governments will help prevent future problems.

Vietnam has hardly any NGOs with concern for Mekong-related environmental issues. Due to the political instability of Vietnam's central highlands, through which the Sesan River runs, NGOs performing studies and making policy proposals find their access is limited. An international NGO that started a study on the environmental and social impacts of the Yali Falls Dam received a warning from the Vietnamese government about further involvement in the issue. But in the Mekong delta, which will likely be affected by upper Mekong development, people conceive development on the upper reaches as something happening in a geographically distant country.

Although in some of these countries civil society can to an extent influence governments in working toward solutions, there are situations in which the NGOs of one country alone are powerless when it comes to cross-border issues. More deserving of our attention with regard to efforts toward solutions is that civil society in the Mekong basin has begun forming cross-border networks tailored to individual issues. About five years ago Thai and Cambodian NGOs took the lead in creating a network of Mekong-basin small-scale fisherfolk, whose livelihoods depend on fish that migrate through the Mekong's mainstream and tributaries. These people are most affected by development projects such as dams and dredging. In this network, the fishers and civil society groups that support them work together to study the impacts of development on fisheries and to make policy proposals. At the November 2002 first Greater Mekong Subregion Summit held in Cambodia, over 80 people from citizens' groups with concern for natural resource management in Mekong-basin countries appealed to policymakers about the importance of the Mekong environment to basin inhabitants and about the negative impacts of development. Such networks enjoy the participation of not only citizens' groups in Mekong-basin countries, but also many NGOs from aid donor countries, which has built the foundation for basin countries and international civil society to understand the concerns and cooperate with the initiatives of local people to deal with cross-border environmental problems.

## 6.   Summation: Civil Society Participation Needed in Decision-making by Regional Agencies

The intents of countries to develop the Mekong's main channel and tributaries are anticipated to further aggravate cross-border environmental problems. These problems came to light only after the affected Mekong basin inhabitants made their case to central governments and international society with the help of local people's organizations, NGOs, and other citizens' groups, or the help of local governments. And when people cannot make direct statements due to politics, citizens' groups in other basin countries have raised the issues for them. Nevertheless, region-level agencies are not set up to incorporate the opinions of civil society into policies and actions meant to properly solve problems or prevent them.

Because of the way the MRC is organized, talks with Mekong-basin people are left to the National Mekong Committees under each government. These national committees cannot respond to citizen requests because they are very short of personnel and funding. Although the ADB has institutionalized procedures for listening to the views of civil society owing to its accountability to the citizens of donor countries, the bank has in actuality received strong criticism and distrust from civil society for projects it financed such as the Samut Prakarn wastewater treatment project in Thailand and the Theun Hinboun Dam project in Laos. ASEAN has no way to respond as a regional agency, and considers the opinions of regional people to be purely a matter of domestic concern in each country.

In this respect the actions of Mekong-basin countries' local governments are in the spotlight. For example, local governments along the borders between Yunnan Province and Vietnam, and between Yunnan and Laos, are working together on environmental problems such as forest fires. Even in the case of the Yali Falls Dam, local governments can to an extent provide for solutions without leaving the matter up to region-level international agencies.

In consideration of the foregoing points, measures such as the following are needed to solve cross-border environmental problems in the Mekong River basin.

(1) Forums for discussing cross-border environmental problems with the participation of all Mekong-basin countries: There is a need to create a framework under one of the three regional agencies, or by two or three of them complementing each other, which would have the participation of upper- and lower-Mekong countries, and make it possible to discuss cross-border environmental problems and respond to them effectively.

(2) A mechanism for communicating information: Hardly any information flows from countries causing cross-border environmental damage to countries subject to the impacts. Basin countries must create an information-provision mechanism which assumes that development will cause environmental damage across national borders.

(3) Transboundary environmental assessments: None of the three specific instances treated in this chapter had been included in advance environmental assessments of cross-border impacts. There is a need to develop and appropriately implement a method for transboundary EIAs in this region, with reference to predecessors such as Europe's Espoo Convention.[21]

(4) Participation of civil society and local governments in decision-making: Local governments, as well as people's organizations and NGOs, which are close to the parties affected by environmental impacts, play important roles in discovering problems quickly and achieving effective solutions. Mekong-basin countries and the relevant region-level international agencies must allow cross-border citizen networks and local governments to participate in the forums proposed in (1), and provide for the incorporation of their views and information in decision-making.

Needless to say, political will is indispensable to bringing about such proposals. The first step toward solving these transboundary environmental problems is for the governments and agencies of Mekong-basin countries and aid donor countries to fully understand the seriousness of the transboundary environmental impacts spreading throughout Mekong-basin countries, and to recognize that the cross-border cooperation of civil society is essential to solving these problems.

Matsumoto Satoru

## Essay:   Environmental Damage under the Military Government in Burma (Myanmar)

Although human rights violations in Burma are well known, the negative impacts of the military government and multinational corporations on Burma's environment tend to be overlooked. Burma is well-endowed with forests, minerals, natural gas, petroleum, and other natural resources. The country is also rich in biological diversity, with tigers, rhinoceroses, elephants, and other animals finding habitat in the expansive forests covering the land. Yet, environmental damage is proceeding with astonishing rapidity because the military government has allowed the logging of vast forests (especially teak) and the implementation of large-scale development projects in an effort to boost revenues by exporting natural resources.

The Yadana and Yetagun natural gas pipeline construction projects in southern Burma by three Western oil companies are notorious because military units deployed in the area ostensibly to provide security for the pipelines have committed systematic violations of local people's human rights such as forced labor and forced relocation. The projects, however, have also had devastating impacts on the surrounding environment. The pipelines cross an area named by the WWF as one of the 200 most valuable ecosystems on Earth. Forests along the pipelines serve as a vital link between the faunas of both Indochina and the Himalayan Mountains and those of the Malay Peninsula, and the area is inhabited by elephants, tigers, tapirs, and other endangered species. Pipeline construction sundered animal migration pathways, threatened the ecological stability of the entire forest, and cut down many large trees. There were also reports of illegal logging by military units sent to guard the pipelines. Soil erosion caused by the region's heavy rainfall in deforested areas silted the rivers.

Severe degradation of Burma's forests is yet another mark of the regime's exploitation of natural resources. Half of Burma is covered by forests. Burma has at least 70% of the world's remaining teak, and it supplies about 80% of the teak on the world market. In 1988 the government started selling to foreign companies extensive logging concessions in areas inhabited by ethnic peoples, thereby doubling the logging rate that had until that time been the lowest in Southeast Asia. Experts say that nearly all the world's teak likely will disappear in a generation. The traditional logging methods of native peoples had had little environmental impact, but the foreign companies practiced highly damaging clear-cutting and over-cutting. The combination of heavy rains and steep terrain caused erosion of the exposed soil. As a result, less water is available in the dry season.

It is well known that the construction and maintenance of large dams lead to environmental damage. Many dams are planned or already under construction in Burma, and the number is expected to grow because the government has a policy of aggressively pursuing hydropower development to make up for the country's power deficiency. One of these dams is the Ta Sang Dam planned on the Salween River in Shan State. In addition to the environmental impacts that dams generally have, the Ta Sang could result in erosion of the downstream riverbed, logging of forests in the reservoir before it is flooded, mosquito breeding in the reservoir, and devastation of the forest by local people who would lose their livelihoods because of the dam construction. Burma has few regulatory controls for environmental protection when building dams, and those that exist are not properly enforced. Despite a growing international trend towards smaller local energy projects, it does not appear that Burma is encouraging such alternatives.

Burma is blessed with minerals and precious stones, and has mining operations of various sizes. Mining often pollutes extensively with toxic byproducts such as waste rock and tailings. Burma's largest copper mine, Monywa in Mandalay Division, uses an extraction method that generates large amounts of hazardous substances. Ore refining equipment quickly deteriorates in Burma's tropical environment, giving rise to concerns about leaks and spills of hazardous substances. The government imposes no legal obligation on mine operators to carry out environmental and social impact studies, and thus operators are virtually free of any responsibility to prevent or minimize impacts on the environment. The environmental damage described here is growing in severity because policies and regulations for environmental protection are either nonexistent or nonfunctional. Unless some changes are made, the future of the country's natural resources appears to be grim.

Akimoto Yuki

# Chapter 3
# Inner Asia: Balancing the Environment with Socioeconomic Development

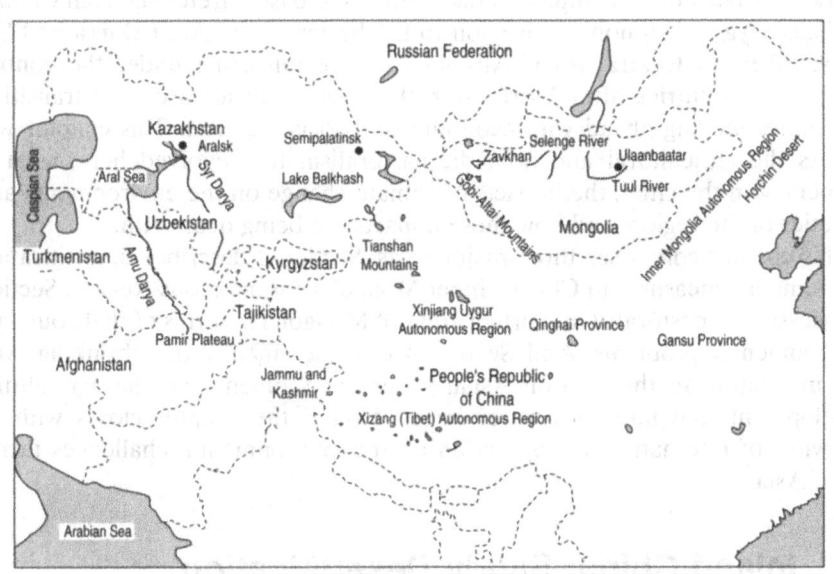

# 1.    Introduction

Inner Asia is the hardest to define of all the regions in Part II. To begin with, "inner" suggests nothing else than the area in question is away from the ocean. For example, it is relatively easy to define "Northeast Asia" as used in this book because it surrounds the Sea of Japan, but there is no such nucleus for Inner Asia. To define this region, let's start with different kinds of ecosystems. Moving from east to west across Eurasia we encounter steppes, deserts, oases, and other kinds of environments (this chapter will not deal with alpine zones), a vast region covering the "Mongolian world" of the highland Mongolian Steppe, and the "Turkmenistan world" comprising the deserts and oases stretching from China's Xinjiang Uygur Autonomous Region to the basins of the Amu Darya and Syr Darya rivers in Central Asia.[1] Although this region came under the control of socialist countries after World War II, those countries are now transition economies moving ahead with economic and social reforms. This chapter will discuss the agricultural and nomadic pastoralism that evolved here, what is happening in the cities, the impacts of climate change on the environments and societies of the region, and how those impacts are being dealt with.

This chapter comprises three major parts. Section 2 describes desertification and remedial measures in China's Inner Mongolia Autonomous Region. Section 3 discusses the pastoralist and urban areas of Mongolia in terms of that country's environmental problems. And Section 4 discusses mainly the shrinking Aral Sea in relation to the Central Asian water environment and the agricultural development that underlies water issues. Finally, the chapter closes with an overview of international cooperation on the environmental challenges facing Inner Asia.

# 2.    Inland China: Sandy Desertification

Although China is a vast country of 9.6 million square km, or about 25 times the size of Japan, 2,674,000 square km, equal to 27.9% of its land area, are desertified. And on the average, desertification engulfs another 10,400 square km of China every year. With each passing decade, more desertification-caused dust storms arise from north China: in the 1950s there were five such events, which increased to eight in the 1960s, 13 in the 1970s, 14 in the 1980s, and 23 in the 1990s. In 2000 there were 12 events, and in 2001 there were 18. Desertification not only causes heavy economic losses in China, but also affects the Korean Peninsula and Japan with dust storms. Since the 1990s the direct economic losses to China by desertification have been 54 billion yuan annually.

The most serious form of desertification is sandy desertification. China has 1,743,000 square km of sandy desertified land, which is 65.2% of all desertified land, and 18.2% of China's land area. Further, although sandy desertification advanced at an average yearly rate of 1,560 square km from the 1950s through the mid-1970s, the pace quickened to 2,100 square km from then through the mid-1980s, and further accelerated to 2,460 square km from 1985 to 1994 and 3,436 square km from 1995 to 1999.[2]

Nearly all of China's sandy desertified land is concentrated in the inland western region, divided among Xinjiang, Inner Mongolia, Tibet, Qinghai Province, and Gansu Province. These three autonomous regions and two provinces have 93.3% of all the sandy desertified land.

A natural phenomenon—sand dunes advancing because of the wind—causes 5.5% of sandy desertification, while the other 94.5% is anthropogenic. These human causes break down into 25.4% excessive clearing of land for agricultural use, 28.3% overgrazing, 31.8% over-gathering of fuelwood and plants, 8.3% inappropriate use of water resources, and 0.7% mineral extraction and road construction. Anthropogenic factors cause almost all of China's sandy desertification.

China combats sandy desertification in a number of ways, varying somewhat from one place to another. Central government measures or those used throughout China include afforestation, returning agricultural land to forest or grassland, sealing off areas and banning grazing by encouraging farmers to keep livestock housed, prohibiting the excessive collection of fuelwood and plants, and legislation.

Deserving special mention is the Sanbei Shelter Forest Project, the biggest, longest-term, and most effective afforestation project to date. Begun in 1978, this national project covers 551 counties of 13 provinces of China's Northwest, North, and Northeast regions, and its goal is to protect existing forests and to afforest 35.6 million ha by 2050. The project is divided into three stages and eight periods. As of the end of 2000 the third period of the first stage had concluded, and 22,040,000 ha had been afforested over 23 years.[3]

Although sandy desertification control efforts by the central and local governments and by people living in such regions are making a certain amount of progress, they are not going as planned. Sandy desertification is proceeding still faster, and efforts to stem it cannot keep pace. Preventing and mitigating desertification will require reassessing the effectiveness of all control measures, and clarifying the measures that administrative authorities should take, as well as the legal responsibility of organizations and individuals. In this respect there are considerable expectations for the Law on the Prevention and Control of Desertification, which became effective on January 1, 2002.

In the Inner Mongolian Autonomous Region's Horchin Desert, where we performed a study, people are combatting sandy desertification by instituting a new landholding system and building "household eco-economic zones." Livelihoods in the Horchin Desert are half farming and half pastoralism. In 1982 the government introduced a contract production system which gave farmers and pastoralists plots of land that had formerly been farmland held in common under the former people's communes, and allowed land use by individuals, but pasture remained land held in common instead of being distributed to individuals. Both farmers and pastoralists indiscriminately grazed their livestock, which they had substantially increased, on these common lands to increase their incomes, but this overgrazing triggered desertification. A new system instituted in 1996 divided and distributed not only farmland, but also these common lands, to farmers and pastoralists for a 30-year period. This gave them full responsibility for the management and conservation of farmland and pastureland and made them use care to practice rational land use. After this change the rate of desertification was considerably slower than it had been under the contract production system, showing that the land contract system plays an important role in slowing desertification.

In addition to this new land contract system, one more way to combat desertification is building model "household eco-economic zones," under which each farming household is provided with the same amount of land regardless of the number of household members. The land is fenced in, several rows of large trees are planted along the fences as windbreaks, and inside of this enclosure are farmland, fields for growing forage, and commercial forests along with irrigation wells. Until the creation of these zones, farmers tilled much land in the Horchin Desert that could not be irrigated, and after a few years in the same location the land would desertify, forcing the farmers to relocate. Repetition of this pattern allowed desertification to continue, but establishment of these zones beginning in 1996 stopped this practice and the desertification it induces because the system keeps farmers in one place for 30 years. One important fact to note is that building a zone entails considerable investment such as for purchasing fencing materials and digging wells. Due to the limited economic capacity of local governments and farmers, many of the zones that have sprouted quickly in recent years are not the kind of "model" household eco-economic zones originally planned. Local governments have insufficient funds for not only building zones, but also for implementing their own policy measures for combatting desertification, and for that reason efforts do not always progress according to plan. Combatting desertification in the Horchin Desert, which is one of the poorest regions in China, will require active and continuing funding assistance from not only the central government and autonomous regions, but also from developed countries and other entities.

# 3.   Mongolia

## 3-1   Nomadic Pastoralism and the Environment

### Changes in Mongolia's Nomadic Pastoralism[4]

Mongolia has an area of about 1,570,000 square km, or four times that of Japan, but a population of only about 2.5 million. A landlocked country sandwiched between the two giants of Russia and China, it has very low rainfall, and wide temperature differences not only throughout the year, but even in the space of one day. Owing to these conditions Mongolia is for the most part steppe, desert, and semidesert, leading to the development of nomadic pastoralism that is characterized by keeping several of five types of domestic animals—sheep, horses, cattle, camels, and goats—and moving them to different pastures depending on

PHOTO 1. Northern area of Gobi-Altai Province
(September 2001 photo by Okada Tomokazu)

the season. This is still Mongolia's biggest industry, providing Mongolians with food, clothing, housing materials, and fuel.

Until the early 20th century this was practically the only way of life in Mongolia, but the 1921 People's Revolution brought a socialist system and modernization such as cities and factories with enormous aid from the Soviet Union. Under the collectivization of pastoral regions completed in 1959, nomadic pastoralists put their livestock into pastoral collectives called *negdel* and worked as collective members, but the livestock industry was doing poorly at that time, and domestic herd size stagnated. At the same time, state farms were established mostly in Mongolia's central and northern regions, where large-scale operations produced wheat, vegetables, and livestock feed.

The 1989 democratization movement led to the demise of socialism, and in 1991 the market economy came to Mongolia. Sudden economic reforms coincided with the Soviet Union's collapse and plunged Mongolia's economy and society into chaos that has yet to be completely brought under control. In the early 1990s the *negdel* were broken up and their livestock put into private hands, after which the number of livestock grew rapidly. Many livestock died because of a *dzud*[5] and *gan* (droughts) from 1999 through 2001, resulting in the loss of 5.2 million animals as estimated by the United Nations.

## Changing Weather Patterns

*Dzud* events occur periodically in pastoral regions, but the cycle has shortened in recent years. Records of *dzud* damage and drought over the last 60 years (Table 1) show that from 1940 through 1980 such events generally occurred once in 10 years, but beginning with the winter of 1986–1987 they happened once every three to seven years.

Recently there is also less rainfall, and rainy periods have shifted. Formerly the period from June through mid-July had the most rain, but in recent years not much rain falls at this time, causing small rivers and wetlands to disappear in some regions. These changes are combining to not only reduce livestock numbers, but even cause people to leave for other parts of the country. Since 1999 people have been leaving western Mongolia, which is especially hard hit. An example of the seriousness is Zavkhan Province, which lost about 13,000 people or 13.3% of its population in two years.[6]

## Pasturelands and Their Degradation

Frequent extreme weather events have a suspected link with global warming, and are currently under study as a factor behind the degradation of pasturelands. Yet, various other factors also put pressure on pasturelands.

The degradation of pastoral lands is already serious in Mongolia's neighbors, the Inner Mongolian Autonomous Region and Russia.[7] Nomads themselves, who are intimately familiar with the problem, say that grass is shorter and sparser than before.[8]

Degradation is especially serious near the capital city of Ulaanbaatar, which has over one-third of Mongolia's population and the country's biggest market, and therefore a high concentration of nomads and their livestock from outlying regions.[9] One of the authors (Minato) conducted interviews of 10 nomadic households in this area in 2001, and nine of them said that the amount of grass had lessened over the previous few years.

There are various reasons for this decline, chief among them overgrazing. A rapid increase in the number of livestock translates into a smaller area of pasture per head (Table 2), and one study claims that already Mongolia as a whole is overgrazed.[10] Especially prized among livestock are goats because they yield cashmere, which sells for high prices, and over the decade from 1989 Mongolia doubled its number of goats. This has proved especially damaging to pasturelands because, unlike other animals, goats uproot grass when they eat it.

Another cause of pastureland degradation is the increased dependence on natural pasture grass due to the demise of the *negdel* and state farms, which had produced hay and grown feed crops to sustain livestock over the winter. Table 2 shows that grass harvests declined nearly 40% in 10 years and that feed crop production virtually stopped.

Motor vehicle traffic is another cause. Because paved roads are almost nonexistent in pastoral areas, motor vehicle traffic over the steppe exposes the soil and allows erosion. Another of the many pressures on pastoral lands is pollution by wastes in mining areas.

## Diminishing Forests

According to the United Nations Environment Programme, during the 20 years from 1974 to 1994 Mongolia's forested area increased from 14,550,000 to 16,130,000 ha, but the breakdown tells the real story. The category "forests" comprises the two large categories of "dense forest" and open area, the latter including fire-damaged or clear-cut areas. "Dense forest" actually declined from

TABLE 1. Disasters in Mongolia and Their Damage

| Years | 1944-45 | 54-55 | 56-57 | 67-68 | 76-77 | 86-87 | 93 | 96-97 | 99-2000 | 00-01 |
|---|---|---|---|---|---|---|---|---|---|---|
| Type of disaster | Drought and dzud | Dzud | Dzud | Drought and dzud | Dzud | Dzud | Dzud | Dzud | Dzud | Dzud |
| Livestocks lost (million head) | 8.1 | 1.9 | 1.5 | 2.7 | 2 | 0.8 | 1.6 | 0.6 | 3 | 2.2 |

Source: UN-Mongolia website <http://www.un-mongolia.mn/>.

TABLE 2. Livestock Sector Indicators

| Year | Number of livestock Total (1,000 head) | No. of goats (1,000 head) | Natural meadows and pastures Total (1,000ha) | Per head (ha) | Gross hay harvest Total (1,000 tons) | Per head (kg) | Fodder crop harvest Total (1,000 tons) | Per head (kg) |
|---|---|---|---|---|---|---|---|---|
| 1989 | 24,674.9 | 4,959.1 | 124,157.0 | 5.03 | 1,166.3 | 47.3 | 5,510.0 | 223.30 |
| 1994 | 26,808.1 | 7,241.3 | 117,147.0 | 4.37 | 672.2 | 25.1 | 29.1 | 1.09 |
| 1999 | 33,568.9 | 11,033.9 | 129,091.0 | 3.85 | 715.2 | 21.3 | 5.3 | 0.16 |
| 2001 | 26,075.3 | 9,591.3 | 128,950.8 | 4.95 | 831.5 | 31.9 | 2.7 | 0.10 |

Source: National Statistical Office of Mongolia, Mongolian Statistical Yearbook, various years; State Statistical Office of Mongolia, Agriculture in Mongolia 1971–1995, 1996.

13,910,000 to 12,710,000 ha, meaning that real forested area shrank.[11] Reasons for this include the forest development and logging that gained momentum during the socialist period, and the forest fires that occur every year. Recently illegal logging in particular has been a continuing problem, and there was a major forest fire in the spring of 1996. Such events have likely further diminished the forests.

Although Mongolia's forests cover only about 10% of the country, they are essential for protecting the ecosystems of pastoral lands. There is an urgent need for an effective forest protection policy due to the great difficulty of regenerating lost forests in Mongolia's cold, dry climate.

### Using Traditional Insights and Technologies

Burdened as it is with economic difficulties and poverty, Mongolia puts economic development first to the great peril of its environment. Because nomadic pastoralism still underpins Mongolia's economy, it would be jeopardizing itself by not attending to problems that would undermine that foundation.

Some guidelines for action are mitigating overgrazing by limiting the increase in livestock, and Mongolia needs to create appropriate laws and institutions as other countries are. But Mongolia's traditions also offer some ideas for developing solutions. Through their practice of nomadic pastoralism Mongolians have developed traditional insights and technologies for protecting pasturelands, water, and forests. Of course a return to tradition is not a total solution, and traditions have not been fully passed on through the period of collectivization and the transition to a market economy, but the traditions fostered by nature and people in pastoral regions can sometimes yield results that are superior to what agrarian peoples conceive as "environmental protection." Accordingly, Mongolians might do well to approach their challenges using these traditional insights and technologies.[12]

## 3-2    Urban Environmental Challenges

### The Burgeoning Urban Population

Despite its image as a steppe nation, Mongolia too has a rapidly growing urban population. Although people were not free to move under socialism, the freedom to move was recognized in 1992 under democratization. But the gross domestic product peaked at 214 billion tug in 1989 (real prices) and fell to 166.2 billion tug in 1993. Further, economic potential in terms of per capita GDP based on contemporary exchange rates (converted to US$) fell from $1,700 to $250, or to one-seventh over those five years. As a consequence, the average price of consumer goods skyrocketed an average of 36.6-fold[13] from 1991 to 1995, making it hard for city dwellers to even buy food. This prompted a 2.3% decline in urban population and a 20.1% increase in the provinces between 1990 and 1995.

In 1994 the GDP started growing because of factors including a rise in the international market price for copper, Mongolia's main export. The economic engines were mining, commerce, and services, with especially rapid growth in the urban industries that require a certain critical mass of population. Such changes in the economy manifested themselves in population dynamics, increasing the urban population 17.9% and decreasing that in outlying regions by 4.8% from 1995 to 2000. Mongolia's population in 2001 was 2,443,000, and about 60% or 1,377,000 people live in the cities.

Population growth in Ulaanbaatar has been especially pronounced. City statistics indicate that the population in 1992 when democratization occurred was 589,000, which increased 37.9% to 813,000 in 2001. The main causes of this surge in population include increased employment opportunities created by development of the urban economy and the creation and enhancement of institutions for higher education, but another reason is that nomads and their family members go to the cities in search of jobs because their economic base has been lost due to steppe and forest fires, *dzud* events, and drought. Many of these people build traditional homes called *ger* or wooden houses at the bases or on the sides of mountains around Ulaanbaatar's existing urban area, but most of them cannot pay the registration fees to become Ulaanbaatar citizens, and are not counted as city residents. The unofficial residents are currently estimated to number at least 200,000. While Mongolia is more than four times larger than Japan, at least 40% of its population is concentrated into a place that is only 0.3% of the total land area, creating a city on the vast steppe with a growing disparity between rich and poor.

## Air Pollution

As in other countries, environmental problems are emerging in Mongolia's urban areas as the population concentrates in the cities, as energy consumption rises, and as the motor vehicle fleet grows. The most visible manifestation of deteriorating urban environments is the air pollution in Ulaanbaatar, and in other cities including Erdenet, Moron, Ulaangom, and Choybalsan.

Especially severe is winter air pollution in Ulaanbaatar, which is situated in a basin surrounded by mountains. Winter winds from the northwest enter from between mountains, change direction from southwest to east by following the path of the Tuul River in the city, and flow gently into the basin. The coal-fired power plants and factory zone that are the main stationary air pollution sources are located in the southwest, putting them upwind of the city center. This air pollution is aggravated by coal-fired space-heating boilers throughout the city and by coal stoves in the *ger* towns surrounding the city.

Additionally, inversion layers tend to form over Ulaanbaatar in winter because of cold air blowing into the basin and ground-level radiative cooling. Once an inversion layer forms, the warm air mass over the city keeps polluted air in the city for long time periods and smoke from pollution sources keeps worsening the situation. Annual coal consumption in Ulaanbaatar (according to 2001 data) is estimated to be about 3.5 million tons at thermal power plants, about 300,000 tons at the city's 150 heating boiler plants, and about 400,000 tons at the 78,000 *ger* households using coal stoves, while the estimated air pollutants emitted per year are 568,000 tons $SO_2$, 456,000 tons $NO_2$, 1.3 million tons of particulate matter, and 10.9 million tons of $CO_2$.[14]

Another major contributor to air pollution is the rapidly growing motor vehicle fleet.[15] All vehicles are imported, and most are used vehicles four or five years old. One study found that 80% or more of all registered vehicles are at least 10 years old,[16] are poorly maintained and hard to obtain parts for. Studies on motor vehicle exhaust by the Central Laboratory of Environmental Monitoring of Mongolia's Ministry of Nature and Environment (1999, 2000) found that 36% (average of both years) of vehicles exceeded environmental standards for CO, and 10% for HCs, yet vehicles whose emissions do not meet standards are not fined or required to seek repair. Further, the increasing number of vehicles, inadequate roads, and lack of traffic control breed traffic congestion everywhere and assure that air pollution will further worsen.

A study by the Ulaanbaatar Public Health Center on the impacts of air pollution on human health found that in 1994, 134 out of every 1,000 people in Ulaanbaatar suffered from respiratory ailments, and in 1995 it was 175 of every 1,000, while in Mandal Village of Selenge Province the numbers were only 65 and 66 out of 1,000, respectively. Especially for people aged zero to 17 the numbers were 273 and 354 in Ulaanbaatar, but only 50 in Mandal Village both years. Between five and seven times as many people fell victim to air pollution in Ulaanbaatar, and air pollution has a heavy impact on children.[17]

One characteristic of Mongolia's climate is long winters with temperatures descending to between −30°C and −40°C, making coal-fired power production, water heating, and space heating vital to survival. In addition to emitting $CO_2$, coal contains heavy metals, radioactive substances, halogens, and other substances not subject to environmental standards. A variety of actions are urgently needed to mitigate pollution and protect the citizens' health. These include: Reducing the fuel consumption of coal-fired facilities to improve efficiency, installing pollution abatement hardware, requiring motor vehicle owners to repair vehicles which do not satisfy environmental standards, and ensuring smooth traffic flow by means of improvements in roads and traffic control systems.[18]

## Water Pollution

Average 1995–2000 values for pollution of the Tuul River, which passes through southern Ulaanbaatar, produced these results for a location called Zaysan just upstream of where an urban tributary joins the Tuul: Biological oxygen demand was 2 mg/liter and ammonia ranged between 0.2 and 0.5 mg/liter, but at Songino about 20 km downstream and below the city and sewage treatment plant BOD was between 7 and 10 mg/liter and ammonia was about 8 mg/liter. Pollution worsens year by year. Some of the primary causes are believed to be a reduction in runoff due to changes in meteorological conditions, irrigation, logging of forests in headwater areas, and other factors; graywater from *ger* villages without adequate sewage treatment facilities and effluent from factories entering tributaries of the Selbe River; and the increasing disposal of wastes on river banks. From June through August many Ulaanbaatar residents leave the city and live in summer homes (wooden structures or *gers*) built in river valleys, but as these homes have no waste treatment facilities, their wastes seep into the soil or run directly into rivers. Further, as of 2001 Ulaanbaatar had 221,000 head of livestock, whose wastes have the same fate.

To supply drinking water, Ulaanbaatar pumps 170,000 cubic meters of groundwater daily from 150 wells of depths between 30 and 70 meters along the Tuul River. By contrast the city has one sewage treatment plant with a capacity of 270,000 cubic meters. Because *ger* villages lack facilities for drinking water and sewage treatment, residents dispose of all wastes by dumping them with disinfectant in pits, which contaminates all the 2- to 3-meter wells in these villages. Some wells are reported to be unfit for drinking.[19]

Ulaanbaatar's water utility says it is taking no preventive measures now because there are no problems with water quality except for certain polluted areas, but the anticipated continuing concentration of population and industry in the cities,[20] and the increase in wastes due to changing lifestyles could very well lead to pollution of rivers and soil by wastes and hazardous chemical substances, and therefore the contamination of the groundwater serving as drinking water, unless the authorities take quick action on waste disposal sites and environmental protection measures in *ger* villages.

## Waste Management

Waste collection and disposal in Mongolia are performed by publicly created, privately operated management companies commissioned by administrative authorities, and by private companies, but the lack of laws or regulations governing waste management and of environmental protection measures has created sizeable public health problems.

For example, 5,527,000 cubic meters of waste were collected from Ulaanbaatar's offices, apartments, *gers*, and roads in 2001, and discarded in landfills on the city's outskirts.[21] Proprietors and contractors are responsible for disposing of wastes generated at factories, commercial facilities, and construction sites. Yet, it is evident that the system is not functioning well because collection bins in the city are overflowing with uncollected trash, and people dump trash on roadsides, riverbanks, and into pits they dig in vacant lots. Once much waste accumulates, local people often burn it on the spot to dispose of it and stop odors.

Collected wastes are disposed in three landfill sites[22] in the mountains outside of the city. These landfills have no facilities for treatment or to prevent trash scattering or leakage; soil is packed down over wastes after they are burned. Incineration is normally in the open, allowing smoke and fly ash to settle over the city and *ger* villages. The incineration of many kinds of wastes gives rise to concerns about the de novo synthesis of dioxins, and the contamination of soil, crops, and livestock by these and other hazardous substances. Still more concerns about soil contamination arise due to the dumping of oil and other wastes by gas stations and repair facilities, whose numbers have grown with the increase in vehicle fleet size; the use of chemicals such as disinfectants, insecticides, and pesticides, which rises with population; and the generation of medical waste and construction and demolition waste.

A German company plans to build a waste recycling facility, but this project has gotten no further than site proposal.

# 4.    Central Asia's Water Environment and Agriculture: Focus on the Aral Sea[23]

## 4-1    Introduction

The five Central Asian countries of Kazakhstan, Uzbekistan, Turkmenistan, Kyrgyzstan, and Tajikistan gained their independence in 1991. Although the appellation "Central Asia" already existed then, there was hardly any detailed information on what countries there were, what kind of people lived there, and what livelihoods they pursued. Rather than part of Asia, these countries were republics of the former Soviet Union, and information on them came via Moscow instead of directly to us. A decade after their independence these countries have debuted in international society, but the task of gathering information on them has only begun. Many people still tend to see them merely as countries on the Silk Road south of Russia and north of the Tian Shan Mountains.

A total of about 50 million people populate the region covered by these five countries. When considering their environmental and agricultural challenges, one must keep in mind that these countries comprise many ethnic groups. Taking Kazakhstan as an example, 2000 statistics indicate that this country has 132 ethnic groups, with Kazakhstanis and Russians accounting for most of the population, at 53% and 30%, respectively, but other ethnic groups over 1% include Germans, Tartars, Uzbeks, Uygurs, and Ukrainians. Kazakhstanis are originally nomadic pastoralists keeping mainly sheep, goats, camels, and cattle. Livestock are on the decline, but there were once twice as many sheep as people. Kazakhstan's Russians migrated to agricultural regions in this country, mainly the wheat-growing area in the north near the border with Russia, while the Germans and Koreans were forcibly relocated to Central Asia during Stalin's rule.

## 4-2    Environmental Characteristics of Central Asia

Topographically these five Central Asian countries are characterized by a mountain system several thousand meters high on the south, comprising the Tian Shan Mountains and the Pamir Plateau. Elevation rapidly drops as one moves north, with a small area of rich verdure that quickly gives way to vast steppes and deserts. Tajikistan and Kyrgyzstan are mostly mountainous, while the greater portions of Turkmenistan, Uzbekistan, and Kazakhstan are desert. Agriculture and the environment are dominated by the two topographical extremes of deserts and glacier-covered mountains.

The region's climate is inland continental, with very hot summers and bitterly cold winters, and only two or three weeks of moderate weather in spring and autumn. Montane areas receive 600 mm or more of precipitation annually, and are thus forested, but most of the region is desert receiving only 100 to 200 mm. Northern Kazakhstan gets 400 mm and is part of a wheat-growing area extending down from Russia. Areas with under 200 mm of rainfall cannot support agriculture without irrigation, and allow only groundwater-dependent livestock raising.

Soil in the Central Asian steppes and deserts is sandy and saline with very low organic content, making it unsuited to agriculture. A characteristic of this soil is that slight precipitation causes salts to migrate downward, where they accumulate at a certain depth, unlike high-precipitation regions where salts are flushed out of the soil system.

The steppes and deserts feature either short halophytes, or woody plants such as saxaul and tamarix that grow no higher than 1.5 meters, making some of the desert areas more like wastelands than sandy or rocky deserts. Both steppes and deserts are green only two or three weeks in the spring, then dead vegetation from June through the rest of the year.

## 4-3  Environmental Characteristics of Central Asia

It is barely possible to grow wheat with rainfall alone in the wheat-growing district of Kazakhstan's northern steppe, and farmers have developed a variety of techniques to use winter snow as an agricultural water supply. This is the only part of Central Asia where agriculture is possible without irrigation, which is why the Amu Darya and Syr Darya, which are fed by the glaciers of the Tian Shan Mountains and the Pamir Plateau, have since ancient times been vital water sources that allowed the development of oasis agriculture. Another valuable desert water source is the Ili River that emerges from China, but this overview of Central Asian agriculture will focus on the Amu Darya and Syr Darya.

These rivers empty into the Aral Sea, which has no outlet and maintained its water level through evaporation. Beginning in the 1960s the Aral Sea started shrinking rapidly due to the diversion of water for irrigation from the Amu Darya and Syr Darya.

This irrigation project, conceived under the Soviet Union, converted a large desert area into farmland for growing cotton and rice, which changed the Soviet Union from a cotton importer into an exporter by constructing many canals of various sizes to carry water to the fields. Especially well known is the Karakum Canal in Turkmenistan, which at the time of its construction was the world's longest canal at 1,100 km. Water supplied by this canal made it possible for Turkmenistan, which is 90% desert, to grow cotton. Irrigated cropland was well over 1 million ha in 1985. Uzbekistan and Kazakhstan both started expanding irrigated farmland mainly in the basins of the Amu Darya and Syr Darya, increasing Uzbekistan's irrigated farmland 1.7 times between 1950 and 1985, from 2,276,000 ha to 3,930,000 ha.

More irrigation translated into less water in the rivers, raising withdrawal from the Amu Darya from 33 cubic km in 1961 to 41 cubic km in 1970, and that from the Syr Darya from 20.6 to 33 cubic km. After 1970 river water use tended to continue rising despite periods of varying water availability. In 1985 withdrawals reached 66 cubic km from the Amu Darya and 54.4 cubic km from the Syr Darya, which are over twice the withdrawals in 1961. At least 90% of the water was used for irrigation. As a consequence, the farther downstream, the smaller the rivers became.

Of great significance is the fact that annual withdrawals grew faster than the increase in irrigated farmland area, for which there are two reasons. First, while at first farmland was opened in areas near the rivers, desert far from the rivers gradually came under cultivation. This required longer canals, which were unlined and therefore leaked an amount of water commensurate with their length. Hence the farther farmland is from the river, the more water needed to supply it.

Second is salinization, the fate of irrigated agriculture in desert regions. Once salts have concentrated in the topsoil, they must be flushed out with large amounts of water before sowing seeds. In recent years the soil is flushed two or three times, while farmland with high salt concentrations requires five applications of water. River water withdrawal has therefore increased faster than farmland expansion. The problem is aggravated by the increasing lack of maintenance for silted drainage canals, which allows the further accumulation of salts and demands still more flushing. Many cotton fields have been abandoned in Uzbekistan recently due to the lack of excavators to maintain drainage canals, which in turn allowed salts to accumulate to levels that make cultivation impossible.

It is not an overstatement to say that cotton and rice are about the only crops grown in irrigated Central Asian agriculture, which produced 95% of the Soviet Union's cotton. Continuous expanses of cotton fields are still found in the basins of the Amu Darya and Syr Darya. Rice is grown in Kzylorda Province along the mid reaches of the Syr Darya, and development of a rice-growing zone in the Ili River system began in the 1970s. Even since the Soviet Union's collapse and the independence of Central Asian countries there has been little change in the cropping of cotton and rice in large-scale irrigated agriculture.

## 4-4    Aral Sea

### The Shrinking Aral Sea

Diversion of water for irrigation has had the effect of shrinking the Aral Sea, which might disappear altogether. Once the fourth-largest lake in the world, it has come to this point in only 40-odd years. While the direct reason for its impending demise is simple, the underlying factors are not.

Sandwiched between Kazakhstan and Uzbekistan, the Aral Sea is fed only by the Amu Darya and Syr Darya, whose basins lie in the five Central Asian countries and Afghanistan. Its decline was long known among scientists, but did not become widely known to the world public until perestroika was launched in 1985. Between 1934 and 1960 the Aral Sea had an average water level of 53 meters, with variance staying within 1 meter, covered an area of 6.6 million ha, and had a volume of 1,064 cubic km. After 1961 the lake started shrinking, until in 1989 its water level was 39 meters and it had divided into the Small Aral Sea and Large Aral Sea. The areas where the Amu Darya and Syr Darya flowed into the Aral Sea were once prime wetlands with rich biota that served as staging sites for migratory birds, but they have nearly disappeared.

## Fisheries

Originally a saline lake, the Aral Sea until the 1960s had a salt concentration of about 10ä and hosted a great variety of life. Salt concentration started rising with the development of large-scale irrigation, and topped 30ä about 1980. Since 1994 the authors have studied water quality in the Amu Darya, Syr Darya, and Aral Sea, and electrical conductivity (mS/cm) readings indicate that salinity of river water increased with proximity to the lake, showing that water draining from the farmland is carrying away salts. Measurements at the Syr Darya mouth and in the Small Aral Sea yielded readings of about 30, but they are twice as high in the Large Aral Sea. As ocean salinity is 30, the Small Aral Sea still allows fish to live, but the Large Aral Sea barely has any plankton.

Declining water quality and shrinking wetlands allowed spawning zones to disappear, and the Aral Sea's annual fish catch dropped from its original nearly 50,000 tons to nearly zero in 1980, which forced fishers living in the basin to concentrate on developing fisheries in the lakes and marshes along the Amu Darya and Syr Darya. About the only catch from the Aral Sea is flatfish, which yielded a mere 270 tons in 1999. Residents of former coastal fishing villages once occupied themselves with half fishing and half livestock raising, but when the Aral Sea receded they switched to raising livestock, primarily camels and goats.

PHOTO 2. A fishing vessel abandoned on a part of the Aral Sea bed that dried up in the 1970s
(September 2001 photo by Ishida Norio)

Aralsk was the Aral Sea's largest fishing port and the location of a fish processing facility, but it has suffered a disastrous economic and social collapse. In
a bid to maintain the processing facility and make up for the decline in the fish
catch stemming from the lake's shrinkage and water quality deterioration, in the
1970s the Soviet government began shipping fish caught in the distant North Sea
and Sea of Japan by rail to Aralsk to have them canned there. This was of course
not profitable, and the facility went out of business. Owing to the loss of fishing,
the city's lifeblood, its population plummeted from an initial 90,000 to 30,000 in
2000. The same fate befell Muynoq, a once-thriving fishing port in Uzbekistan.

## Desertification

As the Aral Sea dries up, more of its former lake bed turns into desert with each
passing year, and the lake bed's appearance depends on when it was exposed.
Some of the area exposed in the 1960s now has coastal vegetation and heaps of
shells from shellfish once living there, while other areas are turning into sandy
desert without any vegetation. Needless to say, the entire former lake bed is
covered with salt, which in some places blankets the ground in white bands.

Shrinkage of the Aral Sea not only created new deserts, but also had an
inevitable impact on the regional climate, in sum making summers hotter and
winters colder. In particular, the Aral Sea has begun freezing in the winter, while
higher summer temperatures are boosting updrafts and frequently causing large
sandstorms. Crops are damaged when winds pick up salt from the ground and
drop it on farmland. Impacts on human health are also suspected, but detailed
studies have yet to be performed. In addition to new desert patches emerging on
the exposed lake bed, desertification is also proceeding in the surrounding areas
and in some places devastating agriculture.

Wildlife has experienced a hard setback from the loss of the broad wetlands
once existing at the mounts of the Amu Darya and Syr Darya. A study conducted
in 2000 observed a large flock of pelicans in a wetland at Kazakhstan's Lake
Balkhash but barely found one at the mouth of the Syr Darya. Although large
flocks of flamingos were observed in the Aral Sea in 1990, it has been very hard
to find them over the past decade. The rapid decline of aquatic life in turn induces
the loss of birds and large animals.

## Impacts on Local People

These sudden environmental changes have had serious impacts on local people,
who did not have enough time to prepare, and who were for some time unaware
that the cause was 1,000 km or more upstream. Lacking appropriate government
action, people were deprived of their livelihoods, fell into poverty, and suffered
health damage, with an apparent increase in maladies such as anemia and
tuberculosis.

One salient indication of health impacts is the rising infant mortality rate along
the Aral Sea coast. After the Second World War this rate (the number of infants
per 1,000 who die within one year of birth) fell in Western countries, Japan, and

throughout the Soviet Union thanks to social stability, economic growth, and medical advances, but it continued rising in the Aral Sea basin even in the 1970s and thereafter. For instance, in Uzbekistan it increased sharply beginning in the 1960s, when the Aral Sea started to recede, and went as high as 80‰. It jumped especially high in the Karakalpakistan Autonomous Republic on the Aral Sea's north shore. These high rates also appear in the statistics of national government and autonomous republics. Although statistical data for individual communities along the Aral Sea coast have not been released, interviews conducted by the authors at village clinics and other places found villages with rates of over 200‰, which is likely an indicator of the rapid environmental deterioration along the coast. While it might be going too far to assume that the Aral Sea is the sole cause, it is likely that environmental changes, economic deterioration sparked by the collapse of the fisheries, and declining livelihoods all conspired to raise the infant mortality rate. Recent reports state that 45.9% of women and 64.3% of children are anemic, and anemia among pregnant women is thought to seriously affect the health of newborn babies.

Causes of health damage are the chemical fertilizers and pesticides used on the irrigated farmland along the Amu Darya and Syr Darya. These substances end up in the Aral Sea, contaminating the rivers, lake, and groundwater. Accumulated fertilizers and pesticides on the dry lake bed assail the local people during sandstorms. Although the public is sure these substances are harming them, no study results exist among publicly released documents.

Drinking water quality is impaired, with salt concentration particularly high. Soil throughout Central Asia is alkaline and saline, and water is likewise saline with high concentrations of calcium and sodium in both surface and groundwater. Water drained from irrigated land raises water salinity and forces basin inhabitants to seek new sources owing to their dependence on river water and shallow wells for drinking water. Further, shallow groundwater veins are drying up in the coastal zone owing to Aral Sea shrinkage. To supply water for people and livestock the government drilled 2,500 deep wells near the Syr Darya mouth from 1970 on. Generally these wells are highly saline and not fit for drinking, but many villages have no choice but to rely on them.

## 4-5  Internationalization

Agriculture in post-independence Central Asia faces severe economic circumstances that make farming on a large scale nearly impossible to continue. As the market economy was instituted, state-owned and collective farms in all former Soviet states were dismantled and replaced by new forms: farming corporations, family farms, and family groups. However, these are still fluid by nature and changing from year to year. Cultivated area also changes, an example being a state-owned collective that during the Soviet era grew 5,000 ha of rice, but which has now divided into the above three forms and is cultivating 1,500 ha. Each type of farming is pressed to make changes for its survival, but agriculture in Central Asia as a whole is challenged severely by the onslaught of internationalization.

It is also doubtful that these operations can survive with monocultures of cotton or rice. How environmental regeneration happens in the Aral Sea basin will be governed by the scale and quality of agricultural practices adopted in this arid region.

The Amu Darya and Syr Darya were domestic rivers under the Soviet Union but are now international rivers, thereby obliging watercourse states to develop water management systems among themselves. Regenerating the Aral Sea will be impossible if the parties merely insist on their vested interests. Already no water at all enters the Aral Sea from the Amu Darya, leaving no hope for reviving the Large Aral Sea. In another few years it will probably disappear and leave a new desert. The only means left for saving part of the lake would be to rebuild a dam at the boundary (Berg Strait) between the Small Aral Sea, into which the Syr Darya flows, and the Large Aral Sea. First built in 1996, the dam collapsed in 1999 and is slated for rebuilding. Although only half the original lake remains, it is quite possible to revive the Small Aral Sea's biota and recreate a decent living environment for people in the Syr Darya basin.

Large-scale irrigated farming under the Soviet Union began with sloganeering about "greening the desert," but the situation changed with the Soviet Union's collapse and the Aral Sea's disappearance. The simplistic idea that agriculture is possible with land, good weather, and a big river to supply water was totally discredited after several decades. The Aral Sea illustrates the miserable failure of iron-hand development that ignores the relationship between basin inhabitants and their environment.

## 5.  Prospects for International Cooperation

We shall close this chapter by examining the prospects for international environmental cooperation in Inner Asia.

First, in view of the severe environmental damage in this region, immediate consideration must be given to assessing human health damage and how to help the victims. In particular, there are concerns about rising infant mortality rates and other health impacts occurring as the Aral Sea shrinks, but the connection needs more study. This is not only meant as humanitarian assistance because it might also be important for coping with future risks that humanity may face. If in the future many people face such risks by having to live under conditions such as those in the Aral Sea basin, now is the time to seriously consider how to maintain their health and livelihoods.

Second, while environmental damage in this region is indeed grave, we want to stress that sufficient attention be given not only to environmental conservation, but also to development of the regional society and economy. This requirement is of course not limited to this region, but rather a widely recognized and crucial facet of the universal theme "development and environment." For example, the "household eco-economic zones" created in China's Inner

Mongolian Autonomous Region are of great interest because they are designed to both arrest desertification and carry on agriculture and forestry.

When envisioning how to achieve environmental and socioeconomic progress in a region, it is important to obtain the views of local people and tailor action to the circumstances. Interviews and questionnaires[24] conducted by one of the authors (Kusumi) during the summer of 2001 in Aralsk and other places on the Aral Sea's eastern coast found that people responding that environmental conservation should have priority slightly outnumbered those in favor of the economy, agriculture, and the fishing industry. However, locally there is a tendency to blame Aral Sea shrinkage or the Baikonur Cosmodrome for nearly all problems including sandstorms and illness. Because in some ways the Aral Sea has become a draw for eliciting foreign aid, one should read these results with a bit of skepticism. Indeed, the impression gained in fieldwork was that local economic development should have priority. While it is not clear where emphasis should be placed, the efforts of notable local people working diligently on running collective farms and raising fish species with high salt tolerance suggested that helping people who form the nucleus of a community is an effective way to achieve socioeconomic development, and then move on to sustainable use of the environment.[25]

Third is the importance of international cooperation. The foregoing discussions covered domestic environmental problems such as air pollution, water pollution, and solid wastes that other countries likewise face in the process of economic development, and problems with cross-border causes and effects, such as desertification and the shrinkage of international lakes. In the former category the main causes include governmental mismanagement during the socialist era and inadequate steps to cope with growing development pressure during the transition to a market economy. To start with, Central Asian countries cannot do without developing and implementing domestic policies for development and the environment. In this area they can probably learn something from the devastating pollution of East and Southeast Asian countries and how they addressed it. In the latter category, each country has to both take domestic action and cooperate with other countries, but even East and Southeast Asian countries are still at the trial-and-error stage and have yet to accumulate enough experience. Environmental cooperation in Inner Asia will demand not only cooperation with other individual countries, but also much effort in developing mechanisms for international cooperation to address cross-border ecosystem damage.

There are also hopes that Central Asia will look for ways to cooperate internationally on the grassroots level. For example, a group of people interested in Mongolia has held monthly meetings in an attempt to share information among researchers, NGOs, students, and other participants, and to build a cooperative relationship with Mongolian researchers. In this process group members have learned from Chinese examples, especially in the environmental field.

Meanwhile, since Mongolia's switch to a market economy in the 1990s, reductions in the national budget have slashed research funds and curtailed research staffs, making it very hard for Mongolian researchers to continue environmentally related research on their own. In fact, it is almost impossible for them to conduct studies and research without overseas cooperation in both funding and equipment. A fact of note is that there are researchers who have done much work with counterparts in former East bloc countries.

A specific example is a project to combat desertification in the northwestern part of Mongolia's Govi-Altay Province called Hoh-morit District. Until 1992 when Mongolia abandoned socialism and set its course toward a market economy, primarily Mongolian researchers—sometimes conducting joint studies with Soviet scientists—conducted desertification-control projects in this region with Mongolian government funding. Playing the leading role was Baasan Tudev, who lost his job when the Mongolian Academy of Sciences was downsized after the market economy transition, and has since continued research on his own in a limited geographical area. In August and September of 2002 Baasan and others visited Hoh-morit for the first time in a decade, where they found that *dzud* damage had occurred three years running. The resumption of cooperation for Mongolian-led efforts to protect Hof-morit's central town Sayn-ust from blowing desert sand is now under consideration.

Local ideas for future initiatives include education for learning from the local environment, mainly the desert, with Hoh-morit school students, and efforts to protect Sayn-ust from blowing sand, primarily by reestablishing perennials and creating windbreak forests with native species. Japan will have to offer more than just research cooperation, including intellectual support for Mongolian researchers, and helping spread the word about their activities.[26]

Bao Zhiming, Ishida Norio, Kamo Yoshiaki, Kusumi Ariyoshi,
Minato Kunio, Okada Tomokazu, Otsuka Kenji

# Chapter 4
# Country/Region Updates

A site with severe soil contamination in Anshun (Tainan County), Taiwan. A caustic soda plant was established by the Japanese company Kanegafuchi Soda before World War II. After the war, Taiwanese companies operated the facilities. The site is contaminated with very high concentrations of substances including mercury, PCBs, and dioxins, and it should be remediated immediately because of the many aquaculture ponds in the vicinity.
Photo: Kojima Michikazu

# Japan

Since the previous volume, changes in Japan's environmental policy have involved mainly (1) global warming and (2) waste management and resource recycling. This section therefore focuses on these two topics and provides a brief follow-up on Japanese environmental policy over the last three years. Other events during this time period include the signing of the Stockholm Convention on Persistent Organic Pollutants, amendment of the Special Measures Law on the Disposal of PCB Wastes and the Chemical Substance Screening Law, and enactment of the Pollutant Release and Transfer Register Law, Fluorocarbon Recovery and Destruction Law, Soil Contamination Countermeasures Law, and Law to Promote the Restoration of Nature. A number of new environmental challenges also arose, but lack of space prevents the authors from reporting on them here.

## 1.    Assessing Japan's Efforts on Global Warming

### 1-1    A New "Guideline of Measures to Prevent Global Warming"

After the December 1997 adoption of the Kyoto Protocol, Japan's government in June 1998 created the "Guideline of Measures to Prevent Global Warming" ("Old Guideline") to serve as a comprehensive framework for combatting global warming, but after several years it was evident that it would be impossible under the Old Guideline to meet Japan's Kyoto Protocol obligation of a 6% reduction from 1990 levels in greenhouse gases (GHGs). The government accordingly released its "New Guideline" in March 2002 in preparation for ratifying the Kyoto Protocol. In line with this New Guideline, the government in June 2002 amended the Law Concerning the Promotion of Measures to Cope with Global Warming and the Law Concerning the Rational Use of Energy (Energy Conservation Law). Additionally, other laws were passed, such as the Special Measures Law Concerning the Use of New Energy by Electric Utilities (RPS Law).

The New Guideline gives four basic ideas on how Japan should combat global warming: (1) balance between the environment and the economy, (2) a step-by-step approach, (3) promotion of combined effort by all sectors of society, and (4) international cooperation for measures to prevent global warming. The step-by-step approach divides the 12 years from 2002 into three steps: step one is 2002–2004, step two is 2005–2007, and step three is 2008–2012, corresponding to the first commitment period. New Guideline procedures are supposed to quantitatively elucidate the steady headway made toward meeting Japan's obligation under the Kyoto Protocol by implementing the measures that start in step one, and to take whatever additional measures are necessary while assessing the progress achieved by measures coming before steps two and three.

## 1-2 Additional Measures and New Policy Instruments

Japan's total GHG emissions as of 1999 were 1,314 million tons ($CO_2$ equivalent), which even under the Old Guideline would result in emissions of 1,320 million tons in 2010, an increase of about 7% over 1990 baseline year emissions of 1,229 million tons. Japan has to lower total emissions to 1,155 million tons to achieve its 6% reduction, and must provide for additional emission cuts equal to the difference of about 165 million tons.

Accordingly, the New Guideline's chief objectives included, first, pushing $CO_2$ emissions from energy use down to the 1990 level in the first commitment period; second, reducing emissions of non-energy $CO_2$, methane, and $N_2O$ by 0.5% of 1990 emissions in the first commitment period; third, a 2% reduction in GHG emissions through development of innovative technologies and further efforts by all citizens to arrest global warming; fourth, holding increases in emissions of the three GHGs HFCs, PFCs, and $SF_6$ down to 2%; and fifth, using Japan's forests to sequester the approximately 13 million tons of carbon (47,670,000 tons $CO_2$ equivalent, about 3.9% of total baseline year emissions) agreed to at COP 7. Further, the New Guideline also strengthens measures in each sector, aiming for reductions from the baseline year of 7% in the industrial sector, 2% in the residential/commercial sector, and 17% in the transportation sector. Included for the "voluntary action plans," which constitute the core of measures in the industrial sector, are: (1) government councils and other bodies check the extent of progress and ensure effectiveness; (2) industries without action plans are urged to quickly develop specific plans by, for example, setting numerical targets, and to release them to the public; and (3) for additional measures the guideline supports the third-party certification and registration system that the Japan Business Federation plans to institute, and will provide for the greater transparency and reliability of voluntary action plans.

As far as one can see from progress to date and the future outlook, it will be virtually impossible for the New Guideline's step-by-step approach, at least in step one, to attain the objectives outlined above, making it imperative to institute additional measures in steps two and three. Hence it will now be increasingly important to consider the appropriate form of the policy mix combining new additional measures such as environmental taxes like the "global warming tax" proposed by the Environment Ministry, emissions trading, and agreements.

# 2. Waste Management and Resource Recycling

## 2-1 Environmental Legislation

At the end of May 2000 the Basic Law for Establishing a Recycling-based Society (Recycling Society Law) was passed after unusually short deliberation, slightly more than a month after the government's bill had been sent to the Diet, which at the same time passed a number of other recycling-related laws: the Wastes Disposal and Public Cleansing Law (a revision), Law for the Promotion of

Utilization of Recycled Resources, Construction and Demolition Waste Recycling Law, Food Waste Recycling Law, and Green Purchasing Law. These laws more or less filled out Japan's lineup of laws on waste management and resource recycling when added to laws that predated this group, such as the Basic Environment Law, (promulgated and took effect in November 1993), Container and Packaging Waste Recycling Law (promulgated in June 1995, took partial effect in April 1997, fully enforced in April 2000), and the Home Electrical Appliance Recycling Law (promulgated in June 1998, took effect in April 2001). And in July 2002 the Automobile Recycling Law was enacted.

Supposedly this suite of laws is meant to reorient Japan away from being a "mass-disposal society," which has characterized Japan's socioeconomic system until now, and make it into a "cyclical society." Article 2.1 of the Recycling Society Law describes such a society as a "society where the consumption of natural resources is restrained and the environmental burden is reduced as far as possible by keeping products out of the waste stream, promoting appropriate recycling of products when they have become recyclable resources, and ensuring appropriate disposal of recyclable resources not recycled." In contrast with the Wastes Disposal and Public Cleansing Law, which served as the basis for Japan's waste-management policy until the new suite of laws appeared, this policy doctrine is laudable for its inclusion of new provisions such as: (1) wastes were originally considered either "wastes"[1] or "unwanted items," but to these were added the concept of "recyclable resources," thereby expanding legal regulation to a broader area, (2) suppressing the generation of wastes comes first, and the "cyclical use" of recyclable resources is encouraged, hence placing emphasis on effort to make things recyclable, and (3) specific purposes are curbing the consumption of natural resources and reducing the environmental burden. Of particular importance is that the Recycling Society Law expressly sets forth this order of priority for managing wastes: (1) long-term use, (2) reuse (using products as is, including repair), (3) material recycling (use as raw materials), (4) thermal recycling (obtaining heat by incineration or other processes), and (5) appropriate disposal as waste. Based on Article 15 of the law, in March 2003 the Cabinet approved the Basic Plan for Establishing a Recycling-Based Society, which includes specific numerical targets for the management of different waste types and resource recycling, and administrative policy measures to comply with the plan are now gaining momentum.

## 2-2  Full Implementation of Extended Producer Responsibility

But is this suite of new laws really helping Japan extricate itself from the mass-disposal society? Not a few issues need to be examined anew. Take, for example, the provision in Article 2.1 of the Recycling Society Law that products are to be kept from becoming wastes. This provision likely had recourse to, but differs from, the provisions in Germany's Ordinance on the Avoidance and Recovery of Packaging Waste and the subsequent Recycling and Waste Management Act, which distinctly say that top priority should be accorded to "avoidance" (*Vermeidung*) of generating wastes. By contrast, Japan's Recycling Society Law has not made a clean break with the policy mindset of curbing waste generation because it still assumes a socioeconomic anatomy that generates huge amounts of wastes. Such being the case, there are concerns and criticism that even if "mass recycling" is encouraged under the "cyclical society" doctrine, this will actually maintain and prolong the mass-disposal system. The primary reason that such concerns and criticism inevitably arise is that in creating this suite of new laws there was still no exhaustive discussion of responsibility and cost allocation for wastes, which has always been a crucial issue in Japan's waste policy.

While on this point the Recycling Society Law is limited in some ways, it has more or less incorporated the idea of extended producer responsibility (EPR), which is of great significance. For instance, Article 11 (Responsibility of Businesses) contains this provision: "[I]t is necessary for the national government, local governments, businesses, and the public to play their respective roles properly in order to ensure the appropriate and problem-free recycling of products and containers that have become recyclable resources, and, if the roles of any particular businesses are deemed important in establishing a cyclical society from the viewpoint of the design and selection of raw materials for their products and containers, and the collection of such products and containers that have become recyclable resources, the role of the businesses themselves, pursuant to the Basic Principles, is to take responsibility for collecting or delivering, or properly recycling, products and containers that have become recyclable resources."

But when seen on the level of the previously enacted Container and Packaging Waste Recycling Law, the subsequent Home Electrical Appliance Recycling Law, and also the Automobile Recycling Law, which is to take effect in 2005, in many ways these EPR provisions need reconsideration with regard to who bears responsibility and pays the costs. Under the Container and Packaging Waste Recycling Law, for instance, the systems have more or less been organized for the "recovery," "takeback," and "recommodification" of the packaging and container waste types designated under the law, but recovery is the responsibility of municipalities, while takeback and recommodification are the responsibility of certain businesses designated under the law. Although the burden on business is somewhat heavier than before, a far greater proportion of the burden is actually on the shoulders of municipalities (in other words, taxpayers). Consequently, there is hardly any effective economic incentive for businesses to curb the excessive use of packaging, which is a problem of colossal proportions at the disposal end of the waste stream. Since the law took effect there have been gains in recovery and recommodification amounts, but if one takes PET bottles as an example, production and sales volumes have eclipsed those gains, and the amount of PET bottles entering the waste stream surged from 167,808 tons in 1996 to 268,700 tons in 2000. In the case of the Household Appliance Recycling Law, businesses are for the time being legally obligated to recover and recommodify four kinds of scrapped appliances: televisions, refrigerators, air conditioners, and washing machines, but the system for covering the costs of recommodification and of transport and collection at designated takeback locations makes the consumers pay when they scrap the products. As predicted at the outset, the law ironically aggravated illegal dumping. A survey released by the Environment Ministry in early August 2001 just after the law took effect revealed that illegal dumping had clearly increased after the law took effect in more than half of 272 municipalities where it was possible to compare dumping before and after it became effective, and illegal dumping has continued since. Just as with the Container and Packaging Waste Recycling Law, this situation has come about because there was too little common-sense discussion of responsibility and cost allocation after appliances are scrapped. Japan needs to review its system for the purpose of more thoroughly implementing the EPR principle, which since 1994 has been the subject of extensive discussion mainly by the OECD and EU, and is gradually developing into a set of international rules for all countries to use.

## 2-3   Illegal Dumping: The Achilles Heel

Recently the most serious problem Japan faces with regard to waste management and resource recycling is illegal dumping (which includes inappropriate disposal and illegal exports). In particular, this section could not go without mentioning the frequent instances of industrial waste dumping in recent years. Incidents of scandalous proportions in the early 1990s were high-volume dumping in Iwaki City, Fukushima Prefecture, and on Kagawa Prefecture's Teshima Island. Later, even heavier illegal dumping was discovered on the border between Aomori and Iwate prefectures.

Under the Wastes Disposal and Public Cleansing Law, responsibility for the appropriate disposal of industrial wastes in principle rests with waste generators, but the reality admits to problems which cannot be overlooked.

First is the reality of waste "self-management," which has been implemented under the principle of waste-generator responsibility. Self-management has been the pretext for the rampant illegal dumping and inappropriate disposal of industrial wastes. For example, Ministry of Health and Welfare (now the Ministry of Labor and Welfare) data on illegal dumping of industrial wastes in 1997 indicate that 119,030 tons, or 29%, of the total discovered illegal dumping of about 408,000 tons was committed by the waste generators themselves. Based on the number of incidents, 435 of the total 855 were by generators themselves, making the proportion 51%. As this shows, waste generators do not always abide by the principle of taking responsibility for appropriate disposal.

Second, authorized disposal contractors may be commissioned because "self-management" is not always a literal requirement. In many instances the fees are not sufficient to cover a contractor's expenses for appropriate disposal, thereby instead encouraging illegal dumping and other forms of improper disposal. Moreover, this has created a situation in which the responsibility of waste generators for illegal dumping by contractors is not legally questioned, and the buck is passed to disposal contractors.

Third, the government allows public involvement in waste disposal under regulations for "co-disposal" and "required disposal" by cities, towns, and villages, and "regional waste disposal" by prefectures, which also makes the principle of waste generator responsibility for industrial wastes meaningless.

Consequently, this principle is not properly implemented despite being a provision of the Wastes Disposal and Public Cleansing Law, and this has led waste generators to seek the most inexpensive option for treatment and disposal. Illegal dumping has become rife nationwide because it is the most economically rational option. Movement of industrial waste across prefectural borders has also become common as businesses seek the cheapest way to unload their wastes, sparking a wave of disputes and lawsuits throughout Japan. One significant factor underlying this situation is the indulgent nature of regulations for punishment and environmental restoration. Although revisions of the Wastes Disposal and Public Cleansing Law in 1991, 1997, and 2000 provided for somewhat more rigorous penalties, other problems remain.

First, to prevent illegal dumping the 1991 revision created a system to designate especially hazardous wastes as "specially managed wastes," and set up a manifest system for such wastes. The 1997 revision went further by expanding the manifest system to cover all industrial wastes. Nevertheless, the manifest system needs many fixes including ways to stop the forging of manifests themselves, and to that end the 2000 revision attempted an improvement by imposing a greater obligation on waste generators for manifest control, requiring them to confirm the proper management of all industrial wastes from collection and transport to intermediate processing and final disposal. Although this represents substantial progress, keeping tabs on wastes with this management system is still ineffective in cases of self-management and when a waste generator has commissioned an unauthorized contractor, situations which account for a considerable percentage of illegal dumping. Effective measures in this area need to be explored.

Second, although revision of the law brought stricter penalties for illegal dumping, it is not a fundamental remedy for the practice of "coming out ahead by dumping." For example, the 1997 revision stiffened the penalty for illegal dumping to a maximum of three years imprisonment and/or a maximum fine of 10 million yen, and if a corporation, an additional maximum fine of 100 million yen, whereas the penalty had previously been a maximum six months imprisonment and a maximum fine of 500,000 yen. Yet the stiffer penalty is still insufficient to outweigh the economic benefit of illegal dumping. The government must consider more rigorous penalties that include confiscation of unjust profits realized from illegal dumping.

Third is the matter of dealing with illegal dumping after it is discovered, including site restoration. Here too, the 1997 revision provided for the creation of a fund to cover the costs of site restoration when the illegal dumper is unknown or cannot cover the costs, but further study is needed in areas such as the fund's place in the overall picture, its size, and how costs should be allocated. Currently industrial waste generators are merely asked to make voluntary contributions to the fund, while local governments pour their tax revenues into it. The government should consider ways to build the fund, such as surcharges on the types and amounts of industrial wastes, to be levied on waste generators as an industrial waste tax.

Awaji Takehisa, Isono Yayoi, Teranishi Shun'ichi

## Essay 1:    Another Look at Minamata Disease and Methyl Mercury Pollution

About a half-century since the official confirmation of Minamata disease in 1956, the story has arrived at another major turning point. Over these five decades the main issues have been (1) the responsibility of government authorities and Chisso, the company which caused it, for the occurrence and worsening of Minamata disease, and (2) how victims are certified and compensated. On the other hand, hardly any effort has been expended on matters such as elucidating the big picture on Minamata disease and determining what level of methyl mercury is needed to affect health. There is still so much we do not know about Minamata disease, the worst instance of methyl mercury damage ever. Nevertheless, over the last 10-odd years research around the world has found that even a trace amount of methyl mercury ingested by an expectant mother has profound impacts on the motor functions and psychological development of unborn children. For that reason there is a need to reframe the question of what Minamata disease is and to take another look at the last 40 years of Japan's pollution administration and environmental policy, including efforts to fully elucidate the effects of trace pollutants.

### Administrative Authorities Not Off the Hook

Chisso, the company that caused Minamata disease, admitted as much and continues to pay compensation to certified patients with about 160 billion yen in subsidies from the Japanese government. The government, however, refuses to admit to any legal responsibility at all and continues fighting in the courts, having appealed the April 2001 decision of Osaka Appellate Court in the Minamata Disease Kansai Lawsuit. At issue is the government's responsibility for doing nothing to stop the pollution even after the official confirmation of Minamata disease, allowing the damage to spread from Minamata Bay throughout the Shiranui Sea, and neglecting to help the victims despite the devastating harm. Especially administrative officials neglected the effort to accurately assess the damage, took no action, and ignored it, which led to a second occurrence of Minamata disease in Niigata Prefecture in 1965. It was not until 1968 that the government officially recognized that Minamata disease is caused by pollution. Administrative authorities have yet to accept their responsibility for allowing Minamata disease to happen and spread, and this is a major roadblock toward a true resolution for Minamata disease.

### Medical Basis for Certification Totally Discredited

At the same time, the certification system for helping Minamata disease victims developed many problems. This system's origin is the Minamata Disease Screening Council created in 1959 to decide who would sign agreements to receive solatia from Chisso. Subsequently Kumamoto Prefecture and Kagoshima Prefecture performed the actual certifications under patient certification systems based on, from 1970, the Law on Special Measures Concerning Redress for Pollution-Related Health Damage, and, from 1973, the Pollution-Related Health Damage Compensation Law. At the outset of the victims' struggle the focus was on getting Chisso to admit it had caused Minamata disease, but after the March 1973 decision in the first Minamata disease lawsuit, and the signing of the agreement based on it, the focus shifted to getting the government, which handled certifications, to certify victims as patients because "administrative certification" was a major barrier to compensation for victims. Victims continued trying to break down this barrier by demanding administrative appeal examinations and filing lawsuits. But the biggest barrier they face was the "acquired Minamata disease determination conditions" created by Minamata disease experts in 1977 because these conditions were used to refuse certification to more than 13,000 victims, claiming that they did not have Minamata disease. Although the conditions supposedly had a medical basis, a 1999 meeting of the Japanese Society of Psychiatry and Neurology concluded that the conditions had no medical basis whatsoever, and were in fact specious, which totally overturned the existing medical basis for Minamata disease certification.

## A New Disease Definition and Fundamental Changes in the Certification System

Controversy swirled around the definition of Minamata disease. It became clear that the primary symptoms stemmed from impairment of the cerebral cortex and central nervous system, which cast serious doubt on the health examination methods and certification requirements employed until that time. In performing Minamata disease examinations, physicians check for sensory disturbances, visual field constriction, hearing loss, ataxia, and other symptoms, and combine these with patients' descriptions in making determinations, but distal symptoms cause changes in patients' descriptions, and if descriptions differed in the second and third examinations, physicians would suspect feigned illness. However, because brain and central nervous system impairment means the parts of the brain controlling sensation and other functions are damaged, it stands to reason that victims report different feelings or none at all. The examinations performed over decades should therefore be reviewed. Especially evident is the harm done by the certification system under the Pollution Victim Relief Program in many fishing villages in Minamata and along the Shiranui Sea coast. For example, even though victims engaged in fishing to the same extent, ate as much fish as others, and suffer from the same symptoms such as headaches, numbness, and sensory disturbances, they will be treated in basically four different ways: some will be certified and receive 16 million yen, some will receive 2.6 million yen for medical care service in a settlement, some will get only medical care expenses for health care service, and others will be turned down and get nothing. Victims do not find this acceptable. Seriously ill victims in particular are refused on the grounds they have a second disease or complications, making bad feelings common within families or between neighbors. Further, the certification system requires victims to apply personally, which has left many of them without help because they lacked this information, or did not apply because they feared discrimination and prejudice. It is time for a fundamental review of the Pollution Victim Relief Program, which was at first meant to provide victims with quick relief.

## Long-term Trace Pollutants and Impacts on Fetuses

A major issue at the 6th International Conference on Mercury as a Global Pollutant, held in Minamata in October 2001, was the most basic indicator of methyl mercury contamination: At what level of mercury in the hair do impacts on human health emerge? Research in Denmark's Faeroe Islands, in New Zealand, and in other places found that even hair mercury levels below 10 ppm result in damage to psychological development and motor functions in fetuses and children. These findings put pressure on Japan to totally reconsider Minamata disease and review mercury restrictions. No studies were performed from 1955 to 1959, when many deaths occurred in the Minamata area, and even in 1969 and thereafter when mercury pollution lessened slightly, hair mercury was 920 ppm maximum and about 200 ppm in fishers. It ranged between 40 and 100 ppm or even higher in Hondo and Ushibuka cities, and even in Kumamoto, which are far from Minamata, showing that mercury pollution was both very heavy and widespread. What is more, hardly any follow-up studies were performed in these areas. In 1956 fish in Minamata Bay and the Shiranui Sea contained mercury of over 300 ppm, and those fish made their way through the food distribution system in many ways, including through itinerant fishmongers and processed foods. Unless investigators abandon the idea of a victim population of about 100,000 people and adopt methods assuming a population of over 2 million, they will be unable to find all the people affected directly.

In 1997 the US Environmental Protection Agency submitted a report to Congress detailing the risks of cancer and other health damage. The report recommended that the tolerable daily intake of methyl mercury especially for pregnant women should be below 0.1 $\mu$g, and asked people to limit their fish consumption in response to the degree of contamination. This translated into fish consumption of under 12 ounces (about 340 g) per week for a 50-kg person, which in Japan would be 260 g because fish contamination with methyl mercury is high at 0.13 ppm on the average. Yet such a recommendation has yet to be issued in the vicinity of Minamata Bay or along the Shiranui Sea coast. The government must implement this USEPA standard for the people in that region, especially for fishers, review safety standards for fish consumption, and provide meticulous administrative guidance for the general public and for fishers and their families because they eat much fish. Additionally, symptoms are worsening among many patients, especially congenital Minamata disease patients and Minamata disease victims of the same generation. It is crucial that we squarely face the reality of Minamata disease as a general illness, and work on a solution that includes psychiatric care in view of the heavy discrimination and prejudice against Minamata disease victims.

Tani Yoichi

# Republic of Korea

## 1. Environmental Policy of the Kim Dae-jung Government

In February 2003 the Kim Dae-jung government was succeeded by that of the new president, Roh Moo-hyun. While it will be interesting to see what environmental initiatives Roh's administration will implement, at this time it is vital to review the environmental policies of former President Kim during his five years in office and note what policy challenges face the new administration. This section will therefore examine a number of Kim's policies (air, water, wastes, and soil) while assessing the current state of South Korean environmental policy and surveying future challenges.

Five years before the change in administration President Kim had offered a new general framework for environmental policy whose objective was "achieving a sustainable community of life in which people and nature are in symbiosis." The basic principles by which this objective would be achieved included implementing environmental policy with citizen participation, full application of the precautionary principle, switching from a policy concerned mainly with supply to a policy that manages demand, harmonizing the environment and the economy, emphasis on the roles of local governments, and full application of the polluter pays principle. Policies in individual areas were then implemented in accordance with these principles. A new assessment is needed to determine the extent to which the government carried out this bold objective and basic policy principles, and how much was actually achieved.

Cutting right to the conclusion, it appears that environmental policy over the last five years has made substantial progress in creating a superficial and outwardly well-structured legal system and in setting up a policy framework based on it, but policy implementation leaves much to be desired, with results yet to be seen, or the situation in fact worsen than before.

In August 2000 the government instituted the Sustainable Development Committee as an advisory body for the president to bring about citizen participation in the environmental policy decision-making process. Comprising representatives of the government, industry, and citizens' organizations, the committee was meant to reconcile conflicts of interest over large government projects that could have serious environmental impacts, and make recommendations to the president about policy orientation, but it was incapable of achieving this and for all practical purposes was a failure. Internal divisions among committee members over the Semangum land reclamation project, a string of resignations by non-government members, and the rigid approach of development-oriented government agencies speak eloquently of this failure.

Meanwhile pollution of the air, water, and soil grows worse with each passing year. Despite repeated tightening of automobile emission regulations, air pollution in the Seoul region has exceeded acceptable limits, while the special laws created individually to improve four major rivers has failed to stop the worsening quality of drinking water sources. Although South Korea has built a national measurement network to study soil contamination, the state of contamination is still not well known. Moreover, a series of real estate bubbles underlies an unstoppable wave of chaotic land development projects around the country.

## 2.    Air Pollution

Forty-six percent of South Korea's population lives in the Seoul area, and the trend toward concentration of the population in this area continues unabated. A nearly equal percentage, 46%, of the country's automobiles are also found here. These high densities of people and automobiles make Seoul's air pollution two to four times worse than that in other industrialized countries. The concentrations of NOx and $PM_{10}$ particulates in particular have not improved, and in fact are increasing year by year.[1] Further, over 90% of South Korea's ozone alerts are in the Seoul area.

Heavy air pollution imposes onerous social costs. According to government calculations, the social costs of urban air pollution (the damage to health and assets by air pollution) amount to 45 trillion won annually, and almost half of Seoul's residents have suffered respiratory system damage because of air pollution, which has surpassed what people can endure.

In a bid to improve the situation, the Ministry of Environment (MOE) developed Special Measures to Improve Air Quality in the Capital Region, marking the start of serious efforts to stem air pollution in and around Seoul. These special measures are aimed especially at NOx and $PM_{10}$, which are mainly from automobile exhaust. As such, remedial measures for motor vehicle exhaust are the focus of these measures, whose goal is to reduce NOx and $PM_{10}$ emissions to below half their 2000 levels no later than 2012. Policy measures for attaining this goal are stricter controls on the emissions of motor vehicles, especially diesel vehicles, stricter regulation of emissions from new vehicles, promotion of low-emission vehicles such as hybrids and electric vehicles, expanded use of natural gas-powered buses (from 3,000 in 2003 to 20,000 in 2007), promotion of clean energy (increasing the number of areas that use low-sulfur fuel oil from seven to 20), and the institution of areawide total pollutant load control and an emissions trading system.

Excepting the areawide control and emissions trading system, these measures are the same as the individual-vehicle controls under Japan's Motor Vehicle NOx/PM Law, leaving doubts about what degree of success can be expected. In 2003 South Korea's government approved the production and sale of diesel-powered passenger cars, whose addition to the fleet could very well negatively impact air quality in major cities. Because diesel fuel is considerably cheaper than gasoline, failure to make price adjustments could lead to a jump in the number of diesel passenger cars.

# 3.    Water Pollution

In 1998 the government announced its "Special Comprehensive Measures to Manage Water Quality of Drinking Water Sources in the Han River System" to improve Han River water quality because it supplies the capital city region. These special measures represent a major departure from the water quality management policy followed previously, and they are characterized mainly by the introduction of watershed management systems. The government later announced special measures for managing the water systems of the Nakton, Kum, and Yeongsan rivers, giving the country formal systems to manage the watersheds of its four major rivers.

Behind the creation of these watershed management systems are the limitations of previously used controls on individual emission sources (the Water Quality Preservation Law), and the heavy dependence of South Korea on rivers for drinking water. The major features of these watershed management systems are the introduction of waterfront zone systems, total emission management systems, water use fee systems, and the creation of a water system management commission for each water system. These commissions levy water use fees on drinking water users in downstream areas and use the proceeds as assistance for people and municipalities in upstream regions, where the government designates waterfront zones and imposes strict limitations on land use within them. Total pollutant loads are also set in upstream regions for the purpose of attaining water quality targets, and then total loads are set for each sub-zone. In this way, limitations are imposed on land use, factory siting, and other economic activities in each sub-zone.

South Korea has therefore established a management system for each of four river systems, rules for fair cost allocation (beneficiaries pay), and total pollutant load controls, but water quality in the upper reaches of the Han River has yet to see any improvement, and in not a few areas it has in fact deteriorated because these management systems alone are not enough to control land use in this region, and it is increasingly evident that these systems alone are not up to the task of arresting additional road construction and haphazard land development. South Korea needs a more drastic reform of its institutions for controlling land use.

# 4.    Resource Recycling and Soil Contamination

In February 2002 the South Korean government substantially revised the Act Relating to Promotion of Resources Saving and Reutilization. The revision's chief feature was that it clearly defined the responsibility of producers for the reuse of wastes, thereby fully applying extended producer responsibility. Especially worth noting is that the law specifies producer responsibility for the recovery and reuse of 18 product types including televisions, refrigerators, air conditioners, washing machines, personal computers, and other consumer appliances, as well as tires and lubricating oils. Especially in the case of consumer appliances, producers must take back scrapped products without charge, thereby creating expectations for the development of new channels for recovery and reuse. Yet major problems remain, such as how high to set the reuse rate, and how to secure recovery channels for scrapped products and reuse facilities.

As society became aware of soil contamination by US military bases, the issue of soil contamination itself suddenly became a matter of public concern. In response to this public concern, the government considerably revised the system for preventing soil contamination and remediating contaminated soil. Primarily this revision imposed much greater responsibility on polluters and on parties responsible for polluting facilities. Also important is that under the revised system the polluters themselves check for soil contamination in advance when a facility that may have contaminated soil is to change hands.

Nevertheless, the state of soil contamination in South Korea is still largely unknown. Further, as the substances designated as soil contaminants are mainly limited to heavy metals, even less is known about contamination by new chemical substances. A major shortcoming is that the national government must take responsibility when the polluter cannot be identified, or when it cannot pay remediation costs.

# 5.    Future Challenges

This has been a brief survey of environmental policy under the Kim Dae-jung Government. All the policy areas covered here share the problem of how to implement their policy frameworks in better ways. Despite their limitations, these policies are based on several principles including the precautionary principle, the polluter pays principle, and citizen participation. It will be no small job to improve the operation of the systems created and elicit positive results.

Regrettably, however, the implementation of policies by the new administration to date suggests that environmental policy could take a big step backward. Its "Government Handover Commission" did not establish any organization to handle environmental policy because, in view of harsh circumstances influenced by the slow economy and international instability, the new government could not afford to allocate funding and people to environmental policy due to factors including fiscal deficit concerns, diplomatic maneuvering with North Korea, and rising military tensions across the entire peninsula. Further, environment-related government expenditures over the last five years are only 2.3% of the entire budget, and the MOE budget was a mere 1% of the whole, a situation that will likely remain unchanged. This low funding allocation is far below that in the Second Mid-term Comprehensive Plan for Improving the Environment, which started in 1997.

Moreover, there are no signs at all of improvement in South Korea's compartmentalized administrative apparatus, which is the greatest obstacle to environmental policy progress. MOE alone cannot overcome the barrier posed by the requirement for discussions with other ministries and agencies. As such, the new administration's biggest policy challenge is perhaps creating a new administrative agency that brings together the environment, transportation, resources, and energy.

Jung Sung-Chun

# People's Republic of China

## 1.    The Chinese "Threat" and Statistics

Whether it is joining the WTO, the flood of Chinese goods on the market, the root cause of deflation, growing energy demand, cross-border pollution, sandstorms, global warming, militarization, or any one of many other issues, hardly a day passes without China in the international news. It is almost as if a monster called "the Chinese threat" begins taking over the world as soon as it appears on the international stage. This sense of a threat probably comes from the size and speed one perceives when looking at China superficially, but the world public does not often accurately comprehend the actual numbers or understand the Chinese people's livelihoods and way of life, which give rise to those figures.

Let us consider global warming as an example. It is widely known that the US forcefully asserts that China is reluctant to reduce its greenhouse gas emissions, and Japan's politicians and bureaucrats parrot this view.

While China does in fact have the world's second-highest emissions after the US, it has already ratified the Kyoto Protocol, and as far as official statistics for the past few years indicate, US criticism is somewhat off the mark because China's coal consumption is declining (about 1.4 billion tons in 1996, but about 1 billion tons in 2000), and its $CO_2$ emissions have dropped rapidly since 1997.

There is criticism that China's statistics are deficient, and in fact it contains an element of truth. Undeniably, statistics do not reflect illegal business transactions, especially those by small privately operated and unregistered coal mines, as well as the considerable amount of coal produced by small mines still operating out of necessity even though they were shut down by the government for consolidation or environmental purposes.

But in order to understand China, one of the authors (Asuka) desires to emphasize the need to consider a structural problem that makes statistics uncertain, and in the case of coal in China, the crucial factor is none other than poverty. According to the measure of poverty used by the UN and other agencies, which is under $1 a day per capita, even now China has about 100 million people living in poverty, most of them in rural areas. Excess labor in those areas is several hundred million people, a number that will surely rise due to WTO membership, water shortages, and environmental degradation. To such people, virtually the only way to secure a cash income is coal mining.

Mine workers get extremely low wages, and accidents are frequent under their poor working conditions. For example, in July 2001 there was an explosion at a small privately operated coal mine in Xuzhou, about 600km from Shanghai, in which 92 people died. Some of them were farmers, and some were coal miners who had found work there after losing their previous jobs due to restructuring at state-owned mines.[1]

A Chinese business magazine featured a cover story on coal mining,[2] relating that miners are paid only about 20 yuan for a day of 10-odd hours, and that a death benefit of only 10,000 to 20,000 yuan is paid when a miner perishes in an accident. It is not uncommon for accidents to be kept secret. The article further observed that illegal operation is possible due to the involvement of local governments, and that the approximately 10,000 coal mining deaths per year mean China's number of deaths per unit amount of coal produced is 200 times that of the US and eight times that of India.

Owing to the lack of other work opportunities, farmers continue working in the mines even if they have accidents or have lost family members. Although mines may have been closed by the government and operating illegally, many farmers have no other way to support themselves and their families. And as their energy source they use coal that is unbelievably cheap thanks to the many lives sacrificed. That cheap energy in turn allows the export of inexpensive Chinese goods.

China probably desires to conceal this state of affairs from the world public, and also to avoid facing these facts itself. People in other countries give little thought to why the goods they benefit from are so inexpensive, and rarely feel any responsibility. Yet, unless people share an awareness of such problems, true international cooperation will be impossible on the environment or any other area. As such, world citizens most need to transcend superficial arguments about the "Chinese threat" and statistical indicators, and instead grapple with the true underlying causes.

## 2.  Consolidating the Achievements of Environmental Policy

The year 2000, the final year of the ninth Five-Year Plan, was the make-or-break year for China's environmental policy, especially on whether pollution could be controlled because a State Council decision of 1996 had declared that small industrial pollution sources in 15 heavily polluting industries would be selectively closed, and that 2000 would be the deadline for meeting definite targets established by the decision for national industrial emission standards and for air and water quality standards in China's major cities.

According to the *China Environment Yearbook 2001*, emissions by major mining and manufacturing companies decreased, with total emissions dropping from 20.6 billion tons to 15.3 billion tons, COD load from 7,035,000 tons to 6,624,000 tons, particulates from 5,620,000 to 4,040,000 tons, and flue dust from 7,580,000 tons to 5,170,000 tons. Additionally, compliance with effluent standards increased from 59.1 to 82.0%. These figures show that China achieved a measure of progress across the board. However, nearly 20% of companies had not attained the emission standards by the end of 2000, and results by province showed that less than half of the companies in some provinces had complied. Industrial pollution control efforts alone were insufficient, and many companies throughout China repeatedly violated pollution regulations. Sampling surveys conducted in some provinces in 2001 found that the percentages of companies not complying with emission standards came to 17% in Zhejiang Province, 30% in Hebei Province, 40% in Henan Province, and under 50% in the priority areas of Shandong Province, or that those companies continued production in violation of selective closing orders.[3]

On the conservation of ecosystems, which is another major challenge for China's environmental policy, the Ninth Five-Year Plan declared a goal of "basically bringing the trend toward worsening environmental damage under control" by 2000, but in fact there is no end to disasters of crisis proportions. For example, articles in *China Environment News* from 2000 through mid-2002 cover stories such as large sandstorms occurring mainly in northern China, droughts that come nearly every year, drying of rivers even in the south despite this region receiving more rainfall than the north, and red tides along the coast. Although climate change is very likely one of the major factors underlying such disasters, attention is also focused on anthropogenic causes including inappropriate use of water resources and land, which are issues that are beginning to appear on the government's agenda. It remains to be seen if any substantial results accrue from efforts including development of the National Environmental Protection Blueprint, passage of the Law on Prevention and Control of Desertification, returning agricultural land to forest or grassland (a suite of policy measures for returning farmland that should not be tilled for environmental reasons back to forests or grasslands), and implementation of the Bo Hai Conservation Action Plan.

Further, there are concerns about more environmental damage from large-scale investment and construction projects proceeding under the rubric of "environmental construction," such as infrastructure investment in inland regions under the Great Western Development Initiative, and water projects such as the South-North Water Diversion Project, which is a key component of that initiative. To address these concerns, in October 2002 the National People's Congress passed an environmental assessment law that includes components for plan assessment, and attention is now focused on what role the law will play at the project implementation stage.

Although China has achieved rapid economic growth since instituting reforms, it is still a developing country with wide regional differences. As such, it is realistic to assume that for the time being China will provide for environmental conservation while proceeding with economic development. And now when the government is hammering out a variety of plans and policy measures for conservation, it must work at consolidating the achievements of its plans and policies. Only then will it realistically be able to achieve the rather conservative environmental protection target in the Tenth Five-Year Plan (2001–2005): "Alleviate environmental pollution to an extent, and take the first step toward bringing the trend of worsening environmental damage under control."[4]

# 3.    The "Green Revolution" in Chinese Society

In addition to its previous symbolic meanings, the color green has in recent years come to stand for the environment and ecology, and it is especially visible in China. For instance, Mao Zedong's phrase "Serving the People" is originally written in red to symbolize the socialist and communist revolution, but at the east entrance to the State Environmental Protection Administration it is written in green, which is also becoming the color of choice for conference venue curtains, neckerchiefs, and other items which have some connection to the environment. Also part of this trend is a spate of books and articles constituting a category called "green literature," although many of these are in fact nonfiction works written by journalists.

Many environmental NGOs formed since the mid-1990s also use "green" in their names. The report submitted to the Johannesburg Summit by China's government says there are over 2,000 officially registered environmental NGOs, although it is not clear what organizations were actually counted. Some environmental NGOs are appraised as "grassroots" both in China and abroad, and for their participation in the Johannesburg Summit, a guide entitled "Introduction to Grassroots Environmental NGOs in China" was prepared. It listed 26 organizations which have mutually recognized one another as grassroots. It is no longer unusual to find such environmental NGOs in China's Northeast and inland regions, and an increasing number of them are tackling specific regional problems. More and more of their chief members are professionals who use their skills to further these organizations' aims.

Especially garnering attention is the Center for Legal Assistance to Pollution Victims at the China University of Political Science & Law (see Essay 2). Active since November 1999, the center offers a free legal counseling and support service to help victims of industrial and urban pollution, which are not often targeted by citizen activists despite the severity. As of August 2002 the center had taken a total of over 3,800 calls, whose breakdown indicates that noise complaints accounted for over 30%, air pollution between 20 and 30%, water pollution around 10%, and several percent each for calls related to solid wastes, radioactivity/RF radiation, insolation/light pollution, and others. The center has actually been able to help victims in about 30 cases using legal and administrative means or the media, but it has yet to file any lawsuits for health damage or against large companies. Compensation has been gained in court victories in only a few cases because apparently courts and local governments, in fear of the possible repercussions, often prevent outward victories by victims. In almost all these cases, however, the victims have won compensation or some other kind of real improvement in their situations. Further, the center's efforts have attracted the interest of people in government administration, legislation, and the media, and of socially influential people. In particular, the center has indirectly supported pollution victims by hosting training seminars on environmental law for people in the judiciary with the object of environmental consciousness raising.

The Chinese government undeniably would like to incorporate environmental NGOs into the system by having them advocate China's interests from a stance other than that of the government. Nevertheless, it is hard to say whether it is better to reject compromise and be active in China without maintaining an office there, like Friends of the Earth, or to believe it is meaningful just to have a local office and do as Greenpeace did when it opened a Beijing branch of its Hong Kong office in April 2001. But the world public must understand that as Chinese NGOs do not have the luxury of such a choice, they are doing what they can under the constraints of the system.

In 2001 three advisors from environmental NGOs were chosen to help green the Olympic games scheduled for 2008, but there are concerns about whether "green Olympics" will become a reality because it appears that environmental policy began to regress once Beijing had been chosen. Nevertheless, environmental NGOs are making the best use of the "green Olympics" opportunity to raise the public's environmental consciousness so that the situation is better than before the games.

Asuka Jusen, Otsuka Kenji, Aikawa Yasushi

## Essay 2:  Legal Assistance for Pollution Victims in China's Western Region

One of the main activities of the Center for Legal Assistance to Pollution Victims at the China University of Political Science & Law is to provide assistance with lawsuits in court, and to help pollution victims and uphold their rights by exhorting administrative authorities and the media to act. In a bid to advance its activities in China's western region, which is in the spotlight owing to the government's Great Western Development Initiative, the center in May 2001 obtained help from the British Embassy in China for Protecting Environmental Rights through Legal Means in Western China.

This project took on a number of environmental cases in which poor litigants could not pay their own court costs, attorney fees, expert opinion fees, and other costs. The center organized attorneys, filed lawsuits, and had the media follow and report on the progress of each case. Its aim is to find redress for the victims, to raise the public's consciousness toward the environment, the law, and their rights, and to encourage the enforcement of and compliance with environmental laws in the west. Further, based on the resolution of each case, the center sponsors study and discussion by lawyers, judges, administrative officials, and academics of issues regarding the defense of citizens' environmental rights in China, and makes proposals to legislative bodies on environmental legislation and the means to implement it.

After launching Protecting Environmental Rights through Legal Means in Western China in July 2001, the center accepted requests from the western region for 18 months, then from among them chose eight cases in which to lend assistance. Lawsuits have been filed in five of these cases, evidence is being gathered in two cases, and administrative mediation is in progress in one case. Cases are not only pollution, but also involve resource damage. One case is a class-action suit whose plaintiffs are over 500 households of three villages. Several media are giving the center favorable coverage on its efforts in these cases.

In June 2002 the center held a seminar in Xi'an for environmental law scholars, judges of the Supreme People's Court and of western region courts, administrative officials responsible for enforcing environmental laws in the State Environmental Protection Administration and in the west, lawyers involved in environmental cases, environmental monitoring technicians, and others, who used actual cases as the basis for discussions on three important issues: evidence-gathering and fact-finding in environmental lawsuits, defendants' scope of adducing proof and its application, and legislation on environmental damage compensation. The seminar was useful for helping to facilitate environmental legislation.

Four problems came to light in the course of the project: (1) Some local government leaders think that economic development should take precedence in the west, and that "pollute and then clean up afterwards" should be the procedure followed. Farmers are the victims of industrial pollution, but they tend to be hesitant about filing lawsuits out of deference to local governments. (2) In environmental lawsuits there are serious limitations on investigative evidence-gathering by parties to a suit. When the factualness of expert opinion reports submitted by local government monitoring and measuring organizations is in doubt, parties may commission experts to perform tests and measurements, but courts will not accept their reports. Further, owing to the extreme difficulty of preserving evidence of pollution-caused damage, winning is very difficult for parties who make demands such as pollution mitigation or stopping pollution emissions. (3) There is deficient understanding about the principle of imputing the burden of proof in environmental lawsuits, and its application. In some cases this imposes an excessive burden of proof on pollution victims. (4) Court hearings in cases involving compensation for pollution damage sometimes lack fairness. Socioeconomic impacts are often taken into consideration in resolving cases.

Xu Kezhu

# Taiwan

## 1.    Construction of the Fourth Nuclear Power Plant

Postwar Taiwan's energy policy has been the underpinning of the island's devel-opment-oriented industrial policy, but as concern for the environment among the Taiwanese people has grown and with advances in democracy, that energy policy has become increasingly inadequate, and it now stands at a crossroads. Especially the way energy is used and the development of energy-related industries are linked directly and indirectly to the environment, and are becoming the focus of anti-development campaigns.

On March 18, 2000 Taiwan had its first change of government in its half-century postwar history when the Democratic Progressive Party (DPP) candidate Chen Shui-bian won the presidential election. Victory by this party, which proclaims a "green administration," is an important first step in Taiwan's democratization, but it will also have a major impact on structural changes in the island's socioeco-nomic system.[1]

The new administration promptly established a Review Committee under the Ministry of Economic Affairs (MOEA) to weigh continuation of the fourth nuclear plant, and the committee delivered its final report several months later after a series of televised weekly meetings.

This was a landmark report because it not only called for stopping construc-tion of the fourth nuclear plant, it also mentioned shutting down the existing plants soon. Minister of Economic Affairs Lin Xin-yi also referred to power deregulation, observing that nuclear power has no economic future, and that Taiwan's need is none other than deregulation. If Taiwan turns its technological expertise in precision instruments and information processing to energy conser-vation policy from a nuclear-phaseout stance, it could bring superior products to the international marketplace. However, political disorder has arisen from per-sistence of the Kuomintang Party (Nationalist Party, or KMT) about continuing construction of the fourth plant.

After falling from power, the KMT raised alarms about a heavy negative impact on Taiwan's industrial investment and economic development owing to possible future energy insecurity caused by cancellation of the fourth nuclear plant, and it also emphasized the reasonableness of the previous government's industrial and energy policies. Further, there are now questions about the new government's administrative prowess stemming from a number of factors includ-ing policy implementation difficulties, river pollution in the south, and construc-tion site accidents. Moreover, the Chen administration is finding it hard to obtain public support for its decision to halt construction of the fourth plant.[2]

At a press conference on October 27, 2000 a Cabinet official announced the decision to suspend construction, which was already at least 30% finished, and cited the following reasons: (1) There will be no electricity shortage until 2007 even if construction is halted. (2) Substitute sources are very likely thanks to energy market deregulation, a distributed energy source policy, and other efforts. (3) There is still no means for the final disposal of nuclear waste. (4) Continuing construction would require a supplementary budget larger than the loss from stopping it. (5) An accident would contaminate the whole island, a situation with which crisis management cannot sufficiently cope. (6) The economic loss and penalties arising from cancellation are smaller than the supplementary budget need to continue construction. And (7) Taiwan's sustainable development requires phasing out nuclear power. President of the Executive Yuan, Chang Jun-hsiung, also used the Chernobyl accident as an example in pointing out the devastating impacts of nuclear accidents, but this failed to quell the backlash from construction proponents.

In the end the Executive Yuan and Legislative Yuan reached a compromise in which they agreed on the long-term goal of a nuclear-free Taiwan, but construction of the fourth nuclear plant would continue, and in February 2001 the Executive Yuan did an about-face and announced that construction would continue.

In sum, the issue of construction continuation had already transcended the bounds of controversy over industrial policy, becoming a political issue that convulsed the new government. The government must still clear many hurdles before realizing its intention of revamping industrial policy to accommodate the nuclear power phaseout and achieve its energy policy shift, which has great significance especially for the environment.

## 2. Coastal and Marine Management

Coastlines are where land and sea meet, and they serve as a valuable place where diverse organisms live and breed, as well as an important scene of human activity. Taiwan's coasts feature oddly shaped rocks created by the waves, and a flourishing marine ecosystem. For example, the Pacific Ocean on Taiwan's east coast is populated with bottle-nosed dolphins (*Tursiops truncatus*), spinner dolphins (*Stenella longirostris*), Risso's dolphins (*Grampus griseus*), Fraser's dolphins (*Lagenodelphis hosei*), and spotted dolphins (*Stenella attenuata*), while Kenting, Ludao Island, and the Bohu area in the south feature coral reefs and deep blue waters. But despite being blessed with such marine resources, Taiwan's neglect of coastal and marine management has allowed inappropriate coastal development around the island, and the consequences are increasingly grave damage to coastal environments and scenery, as well as serious impacts on marine ecosystems.

In Early 1998 Taiwan's government created the Territorial Waters and Propinquity Law and the Exclusive Economic Zones and Continental Shelf Law, which were followed in November 2000 by the Marine Pollution Control Law. Also in 2000 the Executive Yuan issued the *Marine White Paper*. The Ministry of the Interior had for some time been keeping the bill for the Coastal Law before the Legislative Yuan. After the 1998 International Year of the Ocean, initiatives like these evolved into the government's approach on coastal management. Yet, chaotic and unsustainable development has damaged the ecosystem and induced deterioration of the marine environment.[3]

In January 2001 the Greek-flagged freighter *Amorgos* ran aground at Eluanbi, a marine area of Kenting National Park. Although the ship's cargo of oil was at first intact, strong winds worsened damaged to the vessel, which began spilling oil a few days later. This spill was the worst spill in Taiwanese waters in recent years and reached about 2 km of natural shoreline. Upon ascertaining the seriousness of this accident, the Environmental Protection Administration (EPA) began assembling a cleanup team but could not begin full-blown cleanup activities until two weeks after the accident due to the Chinese New Year. Nevertheless, public dissatisfaction with the EPA's response forced both the chief and assistant administrators to resign.

In addition to the fourth nuclear plant issue, the Chen government also faces a plethora of tasks related to coastal and marine management, and is tasked with a long uphill struggle to protect the marine environment even after several years of oil spill cleanup efforts and the design of an oil spill response system.

# 3.    Waste Management and Recycling

Owing to Taiwan's high population density and limited land area, its strategy for municipal solid waste (MSW) disposal is primarily incineration, with landfilling a supplementary means. The government has invested more than NT\$ (New Taiwan dollars) 100 billion in the construction of incinerators and ancillary facilities, an example being the 1991 "Taiwan Area Project for Constructing Waste and Resource Recycling Facilities (Waste Incinerators)," which involved plans for building 21 public incineration facilities. Further, the construction of 15 private-sector incineration facilities was planned under the 1996 "Program for Facilitating BOO/BOT Incinerator Construction." Currently 18 public incinerators and one large private incinerator handle about 50% of all MSW, showing that Taiwan's MSW incineration system is quickly taking shape.

Recycling statistics indicate that in 2001 there were 965,000 tons of total recyclable materials, and that the recycling rate was 12.6%. Daily per capita MSW generation was 0.94 kg, a slight decrease from 1.07 kg in 1999, thereby achieving the reduction intended by policy.

An EPA study on industrial wastes found that in 2000 the amount of general industrial waste generated was about 39.9 million tons, while that of hazardous industrial waste was over 1.6 million tons. Currently almost 8.9 million tons of general industrial waste and 587,000 tons of hazardous industrial waste are not properly disposed, and therefore require incinerators and other disposal facilities. Until revision of the Waste Disposal Act, businesses were responsible for disposing of their own wastes without active government intervention, but after the July 1999 revision the central government agency responsible for the supervision of each industry assumed the responsibility for planning and building facilities to dispose of industrial wastes in each industrial park or science-based industrial park (for example, Hsinchu Science-based Industrial Park). In October 2001 this law was further amended to expressly prescribe regulations by the central government agencies responsible for companies in the industries they supervise. Owing to this revision, a central government agency must take active measures when it is found there is no facility to which a company can commission waste disposal, and the company cannot dispose of its own waste. As a consequence, industrial waste disposal has shifted from generators taking responsibility for their own waste to a system with active government involvement.

Based on the EIA's draft "Industrial Waste Management and Disposal Program," which was approved by the Executive Yuan in 2001, government agencies must complete short-term industrial waste storage facilities by December 2001, and treatment and disposal facilities by December 2003 for the industries the supervise. This plan's aim is to support the proper disposal of industrial wastes and to reduce pollution. By 2006 the industrial waste reuse rate is to be 73%.

Under the draft Industrial Waste Management and Disposal Program, establishing general industrial waste disposal facilities requires that the EPA be responsible for unifying, coordinating, and overseeing the government agency that manages each industry. MOEA's Industrial Development Bureau is responsible for unifying, planning, coordinating, and facilitating the tasks of hazardous waste disposal and creating final disposal landfill sites, which is why the EPA implements the Program for Facilitating BOO/BOT Incinerator Construction to establish final disposal facilities for general industrial waste (including incinerator ash). Twelve counties (cities) are to have completed construction of final disposal facilities by the end of 2003.

The EPA further proposed a plan to quickly establish final disposal facilities for industrial wastes (including incinerator ash), which calls for building large final disposal sites in central and southern Taiwan by the end of 2002 to handle incinerator ash and general industrial waste. In addition to this, MOEA drafted the "Domestic Special Industrial Waste Disposal Plan," under which special industrial waste management centers would be sited in northern, central, and southern Taiwan. As of December 2001 MOEA has built one provisional industrial waste storage facility to help temporarily solve the problem of managing special industrial waste. The medium-term objectives are to complete the first phase of construction on comprehensive management centers by the end of 2003, and to complete the second phase by the end of 2004.

Although building this "modern" waste management system based primarily on incineration is laying the foundation for sound and proper management, depending mainly on incineration as Japan does could bring about the same problems. On the other hand, Taiwan is trying other means that might hold future promise, such as economic means like taxing supermarket shopping bags, and the vigorous use of advanced technologies from around the world through international competitive bidding.

Chen Li-chun, Ueta Kazuhiro

# Republic of the Philippines

## 1.   Environmental Governance

A system for environmental governance has been forming in the Philippines since the 1980s. Local governments, the private sector, NGOs, and a variety of other actors have joined the national government in dealing with environmental problems and developing policy.

On the government level, the Ministry of Natural Resources was reorganized into the Department of Environment and Natural Resources (DENR) when Corazon Aquino assumed the presidency in 1986, and the new department took over primary responsibility for environmental conservation and policy. Although the 1977 Philippine Environmental Code (Presidential Decree No. 151) had created a framework for enacting environmental measures, authority was spread out among several government agencies. This reorganization concentrated authority under the DENR to provide for more efficient implementation. In 1989 the Cabinet approved the Philippine Strategy for Sustainable Development, thereby confirming that economic development would be carried out within the framework of environmental protection and conservation. Behind this clear manifestation of a stance on environmental initiatives by the government was not only the international environmental consciousness that started building in the 1980s, but also the rising tide of NGO and citizens' movements in the Philippines itself.

Until 1986 when president Marcos was driven from power, movements tended to be overwhelmingly dissident and political, but after the ascension of Aquino they increasingly focused on single issues or on self-realization. Environment-related citizens' movements began in the 1980s. The Haribon Foundation was at first concerned exclusively with wild birds but later expanded its work to take in environmental protection in general, and in the late 1980s Green Forum Philippines was founded as an organization to link environmental NGOs. Citizens' movements and NGOs in geographical areas directly affected by environmental damage have been increasingly influencing environmental policy. Examples of their clout are the movement rejecting construction of a geothermal generating plant at Ilosin, Sorsogon Province in 1989, and a petition movement to ban commercial logging in Palawan Province. In response to these developments the DENR created an NGO desk. Many NGOs joined as members of the Philippine Council for Sustainable Development,[1] whose creation by the government in 1992 was timed to coincide with the Earth Summit.

As central government authority was transferred to local governments under the Local Government Code of 1991, local governments gained a measure of discretion in the implementation of policy on development and environmental protection. The code required that NGOs be allowed to participate in decision-making processes such as that of the Regional Development Council, and that NGO views be incorporated into environmental policy.

Since the Aquino administration, the government of the Philippines has been cultivating the conditions to have not only the central government, but also citizens' organizations, NGOs, and local governments involved in the decisions and implementation of environmental policy. There has been considerable progress when viewed over a long time period, but political and economic factors have not always permitted desirable results. A few examples are examined below.

## 2. Waste in the National Capital Region

In January 2001 Gloria Macapagal-Arroyo took over the presidency from Joseph Estrada, who had been impeached by the Senate for involvement in an illegal gambling racket. One of the problems she faced was how to deal with municipal solid waste in the National Capital Region.[2] Although the Manila region generates about 6,000 tons of waste each day, current collection capacity is only 3,280 tons, or about 55%. What is more, by 2010 the Philippines as a whole is expected to generate 40% more waste. Citing the generation of dioxins and other substances, the Philippine Clean Air Act of 1999 made the use of waste incinerators illegal, and made taking waste to a collection point the only option. Yet, complaints about water pollution from people living near disposal sites induced closures of sites in the vicinity of Manila, including Carmona in Cavite Province and San Mateo in Rizal Province. Further only one of the 15 new candidate sites— Magallanes in Cavite Province—has obtained an Environmental Compliance Certificate, while the others have no realistic prospects for opening. Collected waste is either temporarily taken to Pier 18 at Manila Port, or is illegally dumped in vacant lots or rivers. Uncollected waste can be found abandoned everywhere in the cities.

In July 2000 a landfill at Payatas collapsed, killing up to 300 people by some accounts. Although closed temporarily, this landfill was reopened in response to Manila's waste crisis, and to the wishes of Payatas citizens themselves because they make their living by scavenging.

The first law that President Arroyo signed after taking office was the Solid Waste Management Act of 2001, which incorporated NGO opinion to a considerable extent, including provisions for waste reduction such as a ban on disposable packaging, waste sorting, and facilitation of the recycling market. Owing to lack of a budget, however, cities in the National Capital Region are not following through with their responsibilities for sorting and collection. This situation was aggravated in 2001 when cities in the National Capital Region were given authority for disposal in addition to collection. Devolvement of this authority without a system for verifying legal compliance spawned collusion by waste disposal contractors with politicians and bureaucrats, and encouraged illegal dumping in rivers and harbors. In May 2002 DENR Secretary Heherson T. Alvarez responded by issuing business suspension orders to illegal dumpers, and even warned that criminal charges for pollution would be filed.

Over the long term the Philippines must take action to cut waste volume and hold down needless consumption, but there are no prospects for a fundamental solution under current circumstances with incinerators shut down even though proper landfill sites have yet to be found.

## 3.    Revival of the Mining Industry

Under government policy the private sector is in the driver's seat, especially transnational corporations, and development has priority over the environment. The Philippines is said to have some of the world's largest mineral reserves, including gold and copper. Exports of mineral resources accounted for about 25% of total exports at their peak in 1973, but since the 1980s the industry has been in decline owing to low international prices and heavier international competition, and in 1997 it fell so low that mineral exports were only 3.2% of total exports.

In a bid to cut costs as deregulation proceeds, mining companies reduce their work forces and elect to conduct inexpensive open pit mining. Ten of 14 mines operating in 1992 used this technique, which reduces costs at least 10% but has a bigger environmental impact.

To revive the languishing mining industry, the government established the Philippine Mining Act of 1995, whose major feature is its provision for Financial and Technical Assistance Agreements (FTAAs) that allow large companies which can invest at least $5 million to prospect and operate in vast areas for 25 years, with another 25-year extension possible. Foreign-capital companies are allowed to finance 100% of an operation. The idea is to have foreign companies with the needed financing and technology capabilities energize the Philippine mining industry. Although the law requires that a certain portion of costs be used for environmental conservation, it is doubtful this requirement is actually observed because of the conspicuous inhumane treatment of people in mining areas beginning at the prospecting stage.

An example is Western Mining Company (WMC), an Australian company that entered into an FTAA with the Philippine government in 1995, and is prospecting in a vast region of 99,400 ha lying across North Cotabato, South Cotabato, and Sultan Kudarat provinces in Mindanao with the intention of beginning operations in 2004 or 2005. Mining companies must obtain consent from the residents of areas where they prospect and operate, and WMC obtained consent by offering the citizens, through heads of local governments and meetings with indigenous peoples, programs to help the community such as the construction of schools and hospitals, repairing roads, building tapwater systems, and regional electrification. But owing to inadequate explanations of the operation, not a few people gave their signatures without a correct understanding of what they were signing, or were unaware that mining development was starting until trucks rumbled into their villages. Further, resistance and noncooperation from residents elicited an attack by the military in 1996, costing the lives of six B'laan tribe members, and orders were issued expelling 120 families.

At a time when considerable authority has been granted to the private sector in the mining industry, the livelihoods of local people and the environment are taking a back seat to development.

## 4. Collaboration with Business

DENR Secretary Alvarez, who was appointed to his post by President Arroyo, was one of the Congress members who proposed the 1999 Clean Air Act and the 2001 Solid Waste Management Act before assuming his current post. He has consistently been involved in environmental issues, and as such he has continued actively promoting environmental policy since becoming DENR secretary. For example, in an attempt to give tangible form to the Clean Air Act, which had taken effect in November 2000 but had no real programs going under it, Alvarez made the National Capital Region and the surrounding six provinces into a single airshed[3] called the "Metro Manila Airshed," and created a committee for its management in January 2002. That June he developed the Metro Manila Air Quality Improvement Sector Development Program, for which financing support was obtained from the Asian Development Bank (ADB).

Under Alvarez, the DENR also more actively encourages government policy meant to bring about collaboration with business for promoting environmental policy. In June 2001 he launched the Partnership for Clean Air, which is working on eliminating leaded gasoline and on a clean air program with the involvement of business groups such as the Philippine Chamber of Commerce and Industry, the petroleum industry, and the transport industry, as well as citizens' groups like the Citizen's Coalition Against Pollution, and international agencies including the ADB and the U.S. Agency for International Development.

In June 2002 Alvarez himself called for a Partnership for Sustainable Development, and he is trying to create a system for cooperation with business and citizens' groups. Meanwhile, in March 2002 a major distribution company called Shoe Mart exchanged a memorandum with DENR on the inauguration of an environmental conservation program, and that June McDonald's committed itself to cooperation in a publicity campaign to raise the public's environmental awareness. There are also other instances in which businesses have declared their understanding of and cooperation toward environmental policy.

In advancing environmental policy the important question is how much one boosts environmental governance which involves not only administrative authorities, but also has the cooperation of business, NGOs, and citizens' groups. In that sense recent trends in the Philippines are encouraging, but seen from another angle this arrangement also makes it quite possible that environmental measures will be subordinated to the market principle, which assumes deregulation. That is to say, it is possible that environmental policy will be confined within the framework of profit-seeking, and the basic issues disregarded. In fact some NGOs are critical of Alvarez, claiming that he is a "traditional politician" who wants to preserve the conventional politico-economic system. Also, some NGOs say that collaboration by NGOs and citizens' groups with the government presents the risk that they will be co-opted and their capacity for criticism weakened.

Philippines environmental governance has taken form rapidly during recent years, but whether it can become a significant force for environmental protection and conservation is dependent on the public's environmental awareness and how well trends in environmental governance are watched.

Ota Kazuhiro

# Socialist Republic of Vietnam

The Communist Party of Vietnam (CPVN) has set 2020 as its target for making Vietnam an industrial country, but industrialization and urbanization are bringing ever more serious environmental damage while global climate change causes unforeseen disasters. Accordingly, Vietnam's government is implementing a policy to "socialize environmental protection," which uses the private sector to protect the environment.

## 1.  Forest Protection Policy

In November 1997 the National Assembly approved a "5 million ha afforestation plan" and set a goal of expanding forested area from 28% to 43% of the country's area in 13 years. Under this plan forested area increased to nearly 10,920,000 ha or 32% of the country by the end of 2000.

Over the 10 months from January to October 2001 Vietnam planted 166,021 ha of trees (59,798 ha protected forest and 95,000 ha production forest). The government also provided for the management and conservation of 2,140,000 ha of forest by commissioning these tasks to production units, families, and individuals. Its goal for 2002 was to protect 2 million ha, attempt ecosystem improvements in 500,000 ha, and carry out intensive afforestation on 280,000 ha.

During the three years following approval of the 1997 afforestation plan, over 2,100 forest fires burned more than 26,660 ha. Loss due to forest fires is aggravated by factors including swidden farming, illegal logging, negligence by local authorities and communities in measures to prevent forest fires, and tardy implementation of forest fire prevention guidelines. In the first half of 2001, 227 forest fires burned 1,282 ha (62% of the same period in 2000), and there were 24,393 violations of forest-related laws (645 cases of poaching and 16,280 cases of forest product smuggling), resulting in 35,235 kg of wild animals poached and 21,830 cubic meters of wood illegally logged.

Over the 10 months from January to October 2001 there were 39,417 violations of the Law on Forest Protection and Preservation, comprising 7,075 cases of illegal development of forest products, 1,401 cases of deforestation by cultivation, 988 cases of poaching, 25,431 cases of forest product smuggling, and 22 cases of attacks on forest patrols. From 1993 to 1995 Binh Thuan Province in southern Vietnam lost over 53,000 cubic meters of trees in forest and wildlife reserves because of illegal logging, for which 35 people were sentenced to a maximum of 20 years in prison. In August 2000 the chairman of the province's People's Committee was dismissed for suspected corruption regarding illegal logging, and was convicted of dereliction of his duty to protect the land and forests.

## 2.    Floods and Droughts

Unforeseen situations occur throughout the country, threatening the traditional lives of people who have coexisted with flooding because until now it they have been able to predict flooding to an extent.

In the north, flooding induced a huge landslide in mountainous Lao Cai Province, claiming the lives of 20 victims. That same month in Ha Giang Province, the water level of the Lo River rose to a maximum of 7 meters, its highest level in a decade. In the same province flash foods caused by localized heavy rain killed six people, destroyed over 1,000 homes, and damaged 1,500 ha of rice paddies and 300 ha of orchards, causing total damage of about $700,000. During two days in August 2001 the heaviest rain in 17 years flooded Hanoi to depths of 30 to 50 cm in 120 locations.

In central Vietnam flooding came earlier than usual in December 2000, wiping out 26,660 ha of rice and 15,000 ha of secondary crops. From that time until August 2001 there was no rain at all in Quang Tri, Phu Yen, Dac Lac, and other provinces, while temperatures attained 37 to 38°C. Monthly insolation time set a new record at 250 hours. In Quang Tri Province 1,000 ha of rice paddies were damaged, and 1,000 ha of sugar cane fields in Phu Yen Province were devastated. Nearly all wells and rivers dried up, and sea water eroded more land. Ensuing typhoons and flooding claimed 111 lives by mid-November.

In the south, the rainy season water level rise in the Mekong River started 40 days earlier than usual, bringing the worst flooding in 70 years. By the end of November it claimed between 410 and 450 lives, flooded the homes of 4.5 million people, submerged 150,000 ha of rice paddies, and destroyed 56,000 ha of rice, 4,000 ha of secondary crops, and 90,000 ha of orchards and commercial crops, bringing total damage to about $278 million. Crops in Tien Giang Province were especially hard hit, in addition to the flooding of 370 clinics, 10,000 km of roads, and 12,149 classrooms.

In August 2001 the Mekong's water level again suddenly rose, which flooded the Dong Thap Muoi plain (a wetland zone covering Dong Thap, Long An, and Tien Giang provinces) and the Long Xuyen Square (a wetland zone comprising An Giang and Can Tho provinces), claiming 12 lives. In Long An Province 3,000 homes and 23 km of roads were flooded, in addition to damage to or destruction of clinics and bridges. Between August and November this flooding killed 385 people, flooded at least 200,000 homes, several thousand schools, and several thousand ha of rice paddies and orchards, with total damage coming to about $66.7 million.

## 3.    The Cost of Industrialization and Urbanization

Nearly all the of 1.2 billion cubic meters of industrial effluent and household gray-water is dumped untreated into Vietnam's rivers. Severely polluted rivers are the Red River, Tao River, Da River, Lo River, Saigon River, Ty Vai River, Dong Nai

River, and Cau River. Almost 90% of factories have no effluent treatment equipment, and rivers in industrial zones have large amounts of chemical substances. Pollution is said to be at a crisis stage especially near the paper mills and a plant making superphosphate in northern Vietnam's Viet Tri Industrial Zone. Phenol and metallic substances have been found in the Red River, which flows through Lao Cai Province in the north.

At the end of 2001 the daily amount of waste generated in Vietnam was 19,315 tons, which comprised 10,162 tons of industrial waste, 212 tons of medical waste, and 8,941 of municipal solid waste. Under the policy to "socialize environmental protection," in mid-2002 private contractors began collecting and treating solid wastes on a trial basis in certain areas of Lang Son and Bac Giang provinces in the north, and the central city of Da Nang, and this has become a model for private-sector waste management.

## 4.    Clean Energy and Plans for Nuclear Power

In 2000 Vietnam consumed about 1.5 million tons of fossil fuels, and emitted about 113,700 tons of $CO_2$. Energy consumption is rising about 11% annually. The government restricts lead and aromatic substances in gasoline, and has enacted measures for importing fuels with 1.5% sulfur content instead of 3%. It has also lowered taxes and provided subsidies to encourage the importation and use of transportation that uses clean energy in place of fossil fuels. In 2001 Vietnam banned the importation of automobiles that burn leaded gasoline, and the government decided to use unleaded gasoline domestically starting July 1 of that year.

In 2003 the government concluded a feasibility study for a nuclear power plant construction project, and announced that it intents to have a nuclear plant up and running no later than 2019. Proposed construction sites are in the southern provinces of Ninh Thuan and Binh Thuan.

The International Atomic Energy Agency is cooperating in facilitating the peaceful use of nuclear power in Vietnam. Based on a regional cooperation and assistance project, the agency is helping train personnel and transferring applied nuclear power technologies to agriculture, health care, and industry.

Vietnam has entered into the "Japan-Vietnam Framework for Cooperation on the Use of Nuclear Power for Peaceful Purposes and Economic Development," and its government is asking Japan to transfer technologies.

## 5.    The Private Sector in Environmental Protection

In June 1998 the CPVN Politburo issued "Directive 36/CT-TW on Strengthening Environmental Protection Activities When Industrializing and Modernizing," which included promoting a policy of "socializing environmental protection." In substance it could be regarded as the privatization of environmental protection because private enterprises, mass organizations, collectives, and individuals are the leading actors engaging in forest protection, waste management, water source remediation, and similar activities. Private enterprises are managing waste in some cities, and in model projects throughout the country youth and women's organizations and other groups solicit donations from their communities for building water purification plants, facilities for supplying water to mountainous areas, and other facilities.

Every year from April 29 to May 6 is "Water Purification and Environmental Sanitation Week," which includes campaigns in workplaces and communities to see that people have sanitary living environments. Government Reports say that such efforts have expanded green space in cities and industrial zones, and secured cleaner water supplies in rural areas.

From this discussion it is fairly clear that there is a growing awareness of environmental damage both in the government and among the populace. Areas that Vietnam must tackle now include, first of all, information disclosure. Despite the many decisions made by the government, there is insufficient information on their implementation and results. It is unclear whether official standards and permissible levels are universally applied, and as is often the case in socialist countries, there is a tendency to emphasize only the achievement of numerical quotas, while not releasing any particulars. Second, although environmental policy is closely connected to policies for eradicating poverty and helping minorities in mountainous regions, it remains to be seen how well Vietnam can overcome its existing compartmentalized bureaucracy and build institutions for horizontally organized cooperation. And third, while there are provisions for using the private sector to protect the environment, there are concerns that poor people and impoverished regions will be the victims of environmental damage at a time when the gulf between rich and poor is widening. It would be unpardonable for the government to leave environmental protection up to other entities.

Nakano Ari

## Essay 3:  Waste Management and Endocrine Disruptors in Vietnam[1]

Rapid economic growth is giving Vietnam a headache over managing its waste. In Hanoi the Urban Environmental Company (URENCO) manages the city's solid waste. Starting in the second half of 1997 that waste was taken to the URENCO-operated Tay Mo final disposal site, and at the beginning of 1999 the site accepted 1,060 tons of municipal solid waste and 300 tons of construction/demolition waste, sludge, and other wastes daily. Yet because this 5-ha site was full in less than three years, since 2000 the city's waste has been transported to the Nam Son disposal site 50 km from central Hanoi. Household waste is loaded onto handcarts that ply residential streets, and then go to main streets where the waste is transferred to large trucks that carry the waste to the disposal site. Household

TABLE 1. Analysis of Known and Suspected Endocrine Disruptors in Nam Son and Tay Mo Waste Landfill Sites

(μg/ℓ)

| Sites | 2,6-DCP | 3,4-DCP | 2,3,6-TCP | 2,4,5-TCP | 2,4,6-TCP | 2,3,4,5-TeCP | 2,3,4,6-TeCP | 2,3,5,6-TeCP | PCP | NP | BPA |
|---|---|---|---|---|---|---|---|---|---|---|---|
| NS1 | — | — | — | — | — | — | — | 0.19 | 0.38 | 19.46 | 0.14 |
| NS2 | — | — | 0.33 | — | — | — | — | — | 0.33 | 68.08 | 9.73 |
| NS3 | n.d. | n.d. | n.d. | n.d. | n.d. | n.d. | n.d. | n.d. | n.d. | 49.89 | 18.60 |
| NS4 | — | — | — | — | — | — | — | 0.38 | 0.33 | 25.16 | 0.02 |
| NS5 | — | — | — | — | — | — | — | 0.21 | 0.33 | 52.81 | 6.80 |
| TM1 | — | — | — | — | — | 0.12 | — | 1.66 | 2.29 | 40.33 | 21.52 |
| TM2 | 0.49 | — | 0.72 | — | — | — | — | 2.33 | 2.10 | 16.74 | 0.47 |

Notes: NS: Leachate samples from Nam Son Waste Landfill Site. Numbers shows sampling numbers, and 5 samples were taken.
TM: Leachate samples from Tay Mo Waste Landfill Site. Numbers shows sampling numbers, and 2 samples were taken.
2,6-DCP: 2,6-dichlorophenol (a PCP intermediate)
3,4-DCP: 3,4-dichlorophenol (a PCP intermediate)
2,3,6-TCP: 2,3,6-trichlorophenol (a PCP intermediate)
2,4,5-TCP: 2,4,5-trichlorophenol (a PCP intermediate)
2,4,6-TCP: 2,4,6-trichlorophenol (a PCP intermediate)
2,3,4,5-TeCP: 2,3,4,5-tetrachlorophenol (a PCP intermediate)
2,3,4,6-TeCP: 2,3,4,6-tetrachlorophenol (a PCP intermediate)
2,3,5,6-TeCP: 2,3,5,6-tetrachlorophenol (a PCP intermediate)
PCP: Pentachlorophenol
NP: Nonylphenol
BPA: Bisphenol A
n.d.: Below detection limit; —: not analyzed in this study.

waste is a mix of everything from kitchen waste to plastic, empty bottles and cans, and polyethylene bags like those used at Japanese convenience stores.

To examine the impacts of this waste on the areas around these two final disposal sites, we analyzed their leachate. Tay Mo is an open dump and does not collect its leachate, so we sampled the puddles of water found in various places and used that as the leachate. As Nam Son has a liner and leachate collection pipes, we tested the leachate that they collect (Table 1). We detected the typical endocrine disruptors nonylphenol (NP) and bisphenol A (BPA), as well as the endocrine disruptor pentachlorophenol (PCP) and its intermediate degradation products (compounds whose names begin with numerals in the top row of Table 1). Some of these products are more toxic than their parent materials. Although not included in this study, 2.4-DCP, which is an isomer of 2.6-DCP, is a designated endocrine disruptor. All of the substances in the table were definitely detected in the leachate, albeit not in high concentrations. NP is a final degradation product of the industrial detergents nonylphenolethoxylates, while BPA is found in plastics, coatings on the insides of food cans, and in the heat-sensitive paper used nowadays for receipts at retail stores. PCP is used as a pesticide and wood preservative. All these substances are tightly integrated into our lives and have a broad range of applications, making it quite meaningful that such a number of them were detected.

While it is true that much remains to be learned about such pollution on a worldwide level, remedial action can be taken to an extent in the industrialized countries if the capital investments are made because they have the social infrastructure, but in countries and regions that do not have even that infrastructure, it is likely that pollution will accelerate. It will be very worthwhile to continue environmental monitoring by means of studies like this one.

Tateda Masafumi, Fujita Masanori

# Kingdom of Thailand

## 1.   Changes under the New Constitution

In January 2001 Thailand held its first general election for the Parliament under the new 1997 constitution, and it was a major victory for the Thai Rak Thai Party. Thaksin Shinawatra replaced Chuan Leekpai, head of the Democratic Party, as prime minister, and members of NGOs were elected to the Senate.

Just after assuming his post it appeared that Prime Minister Thaksin would take positive action, untried by previous political leaders, meant to resolve environmental disputes. To begin with, he held a luncheon for members of the Assembly of the Poor, which had carried out large demonstrations in Bangkok with people demanding that the Pak Moon Dam floodgates be opened. Later he was the first prime minister to visit the sites of Thailand's three most controversial projects: the Bo Nok and Hin Krut coal-fired thermal power plant projects in Prachuap Khiri Khan Province, and the Thai-Malaysian gas pipeline project in Songkhla Province.

After the January 2000 election, senators formed groups to deal with particular issues, and committees were created to work on resolving environmental disputes, such as by completing an environmental impact study (EIS) on the Thai-Malaysian pipeline project, and visiting the Samut Prakan wastewater treatment project site.

Additionally, the new constitution expressly provides for the right of citizens to participate in the management of local natural resources, and the central government's obligation to facilitate their participation. It also prescribes the performance of EISs, and for providing local community members with opportunities to gather information and state their opinions in the public hearing process, before the implementation of development projects that could have serious environmental impacts.

Nevertheless, this does not mean that the new constitution and the elections it requires have changed the existing tack toward priority for development projects or that for central government management of natural resources. In fact, the government is even going as far as changing environmental laws and policies in order to promote development projects, and that has further intensified environmental disputes in some regions.

Below, this section will take a brief look at how effective the community participation provision in the new constitution has been with regard to large-scale development projects and the legislative process, and at how senators elected under the new constitution are tackling environmental disputes.

## 2.   Large-Scale Development Projects

In May 2002 Prime Minister Thaksin made the final decisions on the construction of the Thai-Malaysia gas pipeline in Songkhla Province and the two coal-fired thermal power plants in Prachuap Khiri Khan Province. Although the power stations would be postponed,[1] the gas pipeline project would go ahead over a new route, which precipitated a severe backlash among villagers in Songkhla Province.[2]

Focal points of the controversy over these two projects were the lack of transparency in the government's procedures, detrimental impacts on the environment and on community livelihoods, and little economic benefit and efficiency for Thailand.

Lack of transparency in government procedures included negligence toward holding public hearings before project implementation, and toward environmental impact statements (EIS), as required by articles 56 and 59 of the constitution. Indeed, before the implementation of all three projects public hearings were held, and their EISs were approved, but power purchase agreements and intergovernmental decisions on project implementation predated the public hearings and final approval of the EISs.[3] As a consequence, opinions and debate in the public hearings and recommendations in the EISs were hardly incorporated into the projects' basic designs, and there was no assurance they would be adequately considered at the implementation stage.

There are also questions about the environmental impact studies performed by contractors. A study conducted at the Bo Nok site after EIS approval by the inspector-general of the Ministry of Science, Technology and Environment confirmed the presence of coral reefs and rare species including whales that were unmentioned in the EIS, thereby exposing the statement's slipshod nature. In the case of the gas pipeline, the Office of Environmental Policy and Planning approved the EIS before a study of adverse social impacts and the approval of mitigation measures by an independent panel of experts despite the panel's three rejections of approval for the EIS report on the pipeline project, and six rejections of approval for the gas separation plant.

What is more, the government had not provided local communities with sufficient information on the entire industrial development plan. In the gas pipeline project, the National Economic and Social Development Board apparently has plans to connect the pipeline to an industrial zone, and to develop that zone into a large industrial estate by attracting petrochemical plants to use the components separated from LPG at a gas separation plant as feedstock, but no government agency has responded to community demands for overall environmental impact studies including the construction of new factories and industrial estates.

There is ample reason for the people of Songkhla Province to be nervous about unpublicized plans for a petrochemical industrial estate: Villagers lost their means of livelihood and suffered health damage for 10 years on the Eastern Seaboard, especially at the Map Ta Phut Industrial Estate, and as of 2002 the root causes of pollution had yet to be eliminated. Despite this, the government continues to convert farmland to industrial use, plans to expand of the industrial zone, and attract more companies.

The previous volume of this series described how people living near coal-fired power plants in Prachuap Khiri Khan Province mounted a campaign opposing inadequate environmental management at the Mae Moh power plant.[4] Local opposition to the Thai-Malaysia gas pipeline is assuming the same form.

# 3. New Legislation May Have Serious Adverse Impacts on the Environment

## 3-1 Amendment of the Mineral Resources Law

Under the current Mineral Resources Law, all of Thailand's mineral resources belong to the Mineral Resources Department, which has granted concessions to developers. Meanwhile, the law has no provisions for punishing developers even if they damage the environment or bring injury to local people.

An example of the undesirable consequences is Klity Creek in Kanchanaburi Province, where a mine developer discharged mercury-containing effluent into a river and seriously damaged the health of the Karen people living nearby. Nevertheless, the government did not shut down the operation or compensate the victims.

Moreover, an amendment to the Mineral Resources Law, which was proposed five times beginning in 1992 and passed the House of Representatives in 2001, attempted to further promote mining by avoiding environmental conservation and circumventing people's rights. Specifically, the present law limits the area of a mine concession to 300 *rai* (48 ha), but the amendment would widen concession areas to 10,000 *rai* (16 sq km) for underground mine development and 50,000 *rai* (80 sq km) for undersea mine development. Further, companies are allowed to mine 100 meters below the surface without taking into account public and landowner participation even though it would clearly have an impact on the environment and local livelihoods; additionally, landowners cannot seek compensation unless they themselves can offer proof of damage.[5]

The amendment does indeed require developers to obtain authorization from the Royal Forest Department if they desire concessions to mine underground on private land and in protected forests, including wildlife sanctuaries and national parks. EISs are also needed before the department may grant concessions. This provision makes possible to avoid harmful social and environmental impacts if the department takes study results into sufficient consideration when making its decisions, but past experience does not justify such expectations. It appears instead that this amendment aims to promote underground mining, especially large-scale potash mining in Udon Thani and Chaiyaphum provinces.[6]

For these reasons the Senate turned back the bill that the House had passed, and added the requirement that concession recipients must obtain permission from landowners before initiating development. It is also lobbying the Constitutional Court to keep the House from making any more changes to the bill.

## 3-2 Community Forest Bill

Meanwhile, opinion among Senators is divided on the Community Forest Bill, and Senators are clashing with villagers and NGOs. Behind this bill is the national forestry policy announced in 1985, which calls for 25% of the country's total land area to be conservation forest composed of national parks, wildlife sanctuaries, Class 1A watersheds, and environmental protection areas. This prohibits local people from using forests designated as "conservation forests,"[7] and in accordance with this policy the Royal Forest Department has expanded areas so designated.

However, conservation forests were not designated after detailed surveys of local land use or public hearings in local communities. Consequently land that had been occupied by local people was suddenly designated conservation forest, resulting in the relocation of the villagers for illegal encroachment, or their arrests for merely gathering non-timber forest products as usual.

Since 1994, the Northern Farmers' Network (NFN) and other NGOs have taken the lead in exploring ways to allow hilltribes to maintain their traditional systems for resource management, land use, and agriculture that are adapted to highlands. Their efforts yielded the "Community Forest Bill" based on community participation, decentralized decision-making, and use within the ecological carrying capacity. When the new constitution was enacted in 1997, NGOs followed the provisions of Article 170 in collecting the signatures of over 50,000 eligible voters, proposing their Community Forest Bill to the legislature, and asking that it be passed.

But the Royal Forest Department was dead set against community management, and it produced a bill of its own that provided for rehabilitating degraded land under its own management and reforestation under state control. Some environmental organizations supported the department's bill, arguing that because of the difficulty of managing forest dwellers, highland forests should be conserved as places with valuable biodiversity, food and water sources, and wildlife habitat.

Debate on these points led to public hearings throughout the country, changes in the bill, and finally cabinet approval. Yet in the process of changing the text, hardly any of the conclusions and proposals produced by communities and NGOs in public hearings were incorporated. The cabinet-approved bill would allow the government to create and abolish community forests and to intervene in the design of forest conservation programs, thereby actually discouraging new organizations from creating community forests.[8]

These changes having been made, the bill was passed by the House of Representatives in November 2001, giving communities a legal basis for forest resource management, and providing for government funding.[9] It also stiffened the penalties for breaking the law: there is a maximum 15 years imprisonment and a fine of 300,000 baht (US$7,500) for offenses such as encroaching on forest reserves, illegal logging, poaching, and causing forest fires.

However, the bill then went to the Senate, where deliberation resulted in fundamental changes that overturned the basic concept of community forests.[10]

Requirements in this Senate bill for creating a community forest raised the 50-person requirement to at least 100 people and prohibited the expansion of community forest areas.[11] These changes practically closed the way for traditional forest management by small villages and for improving forest management by expanding community forests if necessary. Even more important, it prohibited the establishment of community forests in conservation forests, which is what NGOs had most wanted, and required permission from the Royal Forest Department for gathering forest products in community forests. Apparently many Senators do not care to recognize the forest conservation achievements of people who live in conservation forests.

Mori Akihisa

## Essay 4: Thailand's Land System[12]

The complexity of Thailand's land system gives rise to a variety of land-related issues that will be explained briefly here.

Thailand has always had a potentially large amount of unowned arable land, and there was a traditional practice called *chap chorn* under which farmers are recognized as prior occupants if they open the land and till it for a certain period of time. But as people make inroads into the forests and the land area under cultivation expands extensively, Thailand's government has been working on the development of a modern land system. In 1964 Thailand created a forest conservation law that prohibited local people from entering "conservation forest zones." However, many of the areas designated as conservation forests were lands already occupied by farmers in accordance with traditional practices. Nevertheless, the government considered them to be trespassers and pressured them to leave.

A problem here is that many of the farmers did not have complete land ownership. The land code of 1954 underwent a series of revisions in the 1960s and 1970s, and on each such occasion the farmers had to comply with troublesome procedures for changes in land rights. Moreover, under the land practices they had followed until that time, land is ultimately managed by kin groups and rural communities, making those practices incompatible with the principle of individual possession in modern land ownership. In fact, not a few rural people found themselves embroiled in land inheritance disputes involving legal ownership, a kind of dispute which had not been seen among family members or kin. Further, many farmers were not at all inconvenienced by continuance of traditional land practices, but feared that the modern land system would cause the loss of many benefits that they had enjoyed in the past. The situation described here is one major reason why many farmers have no certificates of rights to their land, or have only a right of prior occupancy or right of use, thereby putting them in legally weak positions.

Endo Gen

# Malaysia

## 1.  Economic Development Trends and Environmental Policy

Over the last few years in Southeast Asia there has been a marked shift of factories to China. In 2001 and 2002 major Japanese companies in Southeast Asia either closed or substantially downscaled 22 manufacturing facilities, eliminating about 17,000 jobs. Even more were lost if those at client companies are included.[1]

In a bid to shift the emphasis of industrial policy from labor-intensive manufacturing to more knowledge-intensive industries, Malaysia's government launched a plan in 1997 for a Multimedia Super Corridor that would become the core of its information industry.[2] Accommodating a conversion to the knowledge industry make it vital to not only attract high-tech manufacturing, but also to create pleasant urban environments that will be enticing to the people who will sustain the creative production of knowledge. While Malaysia has on the one hand pursued an aggressive industrialization policy, it is behind in the expansion and enhancement of urban infrastructure and services, as well as in the development and upgrading of efficient public management capabilities. Since the 1990s the federal government has pushed reform in urban environmental management.

To understand environmental problems in present-day Malaysia, one must spotlight not only the preexisting problems caused by rapid industrialization and urbanization, but also the challenges to environmental policy that arise in conjunction with Malaysia's economic development strategy as the country makes the transition from industrialization to the post-industrial stage.

## 2.  Waste Management Privatization

Almost ten years have passed since Malaysia decided to privatize waste management,[3] but highly negative reactions from state and local governments have kept Parliament from passing the waste management bill. However, five of the 13 states brought in waste management companies for interim periods and commenced wide-area, privatized waste management. Although their systems differ considerably from the federal government's initial plan, they could wait no longer to begin more efficient waste collection and wide-area disposal due to concentrated urbanization and swelling waste volumes, because interregional interests within wide-area disposal zones had to be reconciled, and reorganization of the waste management system was deemed necessary.[4]

The federal government switched to wide-area privatized management for not only municipal solid waste, but also medical and hazardous waste (see Essay 6 on medical waste). Kualiti Alam Sdn. Bhd., which was granted an exclusive 15-year contract by the federal government to manage hazardous waste, began operating in June 1998. Kualiti Alam charges relatively high disposal fees in comparison with Japan. Large companies, many of which are transnational corporations, are obligated by regulations to commission their waste management to Kualiti Alam. But in consideration of the burden of paying high disposal fees, special measures have been enacted for small companies comprising mainly local firms that generate limited amounts of pollutants, allowing them to dispose of their wastes without treatment. Illegal dumping and disputes with people living near factories are common occurrences because companies try to avoid paying disposal costs and releasing hazardous substances on their own factory sites. It is up to state environmental departments to deal with these situations. In some cases departments have been commended by environmental NGOs for enhancing the regulation of illegal dumpers,[5] but encouraging proper waste management will probably require the development of hazardous waste management systems in which local communities and NGOs participate.

## 3.    Water Resource Privatization Issues

Malaysia has privatized both tapwater and sewage treatment.[6] Public water supplies were the province of state water departments, but since the late 1990s water supplies are being privatized and put under the control of public corporations.

Selangor state has about 25% of Malaysia's population and therefore supplies far more water than other states, and has the biggest water leaks. Selangor's non-revenue water (of which nearly all is leaked; the rest is stolen) accounted for 40% of water supplied in 1995, and 36% in 2000. The federal government plans to reduce its non-revenue water rate to 29% by 2005.[7]

In Selangor one private company has handled everything from dam development and operation to supplying water since July 1998. Water supply companies that promote dam building, state governments, and the federal government advocate flood control and stable water supplies to cover increased demand. Meanwhile, environmental and consumer organizations are dubious about forecasts of growing water demand. They oppose dam construction that destroys forests and displaces indigenous and local people, and demand reduction of water leaks and a crackdown on theft. Further, water privatization has sparked more vigorous debate on the public benefit of dam projects.

## 4.    Cross-Border Water Resource Management

With the coming of the 21st century, water resource management is becoming a cross-border issue, and Malaysia is no exception because Malaysia and Singapore are at odds over supplying water. Singapore depends on the neighboring Malaysian state of Johore for about half of its water supply, and Johore sells that water for 0.03 Malaysian ringgits per 1,000 gallons raw water. At the same time,

Johore has Singapore purify water, and for 0.5 Malaysian ringgits per thousand gallons purchases an amount of purified water that is 12% of the volume it sells to Singapore. However, the Johore state government announced that as a new purification plant will come online in July 2003, it will no longer buy purified water from Singapore. In anticipation of rising water demand in Johore, the Malaysian government demanded that Singapore pay more for water, and Prime Minister Mahatir has taken a hard line, saying he will not rule out cutting off Singapore's water supply if it does not comply.[8]

In response, Singapore's government decided to build a "NEWater" facility that treats sewage and produces tapwater, which will make up 15% of its total water supply by 2012. It also built desalination plants, and it is exploring the purchase of water through an undersea pipeline from Indonesia's Riau Island. The earliest date for commencement of this project is 2005, but high cost makes it undesirable.[9] Singapore is finding it increasingly difficult to purchase water as inexpensively as it has. Governments are faced with managing water resources and across national borders.

# 5.   Environmental Regeneration

In the areas of social infrastructure and services that support urban environments, Malaysia is undergoing extensive reforms that have no precedent in developed countries. At the same time, coping with the detrimental legacy of damage by industrialization and urbanization is a major focus.

In 1999 a survey team formed by the federal government's Environment Department and the University of Putra Malaysia performed environmental impact studies on operating and closed solid waste disposal facilities in the capital city of Kuala Lumpur. In 2001 there were 168 operating final disposal sites throughout Malaysia. Nearly all were open-dump landfills, and presented a great danger of soil and groundwater contamination. The Kuala Lumpur survey found that highly toxic leachate from the wastes had run into nearby rivers and ponds, percolated into the soil, and contaminated groundwater; and that some sites were at risk of gas explosions.[10] Proposals based on the results called for removal of contaminated landfill soil and environmental remediation. Waste management companies took over disposal site management from local governments when waste management was privatized, but it remains to be decided who will shoulder remediation costs and how it will be done.

In addition to the hazardous substances in industrial wastes, experts also point out concerns that high-tech factories might be causing soil contamination and damaging the health of factory workers and people living nearby.[11] There are also concerns about contamination at closed factories as the migration of manufacturing to China accelerates. Beginning in January 2003 action on soil contamination in Japan was provided for by the Soil Contamination Countermeasures Law, and Malaysia too will find that it must now control illegal dumping and

direct efforts at soil remediation and community redevelopment in sunset industry areas. Malaysia, like Japan, must now turn its attention to environmental
restoration.

Aoki Yuko

## Essay 5:  Privatizing Medical Waste Management

In the 1990s Malaysia privatized the management of all wastes except radioactive waste.
In 1995 the federal government decided to privatize medical waste management. For that
purpose it divided Malaysia into four regions (North, Central, South, and Sabah/Sarawak),
and in each region selected a management company (the same company was selected in
the North and Sabah/Sarawak). These wide-area divisions are the same as those for the
privatization of municipal solid waste (MSW) management.

Each day Malaysia generates between 0.3 and 0.8 kg of medical waste per bed. Management companies are charged with providing special medical waste containers to hospitals, hauling wastes away in them, and performing final disposal by incineration. The
federal government established service indexes and requires companies to make their
operations more efficient, evaluate them, and perform audits. Under individual 15-year
contracts, companies build new incinerators or improve existing facilities, and manage
medical waste within their respective wide-area divisions. Currently large incineration
facilities in five locations burn medical waste from small and medium-sized hospitals.
Large hospitals are provided with incinerators that are operated and managed by waste
management companies. While there are no regulations or guidelines on medical waste
incineration, companies are required to comply with the 1978 Environmental Quality
(Clean Air) Regulations emission standards and to perform environmental assessments
when building incineration facilities. Hospitals pay waste management fees to the companies pursuant to the polluter pays principal.

Waste management companies are actively incorporating foreign technologies for
installing, operating, and managing facilities. Initiatives by companies include using British
and Australian incineration equipment, introducing technology through the United States-
Asia Environmental Partnership, and having capital participation from a transnational
medical waste management company based in Great Britain.

Privatization of medical waste management is characterized more by efforts toward
streamlining through large-scale integrated operations, and by development as a service
industry, than by raising efficiency through competition. In August 2002 Malaysian MSW
management companies began operations in the Mideast. It is anticipated that they will
expand into other countries after gaining experience at home.

Theng Lee Chong, Aoki Yuko

# Republic of Indonesia

## 1. Environmental Management and Decentralization

In conjunction with the 1997 Asian economic crisis a rapidly growing democratization movement arose in Indonesia, and 1998 saw the downfall of former President Suharto's authoritarian government, which had lasted 32 years. Consequently, the centralized political regime upheld until then under the 1945 constitution was dismantled, and a major transition is underway to a new constitutional structure supporting a democratic and decentralized government. Constitutional amendments made in preparation for the 2004 general elections came to a temporary conclusion with the fourth constitutional amendment of 2002. Although there had been discussion on decentralization in the past,[1] current developments in decentralization or local autonomy are understood in the context of the collapse of the Suharto political regime and the subsequent democratization process.

During the Suharto years authority was excessively concentrated in the executive branch, while the independence of judicial power was not recognized, and even the separation of powers was not an established principle of government. Hence Indonesia lacked the appearance of a modern constitutional state based on the rule of law. Efforts directed at democratization and decentralization began with the horizontal dispersion of authority vested in central power structures such as presidential authority and the military, and then extended to the vertical dispersion of authority to outlying regions of the country. This vertical dispersion can be seen as an attempt to let central political power disperse central government authority to the rest of the country, and to get democratization on track by introducing local self-government. Nevertheless, decentralization at present is proceeding within the unitary state framework maintained as a governing principle by the 1945 constitution.

Local autonomy, however, is not the same in every country, for centralization and decentralization combine in varying relative degrees. The state of local autonomy will feature different relationships between the center and localities—such as "decentralized but fused" or "centralized but separated"—depending on a country's historical background, expected roles and functions, and socioeconomic circumstances. Generally, however, local autonomy systems are a tug of war over power between the center and localities, and grow out of the relationship of tension and fusion between centralization and decentralization. In this respect, the question is to what extent and in what way Indonesia will advance political decentralization in connection with environmental management.

But Indonesia's development stage of decentralization is still far from having a well-defined principle of local autonomy. The Local Autonomy Law No. 22 of 1999 generates confusion in its interpretation, and therefore its revision of law has been questioned soon after its promulgation.[2] Another problem is insufficient discussion on the implementation level about the decentralization of environmental management, which to Indonesia means reorganizing environmental management into an entirely new system with localities playing the leading role despite the weakness of such management even under the centralized system. Yet, while rapid progress in decentralization looks democratic, going about it in a disorderly manner could bring about grave and irreversible environmental damage, as well as chaotic political and social situations.

In fact there is much criticism that since 1998 the collapse of the Suharto regime accelerated environmental damage all over Indonesia, or left Indonesia without any environmental control.[3] Despite the dictatorial and violent nature of the Suharto regime, there existed a certain extent of monitoring over environmental management, but Suharto's disappearance left a power vacuum which actually hastened environmental deterioration, owing to precedence given to profiteering, deficient legal responses, and too few alternative means. This should not be seen as the temporary disorder that tends to arise in the normal process of democratization, but a chaotic state of environmental management brought about by decentralization policy.

However, as it is impossible to turn the clock back on the decentralization that was started in 2001, Indonesia should perhaps develop a new kind of comprehensive sustainable environmental management of its own geared to decentralization. Below this section will explore the challenges of environmental management arising in conjunction with decentralization, primarily with regard to Indonesia's efforts in this area.

# 2.  Decentralization and Environmental Issues

Environmental damage in Indonesia is proceeding very quickly. Even though Indonesia is predisposed toward resource-dependent economic development, the current unsustainable course is jeopardizing its natural resource base for development. For example, a government report writes that since 1998 in Kalimantan there has been a rapid increase in cases in which people with the support of influential local figures, powerful public employees, and the military, as well as law enforcers and members of parliament, mobilize members of the public locally or in other areas to perform illegal logging.[4] Malaysian military personnel and third-country timber buyers are sometimes involved in illegal logging.[5] Seventeen truckloads of illegally logged timber, transported by night with a police escort, were seized in Padang, western Sumatra, but a newspaper reported that authorities denied any involvement.[6] Not only illegal logging, but also illegal fishing, mining, and other environmentally destructive activities are often carried out in an organized and unregulated fashion through corruption and bribes. Especially in the future when acceptance of domestic and foreign investment by local governments is farther encouraged, and when smaller-sized logging concessions are turned over to localities, concerns that the extensive damage of deforestation will expand are sure to become more realistic.

Due to the onerous pressures of poverty and population throughout Indonesia, local governments will find the increase of their revenues as one of their fiscal responsibilities to local assemblies, and there will be increasing danger of soliciting investment by indiscriminately issuing permits for resource exploitation and by promoting environmentally unsustainable development projects. But having been under the despotic Suharto government for so long, local governments have little knowledge and experience of environmental management, and have yet to pass local regulations for environmental protection. As such, their environmental management systems and environmental awareness are underdeveloped. Moreover, poor local governance that has been considerably weakened by the lack of political trust toward the central government raises the possibility of inconsistencies in environmental management among local governments. And because there is very little trust left anywhere in the country toward the judiciary in particular,[7] it is possible that even if the central government issued environmental guidelines and policy, they could not be adequately heard and enforced.

A recent World Bank report predicted that Sumatra's lowland forest would disappear in 2004, as would that of Kalimantan by 2010.[8] Local environmental management in the course of Indonesia's decentralization faces a double crisis. First is the weakness of the central government's system for comprehensive environmental management, which, in conjunction with local governments' deficient capacity to tackle the job, presents huge difficulties for restructuring environmental management and base it primarily in localities. Second is the increasing pressure on the poverty-stricken outer islands except Java for resource development to secure fiscal revenues, which will likely further spur environmental destruction through illegal and reckless development. Many sources describe environmental damage that tends to accelerate in the excessively rapid process of decentralization[9].

## 3.  Environmental Management Principles in the Decentralization Process

The principle of government in the 1945 constitution is the unitary state, which has been maintained even after decentralization and the "Reformasi" movement. Heretofore the development and management of local resources has been Java-led, and it is not an understatement to say that the earnings from natural resources have been absorbed by Java through so-called local governance. In this respect, establishing local autonomy in Indonesia will entail the two elements of devolving political authority on localities and providing localities with economic foundations. Especially important to localities is the question of the extent to which natural resource management and profit from resource use can be harmonized to firm up localities' economic underpinnings. Exploring this question necessitates throwing light on Indonesia's stand on decentralization and on its principles for natural resource management.

## 3-1  National Policy Guidelines, and the National Development Program

National environmental policy is underpinned by a five-year plan called the "National Policy Guidelines 1999–2004 (GBHN)," and the "National Development Program" (PROPENAS 2000–2004) that implements GBHN. The fourth chapter of GBHN, "Natural Resources and Environmental Protection" provides, in relation to local governments, "Central government authority on natural resource management and environmental protection will be transferred selectively to local governments in stages, and the environmental quality of the ecosystem as set forth by law will be protected." While this clearly states that central government authority for environmental management is to be gradually transferred to localities, no details are given. Meanwhile, chapter 9 of PROPENAS discusses sustainable environmental management as a priority in local development, and chapter 10 presents five programs as ways to achieve environmental policy, but there is not enough specific information on local autonomy and environmental management.

## 3-2  People's Consultative Assembly Decree No. XV/MPR/1998

It was Decree No. 15 of the People's Consultative Assembly (MPR) in 1998 during the administration of President Habibie that fixed the direction of decentralization. This decree provided that local autonomy shall involve the fair division of Indonesia's natural resources and a fiscal balance between the central government and localities (Article 1), as well as implementation based on democratic principles and considerations to local characteristics (Article 2). Provisions on natural resources in Articles 3 through 5 state that Indonesians shall give fair consideration to the prosperity of local communities and the nation as a whole, and broadly provide cooperatives and small and medium enterprises with opportunities. They also endow local governments with the authority to manage natural resources while obligating them to protect the environment.

This decree takes the position that decentralization should proceed with attention to democratization and local characteristics, and provides that local governments should have both the authority to manage natural resources and the duty to protect the environment, but no conclusion has been reached on whether this means that the central government should transfer to local governments even its authority over proceeds from the use of natural resources. At issue is the matter of how far the government should go in decentralizing those proceeds.

## 3-3  Second Amendment of the 1945 Constitution

In 1999 Indonesia's constitution introduced the institution of local autonomy as long as it does not contravene the unitary state principle. This amendment accorded self-government rights to provinces (Provinsi), and to the regencies

(Kabupaten) and towns (Kota) into which provinces are sub-divided, established local assemblies on all three levels (DDD, DPRD and DPR, respectively), and created administrative head offices of provincial govenors (Gubernur), regency governors (Bupati), and town mayors. Based on the principles of self-government and the delegation of administrative duties, local governments are to perform administration and management of government tasks, and for that purpose municipalities are granted rule-making authority. Amended Articles 18, 18A, and 18B of the constitution provide that each municipality head shall be elected democratically; the central government shall establish units of local governments that are special and distinct; and the use of natural resources shall be administered with justice and equity according to law. However, the relationship between the central and local governments regarding the use of natural resources is to be established separately by law.

## 3-4　Laws No. 22 and No. 25 of 1999

Decentralization is specifically prescribed by Law No. 22 (Law Regarding Regional Government) and Law No. 25 (Decentralizing Financial Authority) of 1999. Government Regulation No. 25 of 2000 details the political authority of Law No. 22, which defines regional autonomy as "an autonomous region's authority to autonomously regulate and protect the local community's interests based on the community's aspiration and laws," and says that decentralization is "the transfer of central government authority to autonomous regions within Indonesia's unitary state framework," distinguishing it from the concept of transferring central government authority to governors or other local officials.

While the scope of decentralization in Law No. 22 affects all governance-related authority, it excludes foreign affairs, national defense and public security, the courts, fiscal affairs and finance, religion, and other fields. But "other" includes a wide range such as macroscopic national development, balancing the budget, and national government administration. Another is the efficient use of natural resources and environmental conservation. Despite the inclusion of natural resource use, its definition is vague and abstract, which makes these laws vague regarding the line between environmental management by the central and local governments.

Law No. 25 prescribes a fiscal balance between the central and local governments under the unitary state, and provides that fiscal apportionment between the central and local governments, as well as equality among local governments, shall be maintained in a democratic, fair, and transparent manner, and in accordance with the potentials, situations, and requests of the regions. Localities are allowed four revenue sources, of which the state budget is the biggest. Percentages apportioned to localities include 80% of that related to forests, mining, and fisheries, 15% for petroleum, and 30% for natural gas. However, there are only four oil- and gas-producing provinces: Aceh, Riau, West Papua, and East Kalimantan.

## 3-5   Government Regulation No. 25 of 2000

This regulation details how authority is decentralized, and sets forth the authority of local governments. For the central government it exclusively reserves authority in certain environment-related areas including the development of assessment guidelines for natural resources and the environment, establishment of regulations for the management and use of natural resources within the 12-nautical-mile limit, environmental impact analyses of activities that could have broad social or environmental impacts, and formulating environmental standards, pollution guidelines, and nature conservation guidelines.

But environment-related items are not limited to those items provided in article 18 of this regulation and widely affect other items in this regulation related to natural resources, which considerably broadens the scope of exclusive central government control over the environment. For example, this gives the central government exclusive control over pesticides and chemical fertilizers, the conservation and management of marine resources, management of mineral and energy resources, forest resource management standards, forest conservation, and ecosystem protection.

Article 3 enumerates the environment-related authority of the provinces, including environmental impacts affecting multiple regencies and towns, coordinating management of the environment for using marine resources from four to 12 nautical miles, coordinating the monitoring and protection of water resources spanning multiple regencies and towns, wide-area assessments of environmental impacts that could negatively affect multiple regencies and towns, monitoring for conservation across multiple regencies and towns, and crafting local environmental standards based on national environmental standards. There are separate provisions for the ocean, minerals/energy, forests, and other environment-related items. Finally, the regulation grants some decentralized powers to regencies and towns.

## 3-6   Law Concerning Environmental Management

Indonesia's basic law on the environment, the Law Concerning Environmental Management, had already oriented itself toward the decentralization of environmental management in 1997, when it replaced the old Law of No. 4 of 1982. This amended law allows the devolution of authority from the central to local governments as a means of carrying out national policy on environmental management. But while it provides that government ministers or local governors have supervisory responsibilities, this does not go beyond the dispersion of authority.

Under Presidential Decree No. 2 of January 2002 the Environmental Impact Management Agency (BAPEDAL) was integrated into the Ministry of Environment, and under this new organization the Regional Environmental Management Capability and Development Department took over the task of environmental management for the regions. However, the Law Concerning Environmental Management of 1997 already had provisions to make it possible to delegate some environmental management authority to municipalities and to transfer some environmental management to them. For that purpose there are currently 168 regional environmental management agencies called BAPEDALDA as of August 9, 2002.

In conjunction with the integration of BAPEDAL into the Ministry of Environment, Indonesia faces a number of challenges in connection with decentralization: Developing guidelines for the creation of regency and town minimum conditions for administering the environment; support and supervision for carrying out decentralization, including the provision of guidelines as well as guidance, training, orientation, and supervision in the environmental field; and determination of criteria used when local governments issue environment-related permission.

## 4.    Decentralizing Environmental Management

Political decentralization is inseparable from local economic independence. Indonesia depends on its natural resources for development, and of course obtains most of its revenues from oil, gas, and other natural resources. Nearly all the resource-derived revenues from throughout the country have been controlled by the central government under the highly centralized Java-based political system. Debate on fiscal decentralization at this time concerns primarily the management of resources and securing revenues by local governments, enlarging the balanced budget amount allocated by the central government to local governments, and expanding localities' own revenue sources. In this sense, environmental management in the Indonesian process of decentralization is not just a matter of transferring political authority for environmental management to localities, but also a local-level "development and environment" problem that poses the challenge of how to reconcile local environmental protection with securing the fiscal resources needed for regional development, such as coping with poverty, suppressing population growth, and developing social welfare. Below, this section explores the problems involved.

Much is still unclear about local self-government at this point, with domestic organizations and international aid agencies pointing out a variety of problems. One problem often noted is that self-government is allowed with regencies and towns in parallel, while there is no clear relationship between provinces and regencies/towns, and there are doubts about the positive role of the provinces. It is often observed that laws are conflicting, overlapping, and vague. Further, there is no clear connection with community participation in self-government, and it is quite possible that the issuance of permits and other authorization by the central government will devolve to local governments without reorganizing the system. The World Bank recommends that Indonesia make four improvements in natural resource management. First is building technical and provincial capacity, second is strengthening checks and balances, third is bringing decentralization to each region when it is ready, and fourth is phased devolution of authority to regional governments.[10]

This has led to increasing calls for amending Law No. 22 of 1999. People's Consultative Assembly Decree No. IV/MPR/2000 on policy recommendations for implementing regional autonomy determined a policy recommendation for implementing local self-government. While it holds that bringing about local autonomy and achieving fiscal balance between the central and local governments are necessary for regional development, which is the main component of national development, it takes the stance that "justifiable regulation of natural resources" is important, and while taking into consideration the lesson learned from the failure of local autonomy between the central and local governments, it sets forth the official view that Law No. 22 of 1999 should be amended. Also, MPR Decree No. VI of 2002 states that the Local Autonomy Law has not been completely enforced in terms of application, thereby engendering overlapping authority, legal gaps and imbalance between regions, and legal uncertainties. Its recommendations to the president are ensuring conformity with laws on implementing local self-government and the creation and assessment of government regulations. Additionally, a Ministry of Home Affairs report states that although there are flaws in Article 18 and other constitutional provisions, the constitution should be used as a guide in amending Law No. 22.[11]

Examining Indonesia's decentralization in conjunction with environmental management reveals a possible combination of the shortcomings found in both liberalism and democracy, and for that reason there is a huge deficiency with respect to how Indonesia will assure benefits that accrue through environmental management in the future, as opposed to immediate benefits. In other words, despite the insufficient development of principles for integrated management which would keep liberalism from going too far, there is an attempt to solve this problem with excessive expectations for democracy. Achieving environmental governance in both the center and the regions will require establishing a democratic and modern system of governance for the entire country, and, modeled upon that system, a system for local administration. Most desirable for achieving sustainable development would be the phased dispersion and decentralization of Indonesia's centralized environmental administration, and looking for a steady and unhurried mode of development which incorporates the advantages of participatory methods.

Sakumoto Naoyuki

# India

As section 1 will show, India's means of combatting pollution are underpinned by laws and policies which are, however, not well enforced or implemented. A characteristic of India is that in this area the administrative apparatus does not play as active a role as the courts. On the other hand, section 2 will discuss the initiatives of farmers, who have taken a fresh look at indigenous knowledge and are putting it to work in their livelihoods. These efforts are connected with the protests against water privatization and the imposition of intellectual property rights on biological resources, which are becoming increasingly globalized.

## 1.    Combatting Pollution

### 1-1    Pollution-Related Laws

Laws for coping with pollution go back to the application of criminal law and the Indian Police Act to deal with water and noise pollution during the colonial era. The first event to spark environmental legislation after independence was the 1972 United Nations Conference on the Human Environment in Stockholm, which India followed with the 1974 Water Prevention and Control of Pollution Act and the 1981 Air Prevention and Control of Pollution Act. But these were more end-of-pipe laws than attempts to eliminate the causes of pollution.

The next event that triggered environmental action was the Bhopal disaster in 1984, which served as a wake-up call pointing to the need to have comprehensive environmental laws that include environmental impact assessments, and led to creation of the Environmental Protection Act and the Ministry of Environment and Forests in 1986. India's first regulations on hazardous waste management were based on the Environmental Protection Act, although action in this area is behind that for air and water pollution.

In 1991 India enacted the Public Liability Insurance Act for Helping Victims of Hazardous Chemical Accidents, which establishes a compulsory liability insurance system meant to quickly provide victims with the minimum basic compensation, but victims can also file lawsuits to demand more than basic compensation. This law is considered to be one of the world's three major pollution relief institutions, along with Japan's Pollution-Related Health Damage Compensation Program and America's Superfund Law.[1]

As this shows, India has a comparatively well developed system of environmental laws to deal with pollution (Table 1). Each state government can enact laws on matters such as water, which the constitution puts under state control, as well as laws and regulations stricter than those of the federal government. Additionally, state pollution control boards can order penalties or closures of factories that do not comply with air and water emission standards. In fact, however, officials in charge on site have considerable discretion in these matters. Especially since the start of far-reaching economic reforms of 1991, competition

TABLE 1. Main Environment-Related Laws

| Environment and Industry | Environment [Protection] Act, 1986<br>Boilers Act |
| --- | --- |
| Air pollution | Air [Prevention and Control of Pollution] Act, 1981<br>Motor Vehicles Act, 1989 |
| Water pollution | Water [Prevention and Control of Pollution] Act, 1974<br>Water [Prevention and Control of Pollution] Cess Act of 1977 |
| Radiation | Atomic Energy Act, 1982 |
| Mining | Mines and Minerals [Regulation and Development] Act, 1957 |
| Insecticides | Poison Act, 1919<br>Insecticides Act, 1968 |
| Land | Urban Land [Ceiling and Regulation] Act, 1976<br>Coal Mines [Conservation and Development] Amendment Act |
| Factories | Factory Act, 1948<br>Industries [Development and Regulation] Act, 1951<br>Indian Forest Act, 1927 |
| Forests | Wild Life (Protection) Act, 1972<br>Forest [Conservation] Act, 1980 |
| Pollution damage compensation | Public Liability Insurance Act, 1991<br>National Environment Tribunal Act, 1995<br>National Environment Appellate Authority Act, 1997 |
| Energy conservation | Energy Conservation Act, 2001 |

among states to attract investment has grown increasingly severe, with attempts to attract it by relaxing the application of environmental laws in affected areas.

Needless to say, the government has been aware of the need for environmental considerations in its new industrial policy even after beginning economic reforms in 1991, for in 1992 it issued a Policy Statement for Abatement of Pollution. But without specifically naming the responsible administrative agencies, it is not clear how this statement fits into the overall scheme. The environmental action plan developed in December 1993 gives the following seven policies for the priority environmental problems that India should tackle: (1) conservation of biodiversity and the sustainable use of resources, (2) securing unpolluted water sources, (3) pollution control and reduction of wastes, especially hazardous waste, (4) encouraging the adoption of clean technologies, (5) initiatives on urban environmental problems, (6) better scientific understanding of environmental problems, and (7) development of alternative energy sources.

## 1-2    Court Action on Pollution

Although India receives kudos for having created the laws and administrative institutions for implementing these seven policies, India is faced with the task of their enforcement, compliance, and monitoring.[2] Below we shall explore how decisions of the Supreme Court and other courts, are playing a more active role than state pollution control boards in environmental policy implementation and legal compliance.

Public interest litigation and other legal efforts are also playing an important role in environmental policy implementation and legal compliance. Basically such litigation is aimed at active compliance with laws, not winning compensation. In addition to formal lawsuit procedures, this institution allows the general public to invoke the warrant jurisdiction right, sometimes even by sending a letter directly to the Supreme Court. There are now so many environmental lawsuits that every Friday is "Green Friday," when courts deal with environment-related issues.[3]

Supreme Court rulings on air pollution in the capital city of Delhi are instructive. Studies by the World Health Organization and the United Nations Environment Programme from 1980 through 1984 showed this city ranked fourth worldwide in the amount of particulate pollution, and the 1990 study revealed it was still as high as seventh place. Delhi's air is severely polluted by not only carbon monoxide and hydrocarbons from motor vehicles, but also by particulates from its many factories.[4] Based on the Delhi Master Plan, the Supreme Court ordered the closure and relocation of polluting factories and power plants.[5] In 1998 the government required buses, taxis, and three-wheelers used for public transportation to use compressed natural gas (CNG), and stiffened emission standards for non-commercial four- and three-wheeled vehicles (Table 2). The introduction of CNG occurred under disorderly social circumstances including an insufficient CNG supply, not enough gas stations, long lines at gas stations, traffic congestion in the city, and the burning of buses by citizens. The main problem now is perhaps higher public transportation fares due to rising CNG prices. Air pollution as of April 2001 showed improvements in $SO_2$ and $NO_2$, but not in particulates.[6]

As of February 2002 CNG-powered vehicles numbered 3,800 buses, 10,350 passenger cars, and 4,000 taxis,[7] indicating that more time will be needed to carry out the Supreme Court decision. There are many other cases in which decisions have yet to be carried out, making it urgent to improve the technical feasibility of Supreme Court decisions. Further, India needs not only court decisions that involve orders and control, but also political and economic reasonableness that takes into account, for example, the impacts of CNG introduction on the poor who depend on public transportation.

## 2.  The Reinstatement of Indigenous Knowledge

### 2-1  Water

Rural people are highly dependent on groundwater, but more and more villages are experiencing difficulty in securing it. The green revolution came to India in

TABLE 2. Supreme Court Directives on Delhi Air Pollution (1998) and the Status of Their Implementation

| | Supreme Court order | Status of implementation as of 1 July 2001 |
|---|---|---|
| Directives on the fuels used by public transport | The number of buses must increase to 10,000 by April 1, 2001 and must run on CNG. | About half of the required 10,000 buses are in service, with over 88% running on diesel. |
| | All pre-1990 taxis and autorickshaws must be replaced with new vehicles using clean fuels by March 31, 2000. | All pre-1990 taxis and autorickshaws were removed by the deadline. |
| | The Delhi government must provide financial incentives to replace all post-1990 autos and taxis with new vehicles that operate on CNG or clean fuels by March 31, 2000. | Financial incentives are being offered for new vehicles operating on CNG. 42% of the autorickshaws and 12% of taxis are on CNG. |
| | All buses older than 8 years must be scrapped by April 1, 2000 unless they operate on CNG or other clean fuels. The entire city bus fleet (public and private) must be steadily converted to CNG. | 1,200 CNG buses are in operation (1,171 dedicated CNG buses and 29 retrofitted with CNG kits). |
| | The Gas Authority of India must create a network of 80 CNG refueling stations by March 31, 2000. | 73 CNG refueling stations are open (9 mother, 17 online, 39 daughter, and 8 daughter booster). |
| Directives on emission standards for non-commercial vehicles | Non-commercial four-wheeled vehicles (petrol or diesel) will be registered in the NCR only if they meet (a) Euro I equivalent norms in effect from 1st June 1999, and (b) Euro II equivalent norms from 1st April 2000. | About 9 lakh non-commercial four wheeled vehicles are in use. 17% of these conform to both Euro I and Euro II norms (Euro I 6% and Euro II 11%). |
| | All two-wheelers must conform to Bharat Stage I from 1 April 2000. | Appx. 20 lakhs two-wheelers are on the road. 5% of these conform to Bharat Stage 1 norm. |
| | Introduction of low-sulfur (500 ppm maximum) diesel fuel in the NCR from 1 April 2000. | All petrol pumps sell only 500 ppm S diesel. |

Note: Directives on fuel used by public transport were current as of July 1, 2001; those on emissions standards for non-commercial vehicles were current as of March 31, 2001.

Source: Based on Tata Energy Research Institute website: <http://www.teriin.org/news/aug013.htm>.

the second half of the 1960s. Macroscopically, the production of rice, wheat, and other grains increased, but the heavy use of groundwater for irrigation caused salinization and waterlogging throughout the country. In some states this induced farmers to abandon up to 20% of total cultivated land.

In the 1990s multinational beverage companies invested in India and built factories in a number of locations, then began selling various brands of bottled water with their marketing sights set on the urban middle income bracket with a certain amount of buying power.[8] In some villages groundwater levels have dropped and dried up farmland due to excessive groundwater withdrawals by these beverage factories.

Closely watched in this connection are India's traditional knowledge of and techniques for water harvesting. Ancient documents from the third and fourth centuries B.C. show that already people on the Indian subcontinent were harvesting rainwater for irrigation, but that knowledge was gradually lost during the British colonial era. A CSE report[9] describes the traditional water harvesting systems in different parts of India. In 400 pages it exhaustively reports on water harvesting knowledge such as wells and storage ponds, the use of irrigation channels, forests and how communities managed and maintained them, and innovative ways of arranging farmland and pastureland. CSE went on to compile other reports on irrigation using water storage tanks in southern Indian states[10] and on water harvesting practices.[11] This catalyzed efforts by state governments and NGOs that actively support practical water harvesting activities.

## 2-2  Biological Resources

Neem trees can be seen near every field and along every road in India. For thousands of years its products have been used as natural insecticides, and in toothbrushes with medicinal and antibacterial properties. The turmeric plant has also been traditionally used in India for thousands of years to treat wounds and rashes. However, in many cases outsiders have taken such indigenous Indian knowledge to other countries and applied for patents on it. Since 1985 Western and Japanese companies have won a number of patents on uses of the neem tree's pharmacological properties. In Japan neem seedlings and oil are sold as environment- and people-friendly horticultural supplies.

Vandana Shiva, director of the Research Foundation for Science, Technology and Ecology, uses the term biopiracy in her criticism of this patent strategy of the industrialized countries.[12] In 1995 she allied herself with over 200 groups in 35 countries in requesting that the US Patent and Trademark Office (USPTO) cancel a patent by a US chemical maker on neem extract. Similarly, an Indian government research institute asked that the USPTO cancel a patent on turmeric, citing ancient documents in Sanskrit as the basis for prior knowledge and technology. Thanks to these challenges, in August 1997 the USPTO cancelled a turmeric patent after recognizing that it had no novelty, and in May 2000 the European Patent Office cancelled a patent on neem that had been granted on the basis of a US patent. These were the first instances in which patents obtained in the West were cancelled on the basis of prior knowledge in developing countries, and they will have incalculable repercussions.[13] NGOs and research institutes in India and elsewhere are signaling that in connection with prior knowledge they will challenge the validity of patents obtained in the West on other biological resources.

In parallel with protests against biopiracy, Shiva and others have launched a movement called *Navdanya* (Hindi for "nine seeds") to protect the genetic resources of native species of rice and other grains, and assure the sustainability of agriculture. Specific activities include studying existing species and the state of their erosion, documenting and registering native species, and distributing seeds to farmers. Registration of native seeds is an attempt to assure an alternative way of conceiving intellectual property that recognizes systems of group and traditional knowledge in each region, but the right to that knowledge is never exclusive or monopolized; instead, people can use it and exchange it for free. Although the Navdanya movement has yet to become a mass movement covering all of India, like the water harvesting practices described above, it represents efforts toward the reinstatement of traditional knowledge on resource use, and is broadening slowly but vigorously.

Tsujita Yuko, Kanazawa Kentaro

## Essay 6:  Flush Toilets: Part of the Problem

Several years ago when I visited Stockholm to participate in a water symposium, I received an invitation from the King of Sweden to attend a dinner, but instead of going to the dinner I visited a number of toilets throughout the city with a colleague, Anil Agarwal. Of course I didn't understand my job well enough to go as far as opening the tanks. These were composting toilets that held the waste in tanks and diverted urine for agricultural use. Later we visited shops in the city selling such toilets, and found a variety of types such as low-flush, electrified, and urine-diverting. I finally began to make the connection.

I had always thought of flush toilets and sewerage as synonymous with a hygienic, clean environment. But actually they are part of the problem, not the solution. Though flush toilet technology is very simple, it is ecologically unwise because a considerable amount of clean water is used to carry away a small amount of waste.

India's flush toilets are especially wasteful of water, flushing away at least 10 liters of water with each use. Water is carried to the city by huge dams and other facilities, then after use it is piped to sewage treatment facilities that cost about the same, finally polluting even more water in rivers and lakes. Nearly all India's rivers are dead because of urban sewage.

Investing in large-scale river remediation programs like the Ganga Action Plan to solve the sewage treatment problem benefits the flush toilets of the rich, but does nothing to help the poor. The politics and economics of sewerage systems in developing countries are outrageous. Investments in sewerage systems have not been recovered in any city, and for that reason the users—who are almost all rich—count on subsidies. Sewerage therefore leads to the use of subsidies for the pleasant water closets of the rich. Under this human waste management paradigm, the poor are always kept from enjoying any benefits. Moreover, it is virtually impossible for the government to meet its construction target for sewage treatment facilities, and fast-paced urbanization assures that these high-cost facilities will exceed their treatment capacities in no time. Still more investment in sewerage construction fuels the vicious circle.

A 1999 study by the Central Pollution Control Board found that Delhi generates 2.55 billion liters of sewage daily, but that only 885 million liters of that are taken into the sewerage system and treated. The rest flows untreated into storm drains and ends up in rivers. Delhi plans to spend 7.5 billion rupees to triple its sewage treatment capacity by 2025, but even if the city achieves that goal, it will be far below what is needed. What is more, less than 3% of sewage in urban areas other than Delhi is treated, so it is not surprising that rivers are evil-smelling.

India's scientists think about sending men to the moon, but give no thought whatsoever to the lowly toilet. There is no discussion at all about eco-friendly sewage systems. Who will seek an alternative paradigm? As long as sanitary engineers are captives of their group interests and entrenched thinking, no one will be seeking change. Nevertheless, we must change, and we must remind ourselves that rich people's flush toilets are a cause of today's environmental problems.

Sunita Narain

# Part III

# Data and Commentary

# 1. Economic Inequality, Poverty and Human Development

Per capita GDP is an indicator that has generally been used to gauge a country's affluence. If the per capita GDP of Country A is $3,000 and that of Country B is $1,000, the citizens of the former country are judged to be better off. But this reasoning has several problems.

First, GDP does not take income inequality (income distribution) into account. Even if Country A's per capita GDP is $3,000, the wealthy (the upper 10%) might have a per capita GDP of $27,300, leaving the other 90% of the population with a mere $300. But if all the citizens of Country B have the same $1,000 income, it is mistaken to say that Country A is more affluent on the basis of per-capita GDP.

Second, GDP also ignores the problem of poverty. If we set the poverty line at $365 a year and estimate the percentage of people at income levels below that line (the poverty rate), then the poverty rate of Country A is 90% while that of Country B is 0%. Here again the GDP yardstick yields the opposite result.

Third, real human happiness is determined by a variety of economic, social, and environmental factors. Such being the case, it is not enough to measure affluence by focusing on the economic facet alone using GDP. For this reason the United Nations Development Programme (UNDP) has emphasized human development meant to expand the broad range of choices (capabilities) available to people instead of narrowly conceiving development as the growth of income and wealth.[1]

It is instructive to gauge the state of Asia's economy using different indicators of affluence. In particular the Gini index is used to measure income inequality, the human poverty index for developing countries (HPI-1)[2] to measure poverty, and the human development index (HDI)[3] to measure human development.

Data on per capita real GDP over a number of years (Table 1) show that the economic development of Indonesia, Singapore, and Thailand slowed in 1997 and 1998 because of the Asian currency crisis, which also triggered Indonesian political instability that led to the fall of the Suharto government in May 1998. This suggests that in 1997 and 1998 there should have been changes of some kind in the indicators for economic inequality, poverty, and human development in Southeast Asian countries.

If this same span of years is viewed using the Gini index (Table 2), the impact of the Asian currency crisis on inequality is not discernible. The data for Indonesia, for example, suggest that there was less economic inequality in 1999 than in 1996, but this is hard to believe. This same indicator plotted in relationship to per capita GDP (Fig. 1) places Indonesia and Vietnam in the 0.3 to 0.35 range, while Malaysia, the Philippines, and Thailand are in the vicinity of 0.4 to 0.45. But again, it is hard to believe that Indonesia with its continuing political disorder has less economic inequality than Thailand and Malaysia, in fact being at the same level as Japan and South Korea. From this it is evident that statistical data for the Gini index vary widely from country to country, and must be interpreted with care.

TABLE 1. Real GDP per capita (Purchasing Power Parity in US$)

| Publication year | Year covered | China | India | Indonesia | Japan | S. Korea | Malaysia | Philippines | Singapore | Thailand | Vietnam |
|---|---|---|---|---|---|---|---|---|---|---|---|
| 1997 | 1994 | 2,604 | 1,348 | 3,740 | 21,581 | 10,656 | 8,865 | 2,681 | 20,987 | 7,104 | 1,208 |
| 1998 | 1995 | 2,935 | 1,422 | 3,971 | 21,930 | 11,594 | 9,572 | 2,762 | 22,604 | 7,742 | 1,236 |
| — | 1996 | — | — | — | — | — | — | — | — | — | — |
| 1999 | 1997 | 3,130 | 1,670 | 3,490 | 24,070 | 13,590 | 8,140 | 3,520 | 28,460 | 6,690 | 1,630 |
| 2000 | 1998 | 3,105 | 2,077 | 2,651 | 23,257 | 13,478 | 8,137 | 3,555 | 24,210 | 5,456 | 1,689 |
| 2001 | 1999 | 3,617 | 2,248 | 2,857 | 24,898 | 15,712 | 8,209 | 3,805 | 20,767 | 6,132 | 1,860 |

Source: UNDP, Human Development Report, 1997–2001.

TABLE 2. Gini Indexes of Asian Countries

| Year | China | India | Indonesia | Japan | S. Korea | Malaysia | Philippines | Singapore | Thailand | Vietnam |
|---|---|---|---|---|---|---|---|---|---|---|
| 1990 | — | — | 0.289 | — | — | 0.429 | 0.438 | — | 0.438 | 0.350 |
| 1991 | — | — | — | — | 0.299 | — | 0.438 | — | — | — |
| 1992 | — | — | — | — | 0.299 | 0.429 | — | — | 0.462 | — |
| 1993 | — | — | 0.317 | — | 0.294 | — | — | — | — | 0.350 |
| 1994 | — | — | — | 0.297 | 0.294 | — | 0.429 | — | — | — |
| 1995 | — | — | — | — | 0.291 | 0.437 | — | — | — | — |
| 1996 | — | — | 0.365 | — | 0.297 | 0.442 | 0.462 | — | 0.434 | 0.355 |
| 1997 | — | — | — | — | — | 0.442 | 0.460 | — | — | — |
| 1998 | — | — | — | — | 0.290 | 0.442 | 0.468 | — | 0.414 | 0.354 |
| 1999 | — | — | 0.310 | 0.301 | 0.294 | 0.442 | 0.463 | — | 0.435 | 0.351 |
| 2000 | — | — | 0.307 | — | 0.294 | 0.442 | 0.461 | — | 0.435 | 0.352 |

Sources: For Japan, Ministry of Public Management, Home Affairs, Posts and Telecommunications, National Survey of Family Income and Expenditure (1999 Survey). For other countries, World Bank, East Asia Update Special Focus: Poverty Reduction and International Development Goals, April 2001 or http://www.aric.adb.org/.

FIG. 1. Relationship Between the Gini Indexes and Per Capita GDP in Asian Countries

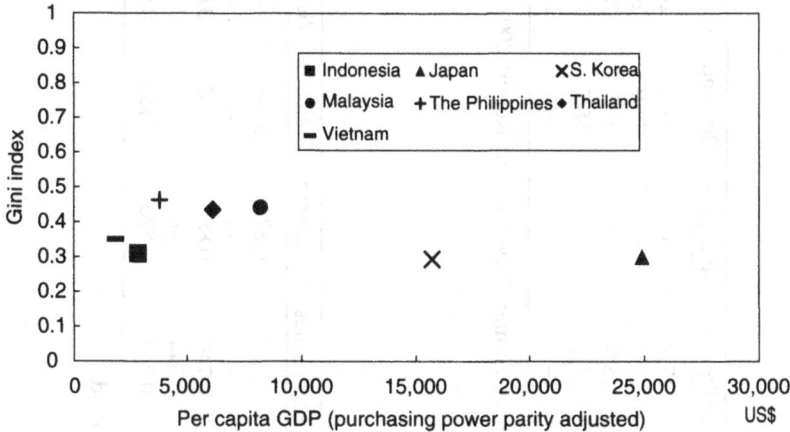

Sources: All Gini index data except for those of Japan are from World Bank, *East Asia Update Special Focus: Poverty Reduction and International Development Goals*, April 2001. Japan data are from Ministry of Internal Affairs and Communications, *National Consumer Survey for 1999* (in Japanese). GDP data are from UNDP, *Human Development Report*, 2001.

HPI-1 (Table 3) is an indicator estimated using the three aspects of longevity, knowledge, and a decent standard of living. The larger a country's value on a scale of 0 to 100, the greater its degree of poverty. HPI-1 increased in Indonesia and Thailand around the time of the currency crisis, which was largely because more people were unable to use health and medical services at that time. However, calculations in the 2001 version of Human Development Report omit the percentage of people who can use health and medical services owing to the unavailability of reliable data. At first glance the increase in HPI-1 values for Indonesia and Thailand from 1995 to 1997 appears to stem from the Asian currency crisis, but in fact it is hard to determine if these values result from the currency crisis or from data of dubious accuracy.

Chronological change in the HDI (Table 4), which like HPI-1 is estimated from the three aspects of longevity, knowledge, and a decent standard of living, indicates that the closer its value is to 1, the more human development has advanced. While the HDI tended to rise from 1995 to 1999, those of Indonesia, Singapore, and Thailand fell in 1998. It is perhaps possible to interpret this as the deleterious effect of the currency crisis on human development, but the high correlation of HDI with GDP calls attention to the significant possibility that it is just a matter of GDP changes being directly manifested in the HDI.

This brief survey leads to the conclusion that these indicators were unable to clearly discern how the Asian currency crisis affected economic inequality, poverty, and human development. Two reasons are the poor reliability of the

TABLE 3. Human Poverty Index (HPI-1) in Asian Countries

| Publication year | Year covered | China | India | Indonesia | Japan | S. Korea | Malaysia | Philippines | Singapore | Thailand | Vietnam |
|---|---|---|---|---|---|---|---|---|---|---|---|
| 1997 | NA | 17.5 | 36.7 | 20.8 | — | — | — | 17.7 | 6.6 | 11.7 | 26.2 |
| 1998 | 1995 | 17.1 | 35.9 | 20.2 | — | — | — | 17.7 | 6.5 | 11.9 | 26.1 |
| 1999 | 1997 | 19.0 | 35.9 | 27.7 | — | — | 14.2 | 16.3 | — | 18.7 | 28.7 |
| 2000 | 1998 | 19.0 | 34.6 | 27.7 | — | — | 14.0 | 16.1 | — | 18.7 | 28.2 |
| 2001 | NA | 15.1 | 34.3 | 21.3 | — | — | 10.9 | 14.7 | — | 14.0 | 29.1 |

Note: NA means the data year was not stated. Due to a change in the calculation method in Human Development Report 2001, data published before and after that report cannot be compared directly.

Source: UNDP, Human Development Report, 1997–2001.

TABLE 4. Human Development Index (HDI) in Asian Countries

| Publication year | Year covered | China | India | Indonesia | Japan | S. Korea | Malaysia | Philippines | Singapore | Thailand | Vietnam |
|---|---|---|---|---|---|---|---|---|---|---|---|
| 1999 | 1995 | 0.679 | — | 0.665 | 0.918 | 0.835 | 0.755 | 0.728 | 0.873 | 0.746 | — |
| — | 1996 | — | — | — | — | — | — | — | — | — | — |
| 1999 | 1997 | 0.701 | 0.545 | 0.681 | 0.924 | 0.852 | 0.768 | 0.740 | 0.888 | 0.753 | 0.664 |
| 2000 | 1998 | 0.706 | 0.563 | 0.670 | 0.924 | 0.854 | 0.772 | 0.744 | 0.881 | 0.745 | 0.671 |
| 2001 | 1999 | 0.718 | 0.571 | 0.677 | 0.928 | 0.875 | 0.774 | 0.749 | 0.876 | 0.757 | 0.682 |

Note: There were no data for 1996. All data were estimated using the 1999 calculation method.

Source: UNDP, Human Development Report, 1999–2001.

data, and the difficulty of comparing countries due to frequent changes in calculation methods. There is a tendency to believe that statistics provide quantitative
data that are highly useful, but in the case of developing countries with imper-
fect systems for generating statistics, great care must be exercised when using
statistical data. Nevertheless, as long as reliable data are available, the above indicators are capable of limning inequality and poverty. Therefore when performing studies on small areas such as settlements and villages, it is well worth the
effort to collect highly reliable data through fieldwork or other means, and use
them to prepare indicators for inequality, poverty, and human development.

Shibasaki Shigemitsu

## 2.    Population and Gender: Reproductive Health and Rights

The UN International Conference on Population and Development (ICPD) held in Cairo, Egypt in 1994 tried a new micro rights-based approach to the population issue by focusing on the health and lives of couples and individuals, especially women, instead of the macro approach of previous population conferences, which focused on population size and quality. Delegates to the ICPD proposed the idea of reproduction health and rights. Gender equality and women's empowerment were also stressed, and introduction of the gender perspective was one factor that encouraged this change in orientation. At the Beijing Fourth World Conference on Women in 1995 reproductive health and rights were clearly defined as a basic human right of women.

While 'sex' refers to the biological characteristics defining humans as female or male, 'gender' refers to the social and cultural attributes and opportunities associated with being female or male. Gender underlies the prevailing idea of 'masculinity' and 'femininity,' and the notion that men work outside and women mind the home. Such gender roles reflect the power relations between men and women in society. Accordingly, the disadvantages and discrimination that women suffer are revealed more clearly through the gender perspective. For example, the lack of self-determination for women about sexuality and fertility underlies high birth rates and unwanted pregnancies.

To briefly review reproductive health and rights, the term means to reframe concerns of each individual's sexuality and fertility from the perspective of health over one's lifetime, and to guarantee such health as a basic human right.[1] Although 'individual' includes both women and men, reproductive health and rights are of far greater importance to women because only women experience pregnancy and childbirth, and because gender discrimination puts women in disadvantageous circumstances in many respects including health, medical care, education, and employment.

In 1999, five years after the ICPD, a special session of the UN General Assembly called ICPD+5 was held to review the degree of progress made on the Cairo consensus, and the session found that many governments and NGOs are endeavoring to promote reproductive health and rights. Although indicators on reproductive health and rights and on population (Table 1) yield statistics that generally indicate improvement, ICPD+5 expressed concerns that a number of issues were more urgent than before, and that the funding needed for program implementation was far short of the goal agreed upon in Cairo.

Tasks of great urgency included (1) decreasing the maternal mortality and morbidity rates, (2) satisfying unmet contraception needs, (3) addressing adolescents' needs, and (4) preventing the spread of HIV/AIDS. About 20% of women's illnesses arise from pregnancy, childbirth, and abortions. Worldwide, at least 585,000 women die each year from complications of pregnancy and childbirth, and about 78,000 of them are caused by complications arising from unsafe abortions. One of the goals

TABLE 1. Reproductive Health and Rights, and Population-Related Indicators

| | 1<br>Total population (millions, 2001) | 2<br>Average population growth rate (%, 2000–05) | 3<br>Total fertility rate (2000–2005) | 4<br>Contraceptive prevelance Modern methods (%) | 5<br>% births with skilled attendants | 6<br>Infant mortality per 1,000 live births | 7<br>Maternal mortality ratio per 100,000 live births | 8<br>% Illiterate (>15 years) M/F | 9<br>GNI per capita ppp$ (2001) |
|---|---|---|---|---|---|---|---|---|---|
| China | 1,285.00 | 0.7 | 1.8 | 83 | 67 | 37 | 60 | 8/23 | 3,550 |
| Indonesia | 214.8 | 1.2 | 2.27 | 55 | 56 | 40 | 470 | 8/17 | 2,660 |
| Rep. of Korea | 47.1 | 0.7 | 1.51 | 67 | 98 | 7 | 20 | 1/03 | 15,530 |
| Malaysia | 22.6 | 1.7 | 2.9 | 30 | 96 | 10 | 39 | 8/16 | 7,640 |
| Philippines | 77.1 | 1.9 | 3.24 | 28 | 56 | 29 | 240 | 4/5 | 3,990 |
| Thailand | 63.6 | 1.1 | 2 | 70 | — | 21 | 44 | 3/6 | 5,950 |
| Vietnam | 79.2 | 1.3 | 2.25 | 56 | 77 | 34 | 95 | 4/8 | 1,860 |
| India | 1,025.1 | 1.5 | 2.97 | 43 | 43 | 65 | 440 | 31/54 | 2,230 |
| Japan | 125.6 | 0.1 | 1.36 | 55.9 | 100 | 3.2 | 6.6 | — | 25,170 |
| Taiwan | 22.2 | 0.8 | 1.5 | — | — | 6.9 | — | 3.7/10.4 | 12,916 |

Note: For Taiwan, items 1, 3, 8, and 9 are from 2000; item 2 is from 1999–2000, and item 6 is the 2001 estimate.

Source: Prepared using United Nations Population Fund, *The State of World Population Report 2001*. For Japan, item 1 is reported by the Statistics Bureau, Management and Coordination Agency, 2000. Items 3 and 6–7 are from the Ministry of Health, Labor and Welfare, 2000. Item 4 is from the National Survey on Family Planning, The Mainichi Shimbun, 2000.

Taiwan statistics are based on Yahoo Taiwan <http://tw.search.yahoo.com>, and CIA, *The World Factbook, 2002* <http://www.odci.gov/cia/publications/factbook/>.

TABLE 2. Countries by Restrictiveness of Abortion Law, According to Region, 1997

| Abortion restrictiveness | Countries/regions | No. of countries (% of world population) |
|---|---|---|
| Without restriction as to reason | US, Cuba, Scandinavia, Germany, France, Italy, Russian Fed., China, Vietnam, Turkey, South Africa, etc. | 49 (41%) |
| Socioeconomic grounds | UK, Finland, India, Japan, Taiwan, Zambia | 6 (20%) |
| Physical health | Argentina, Peru, Uruguay, Poland, Pakistan, Thailand, Morocco, Ethiopia, Mozambique, Rwanda, Republic of Korea, etc. | 23 (10%) |
| Mental health | Jamaica, Portugal, Spain, Switzerland, Australia, Malaysia, Iraq, Israel, Jordan, Gambia, Ghana, etc. | 20 (4%) |
| To save a woman's life | Ireland, Brazil, Chile, Colombia, Guatemala, Bangladesh, Indonesia, Philippines, Iran, Egypt, Kenya, Nigeria, Tanzania, etc. | 54 (25%) |

Notes: 1. Some countries that allow abortion without restriction as to reason do not specify gestational limit.
2. For socioeconomic, physical health, or mental health reasons, some countries include rape, incest, and fetal impairment.
3. When abortion is to save a woman's life, some countries include incest and fetal impairment.
4. Some countries also require spousal or parental authorization regardless of restrictiveness of law.

Source: Anika Rahman, Laura Katzive and Stanley K. Henshaw, "A Global Review of Laws on Induced Abortion, 1985–1997," *International Family Planning Perspectives*, Vol. 24, No. 2, 1998, The Alan Guttmacher Institute.

TABLE 3. Regional HIV/AIDS Statistics and Features, end of 2001

| | People living with HIV/AIDS (Adults and children) | Newly infected with HIV (Adults and children) | Adult prevalence rate (%) | % of HIV positive adults who are women | Young people (15–24 yrs old) | | |
|---|---|---|---|---|---|---|---|
| | | | | | Women | Men | Total |
| Sub-Saharan Africa | 28,100,000 | 3,400,000 | 8.4 | 55 | 5,700,000 | 2,800,000 | 8,600,000 |
| North Africa & Middle East | 440,000 | 80,000 | 0.2 | 40 | 110,000 | 41,000 | 150,000 |
| South and Southeast Asia | 6,100,000 | 800,000 | 0.6 | 35 | 930,000 | 590,000 | 1,500,000 |
| East Asia & Pacific | 100,000 | 270,000 | 0.1 | 20 | 87,000 | 200,000 | 280,000 |
| Latin America | 1,400,000 | 130,000 | 0.5 | 30 | 170,000 | 260,000 | 420,000 |
| Carribian | 420,000 | 60,000 | 2.2 | 50 | 72,000 | 59,000 | 130,000 |
| Eastern Europe & Central Asia | 1,000,000 | 250,000 | 0.5 | 20 | 85,000 | 340,000 | 420,000 |
| Western Europe | 560,000 | 30,000 | 0.3 | 25 | 33,000 | 55,000 | 89,000 |
| North America | 940,000 | 45,000 | 0.6 | 20 | 47,000 | 100,000 | 150,000 |
| Australia & New Zealand | 150,000 | 500 | 0.1 | 10 | — | — | — |
| Total | 40,000,000 | 5,000,000 | 1.2 | 48 | 7,300,000 | 4,500,000 | 11,800,000 |

Notes: 1. Figures on young people are approximate.

2. According to the Ministry of Health, Labor and Welfare's Committee on AIDS Trends, the totals of people in Japan with HIV are 3,145 Japanese (2,736 men, 409 women) and 1,653 foreigners (581 men, 1,072 women). Totals for people with AIDS are, respectively, 1,767 (1,630 men, 137 women) and 621 (422 men, 199 women). All statistics are as of June 30, 2002. Figures for HIV carriers and AIDS sufferers are given as provided by the Committee. HIV infections due to coagulation factor preparations are not included.

Source: Based on the Joint United Nations Program on HIV/AIDS (UNAIDS), Report on the Global HIV/AIDS Epidemic, 2001.

set forth at ICPD+5 was to increase the number of childbirths attended by skilled health personnel such as midwives and physicians to ensure greater safety. Eliminating unsafe abortions makes legalization indispensable, but in many countries it is completely illegal or strictly regulated, and in others where abortion is legal, laws are sometimes not very effective due to the lack of physicians and medical facilities (Table 2). No definite consensus on the legalization of abortion was forthcoming at ICPD+5 owing to strong opposition from Catholic and Islamic countries.

ICPD+5 also focused especially on HIV/AIDS and adolescent health. There is a rapid increase in HIV/AIDS particularly in Sub-Saharan Africa, and it continues to spread in Asia. The 15--24 age group, which is expected to soon be 20% of the world population, accounts for about one-third of the more than 40 million HIV infections (Table 3). Women are more susceptible than men to HIV/AIDS and other sexually transmitted infections (STIs) because of their biological characteristics and because of their subordinate social and sexual status owing to gender discrimination. It is widely known that violence against women, polygamy, harmful practices such as female genital mutilation, the sex trade, and other factors are closely related to HIV/AIDS among women. Additionally, women have limited access to information and medical care. Pandemics of other STIs have also reached a crisis stage.

Japan's maternal mortality and morbidity are far lower than those of other Asian countries, but administrative authorities focus primarily on maternal and child health (MCH), with little emphasis on family planning. The government's 'Healthy Parents and Children 21' initiative (a 2001--2010 national MCH campaign) represents a small measure of progress because it includes an adolescent health program. In Japan abortions are generally performed both safely and legally if they meet the conditions of the Maternal Protection Law, and as long as these conditions are met, women and doctors are immune from prosecution under the Criminal Abortion Law. But this dual legal system does not guarantee women's right of self-determination on their sexuality and fertility. Some women therefore propose abolishing both the Criminal Abortion Law and the Maternal Protection Law, and passing a new law on contraception and abortion, based on the idea of reproductive health and rights. A program covering HIV/AIDS and adolescents is a pressing need for Japan owing to increasing adolescent sexual activity, but government efforts are still inadequate. Gender issues are not given high priority, either. The Basic Law for a Gender-Equal Society was enacted in 1999, but critics say that the very term 'joint participation' used in the law[2] obfuscates the gender discrimination issue.

Even when seen globally, efforts to promote reproductive health and rights with emphasis on the gender perspective have just begun. World population is predicted to reach 8.9 billion in 2050 according to the UN's medium variant, but if women gain the right to choose whether or not to have children, and have only as many children as they want at adequate intervals, it is expected that the world total fertility rate will decline. The international community increasingly recognizes that gender equality and women's empowerment are crucial determinants that will influence population, development, and environmental issues.

Ashino Yuriko

# 3.    Preventing Occupational Injuries and Improving Occupational Safety and Health

The march of industrialization in Asia brings many workers face-to-face with a variety of safety and health risks. Because activities for occupational safety and health directly affect human lives, they are high on the agenda in many industries and workplaces even outside the manufacturing industry. In China about 10,000 workers lose their lives each year in mining accidents alone. Although the government closes very dangerous mines, the poor in those localities continue mining illegally and have accidents because of it. Even primary industries like farming pose many health and safety risks such as machinery accidents, exposure to pesticides and other chemicals, and lifting or moving heavy objects. The author recently visited a farm in southern Vietnam where over the previous two years three people had died owing to poorly maintained used farm machinery overturning on rough terrain and poor agricultural access roads.

The International Labour Organization (ILO) estimates that worldwide there are about 2 million fatalities each year because of work-related injuries. It has implemented a Global Safe Work Programme, which sets four priorities that are also priority issues for safety and health in Asia. This program directs resources preferentially to people who previously had few opportunities to receive appropriate protection and services despite facing safety and health risks. Specifically, the priority areas are: (1) action on mining, construction, agriculture, and other industries with high accident and injury rates, (2) expanded protection for small and medium-sized enterprises (SMEs) and the informal sector, (3) more initiatives on better health and welfare for workers, especially on health problems that have increasing seriousness in the workplace, such as stress, HIV/AIDS, drug dependence, and alcohol abuse, and (4) show employers the economic benefits of safety and health improvements to raise their awareness.

Asia's statistics for occupational accidents and illnesses are not available in many categories according to the country (Table 1). Japan and South Korea regularly maintain statistics and report them, while the considerable unevenness in the statistics from other countries requires ample care in their use when performing analyses and international comparisons. Generally, statistics are often unavailable on occupational accidents and illnesses for SMEs, the self-employed, the informal sector, and agriculture. Whether reliable statistics on occupational accidents and illnesses are regularly generated by a country is itself indicative of that country's degree of progress in addressing such problems.

Governments, workers' and employers' organizations, and NGOs are accordingly making broad efforts for safety and health improvements. Much success has been achieved in worksite-level improvement activities which focus on primary prevention. One question is how government policy measures will support and further worksite efforts. For instance, in 2002 Thailand established a five-year plan for occupational safety and health. As part of this endeavor the government has given priority to, and articulated specific measures for, tasks that include

TABLE 1. Occupational Injuries in Asian Countries

|  |  |  | China | India |
|---|---|---|---|---|
| A | Employment | | | |
| A1 | Employed workers | Thousands, 2000 | 711,500 | 27,941[a] |
|  | Men | Thousands, 2000 | — | 23,515[a] |
|  | Women | Thousands, 2000 | — | 4,426[a] |
| A2 | Working population by industry | Year of study | 2000 | |
|  | 1. Agriculture, forestry, fishing | Thousands | 333,550 | — |
|  | 2. Mining | Thousands | 5,970 | — |
|  | 3. Manufacturing | Thousands | 80,430 | — |
|  | 4. Utilities | Thousands | 240 | — |
|  | 5. Construction | Thousands | 35,520 | — |
|  | 6. Wholesale, retail, hotels, restaurants | Thousands | 46,860 | — |
|  | 7. Transport, storage, communications | Thousands | 20,290 | — |
|  | 8. Finance, insurance, real estate, business services | Thousands | 4,270 | — |
|  | 9. Public employees, social services | Thousands | 20,250 | — |
|  | 10. Others | Thousands | 161,500 | — |
| B | Occupational injuries | | | |
| B1 | Injured persons | Persons, 1996 | 29,036 | — |
|  |  | Persons, 1997 | 26,369 | — |
|  |  | Persons, 1998 | — | — |
|  |  | Persons, 1999 | — | — |
|  |  | Persons, 2000 | — | — |
| B2 | Total occupational deaths | Persons, 1996 | 19,457 | 229 |
|  |  | Persons, 1997 | 17,558 | 242 |
|  |  | Persons, 1998 | — | 211 |
|  |  | Persons, 1999 | — | — |
|  |  | Persons, 2000 | — | — |
| B3 | Occupational deaths by industry | Year of study | 1997 | |
|  | 1. Agriculture, forestry, fishing | Persons | 140 | — |
|  | 2. Mining | Persons | 4,403 | — |
|  | 3. Manufacturing | Persons | 4,384 | — |
|  | 4. Utilities | Persons | — | — |
|  | 5. Construction | Persons | 1,587 | — |
|  | 6. Wholesale, retail, hotels, restaurants | Persons | 345 | — |
|  | 7. Transport, storage, communications | Persons | 552 | — |
|  | 8. Finance, insurance, real estate, business services | Persons | — | — |
|  | 9. Public employees, social services | Persons | — | — |
|  | 10. Others | Persons | 493 | — |

Notes: (a) 1997. (b) 1999. (c) 1998. (d) Excludes hotels. (e) Includes finance, insurance, and real estate, but excludes hotels and restaurants. (f) Includes hotels. (g) Includes hotels and restaurants. (h) Includes finance, insurance, real estate, hotels, restaurants, and social services.

Source: ILO, *Yearbook of Labour Statistics*, 2001.

| Indonesia | Japan | S. Korea | Malaysia | Pakistan | Philippines | Thailand | Vietnam |
|---|---|---|---|---|---|---|---|
| 88,817[b] | 64,460 | 21,061 | 9,321.7 | 35,934[c] | 27,775 | 33,001.0 | 33,664[a] |
| 54,908[b] | 38,180 | 12,353 | 6,096.2 | 30,927[c] | 17,258 | 18,164.9 | — |
| 33,908[b] | 26,300 | 8,708 | 3,235.5 | 5,007[c] | 10,516 | 14,836.1 | — |
| 1999 | 2000 | 2000 | 2000 | 2000 | 2000 | 2000 | 1997 |
| 38,378 | 3,260 | 2,288 | 1,711.8 | 16,980 | 10,401 | 16,095.5 | — |
| 726 | 50 | 18 | 27.3 | 69 | 106 | 38.9 | — |
| 11,516 | 13,210 | 4,243 | 2,125.8 | 3,579 | 2,792 | 4,784.8 | 3,293 |
| 188 | 340 | 63 | 48.1 | 251 | 116 | 172.5 | — |
| 3,415 | 6,530 | 1,581 | 798.9 | 2,251 | 1,430 | 1,280.0 | 977 |
| 17,529 | 14,740[d] | 5,729 | 1,790.1 | 4,984 | 4,587 | 4,801.5[e] | — |
| 4,207 | 4,140 | 1,265 | 422.7 | 1,970 | 2,024 | 951.2 | 856 |
| 634 | 6,160 | 2,085 | 462.0 | 312 | 678 | — | — |
|  |  |  |  |  |  |  |  |
| 12,224 | 15,640[f] | 2,328 | 1,935.1 | 55,231 | 5,636 | 4,864.8[a] | — |
| — | 390 | 1,460 | — | 17 | 4 | 10.2 | — |
|  |  |  |  |  |  |  |  |
| 10,037 | 160,712 | 71,548 | 106,508 | 402 | 50,320 | — | — |
| 8,727 | 154,490 | 66,770 | 86,589 | 376 | — | — | — |
| — | 144,838 | 51,514 | 85,338 | 360 | — | — | — |
| — | 135,836 | 55,405 | 92,074 | 257 | — | 52,025 | — |
| — | 133,948 | — | — | — | — | 50,731 | — |
| 784 | 2,363 | 2,670 | 1,020 | 143 | 240 | 962 | — |
| 1,076 | 2,078 | 2,742 | 1,473 | 90 | — | 1,033 | — |
| — | 1,844 | 2,212 | 1,273 | 89 | — | 790 | — |
| — | 1,992 | 1,412 | 912 | 93 | — | 610 | — |
| — | 1,889 | 1,353 | — | — | — | 616 | — |
| 1997 | 2000 | 1998 | 1995 | 1999 | 1996 | 2000 |  |
| 1,794 | 6,213 | 919 | 20,465 | — | 8,700 | 5,791 | — |
| 587 | 760 | 1,134 | 1,016 | 122 | 370 | 528 | — |
| 4,486 | 37,673 | 22,446 | 62,483 | 135 | 28,220 | 32,155 | — |
| 844 | 80 | 127 | 542 | — | 540 | 417 | — |
| 629 | 33,599 | 13,172 | 4,406 | — | 1,860 | 3,652 | — |
| — | 17,054 | — | 10,187 | — | — | 4,659 | — |
| 286 | 17,242 | 4,657 | 4,826 | — | 3,170 | 1,625 | — |
| — | 1,173 | 1,578 | 4,084 | — | 880 | 1,466 | — |
|  |  |  |  |  |  |  |  |
| — | 3,487 | — | 5,320 | — | — | 183 | — |
| 207 | 16,667 | 8,602[g] | 805 | — | 6,580[h] | 255 | — |

enhancing the legal framework, helping SMEs, developing an effective labor inspection system, improving the system for reporting occupational accidents and illnesses, and providing safety and health support for workers and employers.

In 2001 the ILO created the ILO Occupational Safety and Health Management System Guidelines, and is promoting them in Asian countries. They are designed to be useful as practical guidelines for implementing safety and health activities in a wide variety of workplaces including SMEs. That same year China's government created its own guidelines based on those of the ILO. Thailand and Malaysia are also considering the formulation of their own guidelines. Vietnam is using the ILO guidelines to train SME owners and make substantive improvements in safety and health.

The role of NGOs is more important than ever for making health and safety progress in the informal sector. A good example is a Thai NGO called HomeNet. Since the 1997 economic crisis many workers have lost their jobs and are working at home. At first HomeNet's main purpose was to expand markets for the products of these cottage industries, but later it focused on the chemicals that people took into their home workplaces and the importance of decreasing the safety and health risks of using production machinery. In cooperation with the ILO, the Thai government is conducting participatory safety and health training for home workers' cottage industries around the country. In Cambodia local NGOs are playing an active role in eliminating the most harmful forms of child labor. First, NGO members themselves receive safety and health training, then become child labor monitors who visit workplaces where children often work, and implement whatever measures are possible locally to mitigate safety and heath risks.

From now on it will be crucial that steady community-based efforts for improvement make headway and become networked. An important challenge will be how government policy can lend concrete assistance to these worksite activities and broaden their base.

Kawakami Tsuyoshi

# 4.    Improving Health and Education

Asia has achieved much progress in health and education thanks to socioeconomic development, which is evident from declining infant mortality, and especially from the rising adult literacy rate. But many Asian countries and regions have yet to achieve much in the way of such development.

The relationship between the female adult literacy rate and the infant mortality rate in a number of Asian countries shows there is a strong correlation between them (Fig. 1). Moreover, these countries fall into two distinct groups depending on whether or not they have achieved progress in these areas. At the same time, macro indicators functioning on the level of individual countries are often poor indicators of what is happening at the bottom level of society and among the poor. Even within a single country there are wide differentials between the poor and people who are well off. The challenge is how to approach front-line problems, find measures that can be implemented, and achieve steady progress. Also important is how policy can actually support these local efforts to improve health and education.

Infrastructural improvements related to health issues including maternal and child health, combatting infections, and securing safe drinking water are still of crucial importance. Additionally, Asia has peculiar health problems that demand action tailored to each one, and those with priority include HIV/AIDS, equal medical care services for all citizens, and occupational safety and health (see III-3). HIV/AIDS efforts should focus on the poor because they account for many of the newly infected, but at the same time those already infected need to be helped by eliminating discrimination and enhancing medical care. Malaysia endeavors to stop discrimination with workplace guidelines for dealing with

Fig. 1. Adult Female Literacy Rates and Infant Mortality Rates (2000)

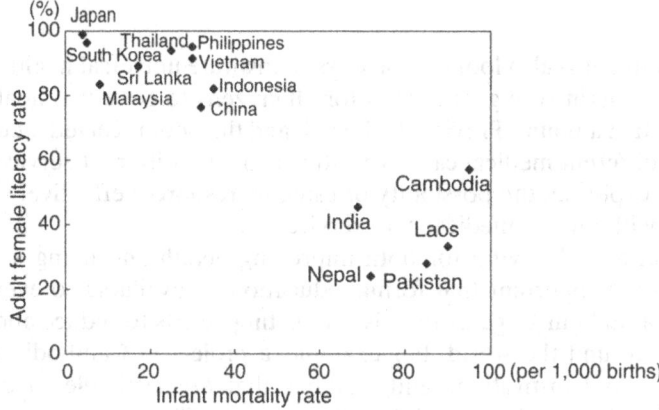

Source: Based on UNDP, *Human Development Report*, 2002.

TABLE 1. Health and Education

| | | China | India |
|---|---|---|---|
| A | Socioeconomic indicators | | |
| A1 | Population | Millions, 2000 | 1275.1 | 1,008.9 |
| A2 | Per capita GDP | US$, 2000 | 3,976 | 2,358 |
| A3 | Share of income or consumption(%) | Survey year | (1998) | (1997) |
| | Poorest 20% | | 5.9 | 8.1 |
| | Richest 20% | | 46.6 | 46.1 |
| B | Health and population indicators | | |
| B1 | Life expectancy at birth, males | Years, 2001 | 69.8 | 60 |
| | Life expectancy at birth, females | Years, 2001 | 72.7 | 61.7 |
| B2 | Under-five mortality rate, males | Per 1,000 live births, 2001 | 34 | 89 |
| | Under-five mortality rate, females | Per 1,000 live births, 2001 | 40 | 98 |
| B3 | Children underweight at birth | %, 1995–2000 | 6 | 26 |
| B4 | Infant mortality rate | Per 1,000 live births, 2000 | 32 | 69 |
| B5 | Maternal mortality | Per 100,000 live births, 1990–99 | 55 | 540 |
| B6 | One-year olds fully immunized against measles | %, 1999 | 99 | 50 |
| B7 | HIV/AIDS prevalence (adults) | %, 15–49 years, 2001 | 0.11 | 0.79 |
| B8 | Physicians | Per 100,000 people, 1990–99 | 162 | 48 |
| B9 | Health expenditures % of GDP | %, 2000 | 5.3 | 4.9 |
| C | Education indicators | | |
| C1 | Net primary education enrollment | %, 1998 | 91 | — |
| C2 | Children reaching grade 5 | %, 1995–97 | 94 | — |
| C3 | Net enrollment in secondary schools | %, 1998 | 50 | 39 |
| C4 | Adult literacy, men | %, ages 15 and above, 2000 | 91.2 | 68.4 |
| | Adult literacy, women | %, ages 15 and above, 2000 | 76.3 | 45.4 |

Sources: A1-A3, B-3-B8, and C1-C4 are from UNDP, *Human Development Report*, 2002. B1, B2, and B9 are from WHO, *World Health Report*, 2002.

HIV/AIDS, and it is also looking for ways to institute universal health insurance. Thailand is experimenting with ways for all citizens to get equal health care by paying a certain amount. Previously the rich and the poor received medical examinations at different medical care institutions and got different services, but now Thailand is exploring the possibility of using its resources effectively to provide all citizens with quality medical care services.

Education for all is vital for both improving health and living a better life, which makes it important that formal education be available to everyone. The International Labour Organization is conducting efforts to reduce and eliminate child labor around the world. For example, a project in Cambodia found that upgrading school institutions and facilities that are available to everyone is indispensable for eliminating child labor. An insufficient government budget translates into low salaries for teachers, and a lack of public funds for purchas-

| Indonesia | Japan | S. Korea | Malaysia | Pakistan | Philippines | Thailand | Vietnam |
|---|---|---|---|---|---|---|---|
| 212.1 | 127.1 | 46.7 | 22.2 | 141.3 | 75.7 | 62.8 | 78.1 |
| 3,043 | 26,755 | 17,380 | 9,068 | 1,928 | 3,971 | 6,402 | 1,996 |
| (1999) | (1993) | (1993) | (1997) | (1996–97) | (1997) | (1998) | (1998) |
| 9.0 | 10.6 | 7.5 | 4.4 | 9.4 | 5.4 | 6.4 | 8.0 |
| 41.1 | 35.6 | 39.3 | 54.3 | 41.1 | 52.3 | 48.4 | 44.5 |
| 64.4 | 77.9 | 71.2 | 69.2 | 61 | 64.2 | 65.7 | 66.9 |
| 67.4 | 84.7 | 78.7 | 74.4 | 61.5 | 71.5 | 72.2 | 71.8 |
| 50 | 5 | 8.0 | 13 | 105 | 46 | 38 | 44 |
| 40 | 4 | 8.0 | 11 | 115 | 33 | 31 | 35 |
| 9 | 7 | — | 9 | 21 | 18 | 7 | 9 |
| 35 | 4 | 5.0 | 8 | 85 | 30 | 25 | 30 |
| 380 | 8 | 20.0 | 41 | — | | | |
| 71 | 94 | 85.0 | 88 | 54 | 79 | 96 | 93 |
| 0.1 | >0.10 | >0.10 | 0.35 | 0.11 | >0.10 | 1.79 | 0.30 |
| 16 | 193 | 136 | 66 | 57 | 123 | 24 | 48 |
| 2.7 | 7.8 | 6.0 | 2.5 | 4.1 | 3.4 | 3.7 | 5.2 |
| — | 100 | 97 | 98 | — | — | 77 | 97 |
| 88 | — | 98 | — | — | — | — | — |
| — | — | — | 93 | — | — | 55 | 49 |
| 91.8 | 99 | 99.1 | 91.4 | 57.5 | 95.5 | 97.1 | 95.5 |
| 82 | 99 | 96.4 | 83.4 | 27.9 | 95.1 | 93.9 | 91.4 |

ing teaching materials imposes burdens on parents. Children cannot attend school if their parents cannot pay, which feeds into a vicious circle that creates the conditions for child labor.

There is also a need to address the inequality in educational opportunity between women and men, which is illustrated by the ratio of male to female literacy rates (Fig. 2). Low female literacy rates in some countries are testament to the need for equal educational opportunities.

Asian countries are exploring a variety of educational activities outside of the schools. To help close the digital divide, the United Nations Development Programme is using a special bus in Malaysia which moves from village to village to teach people about the Internet. China and other countries are experimenting with distance education. Other important efforts that should be made are using workplaces to conduct technical education for workers, and the creation and

FIG. 2. Literacy Rates (Male-to-Female Ratio, 2000)

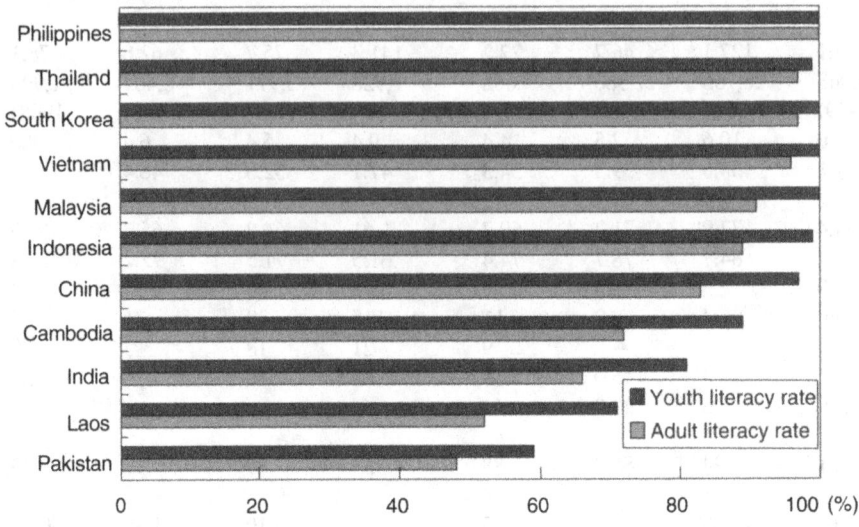

enhancement of night classes for young people so they can study while holding down jobs. Various efforts are under way to provide equal educational opportunities to men and women, and to create educational opportunities that are available to all, while tailoring education to the circumstances of each locale.

Kawakami Tsuyoshi

# 5.    Disputes over Regulating GM Crops and Food

Asia is an important region to the biotechnology industry because it is the focal point of two issues. First is resource access and the sharing of profits because Asia is a gold mine of diverse biological resources.[1] Second is the issue of regulating imports of GM crops and foods from North and South America because Asia is a huge and quickly growing market. Although it is easy to be distracted by the clash over regulations on GM products and labeling between the EU and the US, Asian trends also call for attention because Asia itself will have a major influence including what happens in China, a rapidly growing GM producer, and in India, whose trends in this area are being closely watched.

As in other parts of the world, Asian countries have become dependent on imports of corn and soybeans from the US. Although China and Thailand do barely manage to help supply corn in the region, most countries are dependent on imports from the US for the better part of their soybeans (Table 1). In 2002/03, 40% of the corn and 81% of the soybeans produced by the US were GM crops.[2] Although some countries are increasing their imports from Argentina and Brazil, these two countries are also major GM crop producers. Ninety percent of Argentina's soybeans and 30% of its corn are genetically modified. In Brazil, meanwhile, there is conflict over a ban on GM crops between the federal government and some state governments, while industry is facing off with consumers and environmental organizations.[3] Now that its new administration officially approved the planting of GM soybeans in 2003, however, Brazil has emerged as the fourth-biggest producer of GM crops.[4]

As in EU countries, the consumer movement in Japan showed increasing concern for the effects of GM crops on food safety and the environment, which spurred more discussion of labeling in 1997 and 1998, and led to the revised Law Concerning Standardization and Proper Labeling of Agricultural and Forestry Products, also known as the JAS Law, in April 2000 and to amendments to standards under the Food Sanitation Law the following April. These changes brought about mandatory labeling (5% threshold,[5] with some exceptions, such as soy oil, corn oil, corn syrup, and soy sauce) under control of the Ministry of Agriculture, Forestry and Fisheries, and the former Ministry of Health and Welfare. There is also discussion on requiring animal feed safety inspections (beginning in 2003), which came about in light of the Starlink issue that arose in 2000 and exposed the fragility of the existing GM food regulatory system both in the US and Japan.

A similar trend is observed in South Korea. In February 2001 after Starlink was found in imported corn, the government announced its intention to require labeling for GM soybeans and corn, and planned to implement the requirement with a 3% threshold for unprocessed items in March, for processed items in July, and for potatoes the following March. Despite a six-month probationary period, the requirement is being steadily implemented under the Ministry of Agriculture and Forestry and the Food and Drug Administration. In a bid to require safety screenings that have been voluntary since 2002, the Act on Transboundary Movement of Living Modified Organisms (LMO Act), issued by the Ministry of Com-

TABLE 1.  Asian Soybean and Corn Imports: High Dependence on the US (1,000 t)

Soybeans

| | 2000/01 | | | 2002/03 | | | Self-sufficiency rate in 2002 |
|---|---|---|---|---|---|---|---|
| | Imports from the US | Total imports | % from the US | Imports from the US | Total imports | % from the US | |
| Japan | 3,566 | 4,767 | 74.8 | 3,894 | 5,023 | 77.5 | 5.1 |
| S. Korea | 1,190 | 1,388 | 85.7 | 1,300 | 1,500 | 86.7 | 7.1 |
| China | 6,222 | 13,246 | 47.0 | 6,650 | 16,250 | 40.9 | 50.3 |
| Indonesia | 1,317 | 1,317 | 100.0 | 1,410 | 1,650 | 85.5 | 32.4 |
| Malaysia | 199 | 601 | 33.1 | 270 | 775 | 34.8 | 0.0 |
| Philippines | 350 | 420 | 83.3 | 150 | 250 | 60.0 | 0.4 |
| Taiwan | 1,985 | 2,330 | 85.2 | 1,923 | 2,200 | 87.4 | 0.0 |
| Thailand | 716 | 1,286 | 55.7 | 800 | 1,650 | 48.5 | 13.2 |
| Vietnam | 15 | 515 | 2.9 | 20 | 990 | 2.0 | — |

Corn

| | 2000/01 | | | 2002/03 | | | Self-sufficiency rate in 2002 |
|---|---|---|---|---|---|---|---|
| | Imports from the US | Total imports | % from the US | Imports from the US | Total imports | % from the US | |
| Japan | 14,570 | 16,275 | 89.5 | 13,950 | 15,500 | 90.0 | 0.0 |
| S. Korea | 3,289 | 8,723 | 37.7 | 300 | 8,900 | 3.4 | 0.8 |
| China | 0 | 50 | — | 0 | 50 | — | 102.5 |
| Indonesia | 539 | 1,337 | 40.3 | 100 | 1,200 | 8.3 | 84.1 |
| Malaysia | 18 | 2,613 | 0.7 | 15 | 2,460 | 0.6 | 2.8 |
| Philippines | 79 | 181 | 43.6 | 135 | 215 | 62.8 | 94.3 |
| Taiwan | 5,003 | 5,074 | 98.6 | 4,080 | 4,900 | 83.3 | 1.0 |
| Thailand | 0 | 6 | — | 0 | 10 | — | 100.0 |
| Vietnam | 0 | 210 | — | 0 | 250 | — | 90.4 |

Note:  Market year in each country: October through September in Japan, China, Indonesia, South Korea, Malaysia, and Taiwan. September through August in Thailand (soybeans). January through December in Thailand and the Philippines (soybeans). July through June in the Philippines (corn) and Thailand (corn).
Source:  USDA-FAS, *Gain Report*, for each country.

merce, Industry and Energy, is expected to go into effect in 2004 and require mandatory environmental risk assessments of GM crops and food.[6]

In February 2001 the Department of Health under Taiwan's Executive Yuan announced regulations on the registration and labeling of GM products. Voluntary labeling for non-GM products was confirmed retroactively to the previous month, and mandatory labeling for GM products was gradually enacted starting in January 2003 beginning with unprocessed GMOs; primary processed GMO labeling started in January 2004, and in January 2005 all processed GMOs must be labeled (this regulation will not apply to products which do not contain trans-genes or protein, e.g., corn starch, corn syrup, corn oil, soy oil, and soy sauce). The Legislative Yuan passed a law requiring a 1% threshold, but in October 2001 it confirmed the 5% level sought by the Department of Health.[7]

On the other hand, China is in the spotlight as a GMO producer ranking fifth after the US, Argentina, Canada, and Brazil. Commercial GM crop cultivation is currently limited to insect-resistant cotton, but huge fiscal investments are financing vigorous research and development for GMOs and genomics at universities and research institutes. China also issued and implemented the Regulation on the Safety Administration of Agricultural GMOs (June 2001). In accordance with this regulation, in January 2002 the Ministry of Agriculture issued three Agricultural GMO implementation measures: Safety Evaluation, Import Safety Administration, and Labeling.[8] Although the overall trend is toward stricter regulations on GMOs, it is not easy to say how these measures will actually be implemented because the relevant policies have often changed, and clear and consistent information about these policies is limited.

ASEAN countries are also making progress in regulating GMOs.[9] The Senior Officials Meeting of the ASEAN Ministers on Agriculture and Forestry and the ASEAN Senior Officials on Environment have taken the lead in creating the ASEAN Guidelines on Risk Assessment of Agriculture-Related GMOs (1999) and in holding workshops such as Capacity Development for the Integrated Approaches to Biosafety of GMOs (organized by UN University in 2001). These efforts were based on the 'Singapore Initiative' for harmonizing regulations among ASEAN members, which was agreed upon in 1997. However, legislation is considerably delayed in the three lesser developed countries of Lao PDR, Cambodia, and Myanmar. This is bringing about a widening gap between them and Malaysia, the Philippines, Indonesia, and Thailand, which have devised a variety of regulatory measures while trying to build their biotechnology industries.

In Malaysia, for example, the Ministry of Science, Technology and the Environment (MOSTE) has assumed responsibility for developing a biosafety policy. Underpinning this policy are existing laws such as the Food Law and the Plant Quarantine Law, as well as the National Guidelines on GMO Environmental Releases enacted in 1997 by the Genetic Modification Advisory Committee located in MOSTE, and the Biosafety Law that was submitted to the Cabinet in 2003 and is undergoing revisions following the public consultation process. Meanwhile, regulatory legislation on GMOs including imported agricultural produce had been on the back burner, but a GMO labeling regulation drafted by the Ministry of Health is expected to become law similar to that in Japan, although the threshold is likely to be set at 3%.

The Philippines is home to the International Rice Research Institute, which along with government research institutes, universities, and other institutions is vigorously pursuing biotechnology research. Further, the Philippines was the first ASEAN country to start developing a biotechnology regulation policy, a responsibility that was assumed by the National Biosafety Committee of the Philippines,

TABLE 2. Regulation of GMOs in Asia

| | | Japan | S. Korea | China | Taiwan | Brunei Darussalam |
|---|---|---|---|---|---|---|
| | Commercial cultivation | $\Delta^{(a)}$ | O | O | ? | × |
| | Food distribution | O | O | O | O | $\Delta^{(b)}$ |
| Biosafety legislation | | O | In progress | O | O | × |
| Safety screening guidelines | Food | O | O | O | O | × |
| | Environment | O | O | O | O | × |
| | Laboratories | O | O | O | O | × |
| Field trials | | O | O | O | O | × |
| GMO authorization | | O | O | O | O | × |
| GMO labeling (threshold) | | O (5%) | O (3%) | × | O (5%) | × |

Note: (a) Legally permitted, but not commercially cultivated except on an experimental basis. (b) Unconfirmed, but quite possibly in distribution. (c) Being studied by government agencies.
Source: For ASEAN countries, see: Sakarindr Bhumiratana, *Report on Biosafety Policy Options and Capacity Building Related to Genetically Modified Organisms in the Food Processing Industry of ASEAN*, June 2002.

created in 1990. In accordance with unified guidelines (Administrative Order No. 8: Rules and Regulations for the Importation and Release into the Environment of Plants and Plant Products), which were approved in April 2002, already 10 safety tests have been completed on GMOs for direct use as food, feed, or processing, one of which (Monsanto's insect-resistant Bt corn) is also approved for cultivation within the Philippines. However, legislation on labeling is overdue, and until now labeling has been voluntary.

Indonesia has regulated biosafety since 1997 under the Ministerial Decree on Genetically Engineered Biotechnology Products (Ministry of Agriculture), which was combined with other ministerial decrees to govern the food safety aspect of GM products in 1999. As of February 2001, Indonesia had already approved field tests of insect-resistant cotton developed by the US company Monsanto (limited to seven districts of South Sulawesi). In response to this, 72 organizations including domestic and foreign NGOs filed a lawsuit in June 2001, but the government and the Ministry of Agriculture intend to go ahead. Meanwhile, in March 2002 the government announced its intention to formulate and implement rules for mandatory GMO labeling. A threshold of 5% is expected.

Thailand conducted field tests on long-life tomatoes in 1994 and insect-resistant cotton in 1995, but commercial cultivation has yet to come. Although it has put off signing and ratifying the Biodiversity Convention, it is making efforts for the conservation and use of biological resources, and its focus now is regulating the environmental release of GMOs while allowing GM soybeans and corn into the country for use as food, feed, and processing. The basis of such regulation will likely be the Plant Quarantine Law, which was revised in 1999, as well as biosafety laws and labeling rules (5% threshold), which are currently under development.

Japan and South Korea have many citizens' and environmental organizations that engage in dynamic anti-GMO movements, but such opposition movements can also be found in Thailand and the Philippines where Greenpeace's Southeast Asian chapter is active, and where peasants' organizations and domestic NGOs like the Farmers-Scientists Partnership for Development (MASIPAG) come

| Cambodia | Laos | Myanmar | Indonesia | Malaysia | Philippines | Singapore | Thailand | Vietnam |
|---|---|---|---|---|---|---|---|---|
| × | × | × | ○ | × | × | × | × | × |
| ○ | ○ | △[b] | ○ | ○ | ○ | ○ | ○ | ○ |
| × | In progress | × | ○ | In progress | In progress | × | In progress | In progress |
| × | × | × | ○ | ○ | ○ | × | ○ | × |
| × | × | × | ○ | ○ | ○ | ○ | ○ | × |
| × | × | × | ○ | ○ | ○ | ○ | ○ | × |
| × | × | × | × | × | ○ | △[c] | ○ | × |
| × | × | × | ○ | ○ | ○ | × | ○ | × |
| × | × | × | ○ (5%) | ○ (3%) | In progress | × | ○ (5%) | In progress |

together. The biotechnology regulation policies being developed in these countries are one achievement of such movements, but there are of course strong backlashes from the US and the biotechnology industry. For example, in January 2002 the Office of the US Trade Representative (USTR) sent a high-ranking official to pressure South Korea into raising its 3% threshold for GMO labeling to the same 5% as Japan.[10] Another instance is that in response to stagnating imports of US soybeans and corn by China after that country established GMO rules, in February and March 2002 USTR and the US Department of Agriculture conducted a series of negotiations with China, and on the occasion of President Bush's visit to China, they pressured China to delay its GMO rules and to simplify approval procedures.[11] After the US pointed out a WTO infraction, China protested by saying that its regulations were based on international practice, and were the same as the rules employed by the EU and other Asian countries. Nevertheless, China was obliged to make some concessions, such as relaxing approval procedures and putting off application of the rules.

Monsanto and other major biotechnology companies are waiting to bring out the bioengineered wheat and rice that they have developed. Although wheat is traded in the same way as soybeans, rice is a primary crop and staple food of Asian countries. And in addition to cotton, which is already under commercial cultivation in the region, research and development are proceeding apace on papayas and sweet potatoes (Table 3). There is no reason to expect that impact assessment methods used in North America, which has low biodiversity, can be employed without modification in Asian countries, where the plants for major crops originated. Moreover, GMOs will have an enormous impact on the agricultural structures of Asian countries, whose many small farmers are vital actors in rural economies and societies. It is crucial that instead of leaving GMO regulation to individual countries, these regulations are carefully established and implemented as a joint endeavor by Asian countries and, on a worldwide level, by the international community.

Hisano Shuji

TABLE 3.  Examples of GM Crops with Relevance to Asian Countries

| Crop | Country | Improved trait | Comments | Stage |
|---|---|---|---|---|
| Cotton | India | Pest resistance (Bt) | Approval to grow Bt cotton developed by the company Monsanto granted in March 2002. | Commercialized |
| | Indonesia | Pest resistance (Bt) | 2,700 farmers grow Bt cotton in South Sulawesi. | Commercialized |
| | China | Pest resistance (Bt) | Both locally developed varieties and those by the company Monsanto are grown. | Commercialized |
| Maize | Philippines | Pest resistance (Bt) | Field studies from 2000, multi-site trials from 2001. | Field studies |
| Papaya | Malaysia, Thailand, Philippines, Vietnam, Indonesia | Viral resistance (ringspot virus) | Project undertaken by the Papaya Biotechnology Network of Southeast Asia with support from the company Monsanto and the ISAAA. | Commercialized |
| Rice | India and US | Dwarfing | Gene from *Arabidopsis* transferred into Basmati Rice. | Laboratory studies |
| | Philippines | Micronutrient enrichment (vitamin A precursors, or β-carotene) | Project undertaken by the Golden Rice Network (India, China, Indonesia, Vietnam, Bangladesh, the Philippines and South Africa). Collaborators include the International Rice Research Institute (IRRI), the Rockefeller Foundation and the company Syngenta. | Laboratory studies |

| | | | |
|---|---|---|---|
| S. Korea and US | Abiotic stresses | Salt, drought and cold tolerance. Research undertaken by Cornell University and researchers in South Korea with funding from the Rockefeller Foundation. | Laboratory studies |
| Philippines | Micronutrient enrichment (iron and zinc) | Research undertaken by the Institute of Human Nutrition, at the University of the Philippines, in cooperation with IRRI. | Field studies |
| Philippines and US | Bacterial resistance (bacterial leaf blight) | Gene patented by the University of California, Davis, has been made available free of charge. Field trials conducted by IRRI. | Field studies |
| Sweet potato Vietnam | Pest resistance (Bt) | Research undertaken by the Institute of Biotechnology in Hanoi, Vietnam. Bt strains were donated free of charge by the company Novartis. | Laboratory studies |

Note: Examples of Japanese research and development not included. ISAAA is an international agency whose purpose is facilitating agricultural biotechnology between the North and South. The International Rice Research Institute (IRRI) is located in the Philippines. R&D on GM wheat is conducted mainly in North America.

Source: Excerpts from: Nuffield Council on Bioethics, *The Use of Genetically Modified Crops in Developing Countries: A follow-up Discussion Paper*, June 2003, Appendix 3 (pp. 97–104).

# 6.    Forest Resources on the Decline

Of the greatest importance when considering the state of the forests is that everyone shares clear definitions, assessment methods, and numerical data based on
credible sources. Data in the UN Food and Agriculture Organization's (FAO)
*Global Forest Resources Assessment* (FRA 2000) fulfill these requirements with
a certain degree of reliability. These data were based on reports from countries
around the world, remote sensing data, and other sources, which were vetted by
specialists before appearing in the assessment. This report shall examine Asia's
forests using the FAO's State of the World's Forests 2001,[1] which presents the
results of FRA 2000.

FRA 2000 defines 'forest' as 'tree canopy cover of more than 10 percent and
area of more than 0.5 ha.' Even an area with 10% tree canopy cover and tree
height of under 5 m, such as a recently logged area, is considered forest if that
state is temporary. Both natural forests and plantations are 'forest' no matter how
they are used. Thus, a shelterbelt protecting farmland is also considered forest if
it has a width of 20 m or more and meets the above definition. And with the publication of FRA 2000 even plantations of rubber trees and cork oaks are included
in forests. Excluded are trees for agricultural produce such as orchards, and trees
planted in agroforesty systems.

Let's consider an example. If tree canopy cover remains less than 10% for a
time, that stand of trees is not defined as forest and it is statistically gone, but if
canopy cover drops from 80% (closed forest) to 30% (open forest), it is still forest
by definition. Such changes and decreasing biomass indicate a qualitative deterioration called forest degradation.

Forests not only provide us with wood products and non-wood products (such
as mushrooms), they also recharge aquifers, conserve soil, fix carbon, preserve
biodiversity and genetic resources, provide places for cultural activities and recreation, and have other functions. As such, forest loss means losing these functions and thereby raising serious concerns for the survival of humans and other
organisms.

FRA 2000 says that world forest area is about 3.87 billion ha (Table 1), of which
56% is in tropical and subtropical regions. Plantations account for a mere 5% of
all forests. Thirty percent of the world's land is forested, which means that world
per capita forest is 0.65 ha. Average forest stock per ha is 100 cubic meters according to FRA 2000, so if we assume that all forests could be used, each human
should be limited to consuming a maximum of 65 cubic meters in his or her
lifetime.[2]

Forest area will change owing to the new definition of forest in the 2000 assessment. Forest area in 1990 according to the old and new definitions was about 3.51
billion ha and 3.96 billion ha, respectively. Statistically, this indicates a 450 million
ha increase by counting increased forested area in developed countries and the
addition of rubber plantations and the like to the count. Under the new definition, forest area decreased by 94 million ha from 1990 to 2000, for a −0.24%
annual rate of change.

TABLE 1. Forest Cover in the World

| Region | Forest area, 2000 (million ha) | % of forest area of land | Annual rate of change (%) | Natural forests (million ha) | Forest plantations (million ha) | % of plantations of forest area | % of each area of forest plantations | Wood biomass in forests (million tons) |
|---|---|---|---|---|---|---|---|---|
| Africa | 650 | 22 | -0.8 | 642 | 8 | 1 | 17 | 70,917 |
| Asia | 548 | 18 | -0.1 | 432 | 116 | 21 | 12 | 45,062 |
| Europe | 1,039 | 46 | 0.1 | 1,007 | 32 | 3 | 27 | 61,070 |
| North and Central America | 549 | 26 | -0.1 | 532 | 18 | 3 | 14 | 52,357 |
| Oceania | 198 | 23 | -0.2 | 194 | 3 | 2 | 5 | 12,640 |
| South America | 886 | 51 | -0.4 | 875 | 10 | 1 | 24 | 180,210 |
| Total | 3,869 | 30 | | 3,682 | 187 | 5 | 100 | 422,256 |

Source: FAO, *State of the World's Forests*, 2001.

259

TABLE 2. Forest Changes and Economy in Asia

|  |  | Land area in 1999 (1000 ha) | Population in 1999 (millions) | Population density in 1999 (persons/ha) | GDP per capita in 1997 (US$) |
|---|---|---|---|---|---|
| Total |  | 13,063,900 | 5,978.14 | 0.46 | — |
| South Asia | Bangladesh | 13,017 | 126.947 | 9.75 | 352 |
|  | India | 297,319 | 998.056 | 3.36 | 392 |
|  | Nepal | 14,300 | 23.385 | 1.64 | 216 |
|  | Pakistan | 77,088 | 152.331 | 1.98 | 502 |
|  | Sri Lanka | 6,463 | 18.639 | 2.88 | 770 |
| Continental | Cambodia | 17,652 | 10.945 | 0.62 | 303 |
| Southeast | Laos | 23,080 | 5.297 | 0.23 | 414 |
| Asia | Myanmar | 65,755 | 45.059 | 0.69 | — |
|  | Thailand | 51,089 | 60.856 | 1.19 | 2,821 |
|  | Vietnam | 32,550 | 78.705 | 2.42 | 299 |
| Insular | Brunei Darussalam | 527 | 0.332 | 0.61 | — |
| Southeast | Indonesia | 181,157 | 209.255 | 1.16 | 1,096 |
| Asia | Malaysia | 32,855 | 21.830 | 0.66 | 4,469 |
|  | Philippines | 29,817 | 74.454 | 2.50 | 1,170 |
|  | Singapore | 61 | 3.522 | 57.74 | 32,486 |
| East Asia | China | 932,743 | 1,274.106 | 1.37 | 668 |
|  | DPRK | 12,041 | 23.702 | 1.97 | — |
|  | Japan | 37,652 | 126.505 | 3.36 | 43,574 |
|  | Mongolia | 156,650 | 2.621 | 0.02 | 391 |
|  | S. Korea | 9,873 | 46.480 | 4.71 | 11,028 |

Note: The author recalculated figures of some columns using the original data.
Source: FAO, *State of the World's Forests*, 2001.

Annual change in forest area by region over this decade was –0.8% in Africa and –0.4% in South America, indicating that deforestation is continuing, while the –0.1% rate in Asia shows that the brakes are being put on deforestation in this region as a whole. Europe's forests are growing (Table 1). Asia has a far higher plantation rate than other regions, having over 60% of the world's plantations. It is possible that such afforestation efforts are behind the increase in forest area. By comparison, each year the world's tropical regions lose 14.2 million ha of natural forest and gain 1.9 million ha in plantations by afforestation, for a net forest loss of 12.3 million ha.

A close look at the major countries in South and East Asia (Table 2) shows that deforestation from 1990 to 2000 was pronounced in Nepal, Sri Lanka, Pakistan, the Philippines, Myanmar, Indonesia, and Malaysia. In terms of area, losses were high in Indonesia, Myanmar, Malaysia, and Thailand. Deforestation is still

| Total forests in 1990 in 2000 (1000 ha) | | % of forest cover in 2000 | Forest area per capita (ha) | Forest plantation in 2000 (1000 ha) | Wood volume (m³/ha) | Wood biomass (tons/ha) | Change in forest cover, 1999–2000 (1000 ha) | Annual change rate, 1990–2000 (%) |
|---|---|---|---|---|---|---|---|---|
| 3,963,429 | 3,869,455 | 29.6 | 0.65 | 186,733 | 100 | 109 | −93,974 | −0.24 |
| 1,169 | 1,334 | 10.2 | 0.01 | 625 | 23 | 39 | 165 | 1.41 |
| 63,732 | 64,113 | 21.6 | 0.06 | 32,578 | 43 | 73 | 381 | 0.06 |
| 4,683 | 3,900 | 27.3 | 0.17 | 133 | 100 | 109 | −783 | −1.67 |
| 2,755 | 2,361 | 3.1 | 0.02 | 980 | 22 | 27 | −394 | −1.43 |
| 2,288 | 1,940 | 30.0 | 0.10 | 316 | 34 | 59 | −348 | −1.52 |
| 9,896 | 9,355 | 52.9 | 0.85 | 90 | 40 | 69 | −541 | −0.55 |
| 13,088 | 12,561 | 54.4 | 2.37 | 54 | 29 | 31 | −527 | −0.40 |
| 39,588 | 34,419 | 52.3 | 0.76 | 821 | 33 | 57 | −5,169 | −1.31 |
| 15,886 | 14,762 | 28.9 | 0.24 | 4,920 | 17 | 29 | −1,124 | −0.71 |
| 9,303 | 9,819 | 30.2 | 0.12 | 1,711 | 38 | 66 | 516 | 0.55 |
| 452 | 442 | 83.9 | 1.37 | 3 | 119 | 205 | −10 | −0.22 |
| 118,110 | 104,986 | 58.0 | 0.50 | 9,871 | 79 | 136 | −13,124 | −1.11 |
| 21,661 | 19,292 | 58.7 | 0.88 | 1,750 | 119 | 205 | −2,369 | −1.09 |
| 6,676 | 5,789 | 19.4 | 0.08 | 753 | 66 | 114 | −887 | −1.33 |
| 2 | 2 | 3.3 | 0.00 | — | 119 | 205 | 0 | 0.00 |
| 145,417 | 163,480 | 17.5 | 0.13 | 45,083 | 52 | 61 | 18,063 | 1.24 |
| 8,210 | 8,210 | 68.2 | 0.35 | — | 41 | 25 | 0 | 0.00 |
| 24,047 | 24,081 | 64.0 | 0.19 | 10,682 | 145 | 88 | 34 | 0.01 |
| 11,245 | 10,645 | 6.8 | 4.06 | — | 128 | 80 | −600 | −0.53 |
| 6,299 | 6,248 | 63.3 | 0.13 | — | 58 | 36 | −51 | −0.08 |

proceeding in many of the countries of continental and insular Southeast Asia, which still have comparatively abundant forests.

On the other hand the distribution of plantations is uneven. Of Asia's total 116 million ha, China and India have two-thirds, with 45.08 million ha and 32.57 million ha, respectively. If these plantations had not been planted, Asia's deforestation rate may have been nearly three times greater. Note also that China and India heavily influence forest dynamics in Asia. Specifically, the reason that the pace of deforestation in Asia as a whole appears to be slowing is the misleading effect of these two countries. China is planting trees with the intention of increasing its forest area 7% by 2010.

Countries around the world are planting trees not only to supply wood, but also in many cases to protect soil and water sources, fix carbon, stabilize slopes, and create windbreaks. Industrial afforestation is widely conducted throughout the Asia-Pacific region to supply wood. Hardwood afforestation for industrial wood is 30% eucalyptus, 12% acacia, and 7% teak, while 61% of softwood trees planted are pine.[3] Until the mid-1990s the preferred species were fast-growing trees that matured quickly and could be sold in a few years, but with the supply of large trees from natural forests dwindling, India and Malaysia in recent years have been planting teak and other species that sell for high prices.

A number of interlinked causes are behind deforestation: excessive commercial logging, fuelwood collection, nontraditional swidden farming, agricultural development, underdeveloped land institutions, socioeconomic factors like population pressure, and natural factors such as forest fires. In recent years there is a focus on illegal logging and issues surrounding the timber trade, and remedial measures are under consideration. Initiatives on various levels are necessary if the rate of deforestation is to be slowed. Developing countries in Southeast Asia and elsewhere must develop land institutions and environmental protection systems, carry out afforestation projects using tree species suited to their regions, provide for forest management with community participation, and consider policy measures for effectively using forests and timber to develop regional economies. Also garnering attention are efforts now underway in Indonesia and Malaysia to create national forest certification systems meant to achieve sustainable forest management. The success of these systems will probably hinge on how they are linked to international forest certification.

Tachibana Satoshi

# 7.  Local Perspectives in Protected Area Management

The world's total forested area is 3.87 billion ha, or 30% of the Earth's land area. Fifty-six percent is tropical or subtropical forest, and the other 44% is temperate or boreal forest.[1]

But these valuable forests are quickly dwindling due to plantation development, non-traditional swidden agriculture, commercial and illegal logging, conversion to pastureland and farmland, fuelwood collection, and other causes. Further, drying caused by the 1997 El Niño triggered fires that seriously harmed the world's forests. In particular, grave fire damage occurred in Southeast Asia and South America from anthropogenic impacts such as swidden farming and plantation development. From 1990 to 2000, 14.6 million ha of forest were destroyed, while 5.2 million ha were planted, for a net annual loss of 9.4 million ha. Tropical natural forests especially are being lost at an average annual rate of 14.2 million ha. Especially notable in Asia is the loss of Indonesia's vast forests (Table 1).

There are two conceivable ways of preventing forest loss. One is to restore forests through natural growth and by planting trees. The other is to protect forests by shielding ecosystems from destructive outside impacts through the institution of protected areas. Worldwide there are 12,754 protected areas, and 30,350 areas including those under 1,000 ha, for a total area of about 1.3 billion ha, or 8.81% of the Earth's land area.[2] Only 3,636 of these are marine areas.[3] Asia has the most protected areas after North and Central America (Fig. 1). Of the

FIG. 1. Proportions of World Protected Areas by Size

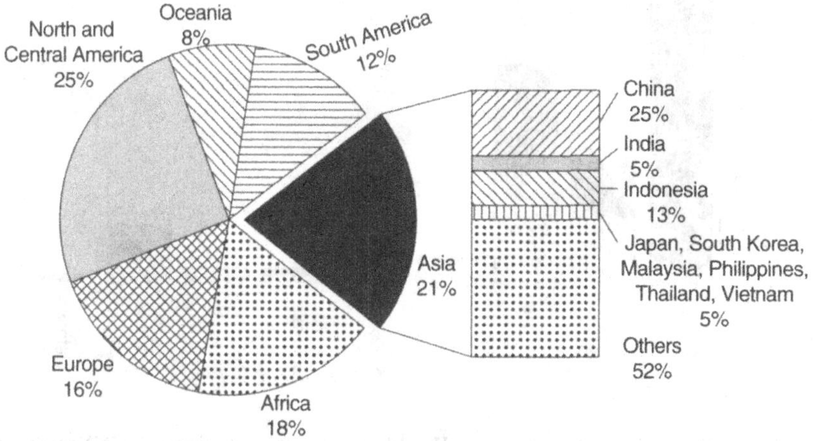

Note:  Covers the protected areas in the World Conservation Union (IUCN) categories I through VI that are 1,000 ha or larger.

Source: IUCN, *1997 United Nations List of Protected Areas*, 1998.

nine Asian countries listed, China, Indonesia, and India excel in terms of both the numbers and sizes of their protected areas, and China has 25% of Asia's total protected land area. In China and Thailand nearly half of the existing forested land is designated as protected, but protected areas account for only about 10% of the forests in Japan, South Korea, and Malaysia. Designation of protected areas proceeded rapidly in China and Indonesia during the 1980s and 1990s, while in India and Thailand it was the 1970s and 1980s. Although protected areas in these countries tend to increase year by year, designation in other countries is confined to certain periods of time (Fig. 2).

FIG. 2. Additions by Decade to Protected Areas in Nine Major Asian Countries

(1 million ha)

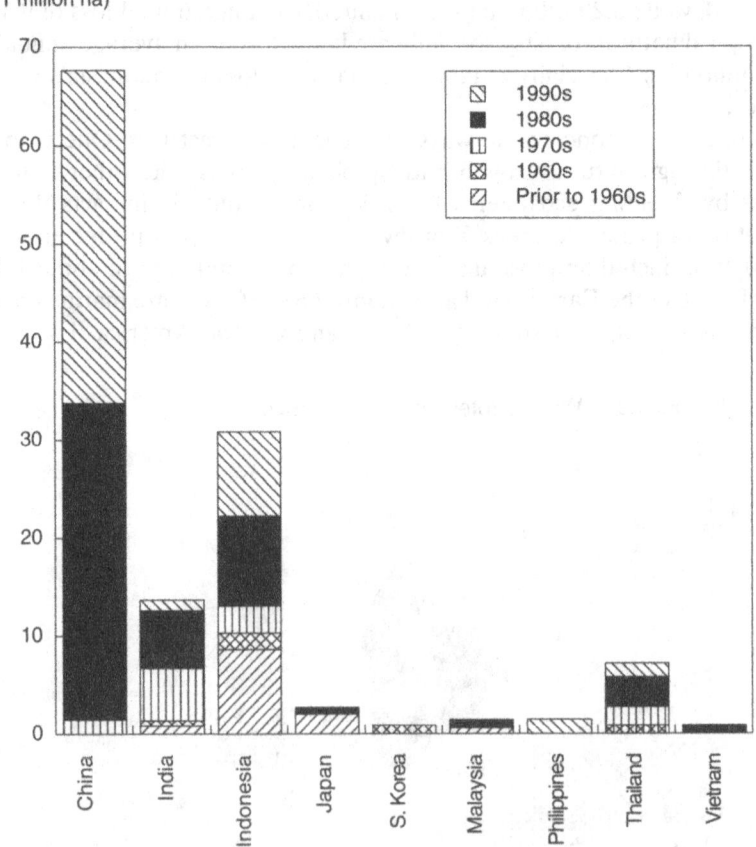

Note: Covers the protected areas in the World Conservation Union (IUCN) categories I through VI that are 1,000 ha or larger. However, does not include areas whose dates of establishment are unknown. Data for 1990s go up to 1997.

Source: Same as Fig. 1.

Protected areas are divided into seven categories according to their purposes: scientific research, wilderness protection, preservation of species and genetic diversity, maintenance of environmental services, protection of specific natural/cultural features, tourism and recreation, education, sustainable use of resources from natural ecosystems, and maintenance of cultural/traditional attributes. Protected area categories are based on the degree of management intervention, and beginning with areas closest to a natural state, they assume this order: strict nature reserves, wilderness areas, national parks, natural monuments, managed resource protected areas, habitat/species management areas, and protected landscapes/seascapes. In the nine Asian countries, wilderness areas, national parks, habitat/species management areas, and managed resource protected areas account for the most protected land, while the other types are small (Table 1). In terms of individual countries, China's largest category by size is wilderness areas while that of Indonesia is national parks. Although the total number of areas is small in Indonesia, each area is very large. In other countries the types of areas with the largest sizes are habitat/species management areas in India and Vietnam, national parks in Japan, Malaysia, and Thailand, and protected landscapes/seascapes in South Korea and the Philippines.

From a global perspective, protection of ecosystems by establishing protected areas preserves biodiversity and in other ways benefits humanity, but a local perspective reveals a number of problems.

First, many people in tropical forest regions depend on forests for their livelihoods, but their access is restricted when those forests are incorporated into the steadily rising number of protected areas. To people who depend on the forests to live, protected status brings disadvantages instead of benefits. Maintaining adequate communication and building relationships of trust between local people and governments are essential to protected area management. For those purposes governments must encourage local participation before creating protected areas and as soon as possible afterwards to increase community incentives and achieve protected area management based on collaboration between both sides.

Second is how to implement management. Plans are developed through cooperation between a government and outside agencies, but in actuality only vague determinations are made on whether plans are properly tailored to their areas. Some plans are therefore ivory-tower schemes that consist of idealistic proposals. One of the most salient examples is the determination of boundaries. Many protected areas of Southeast Asia have boundaries on their maps but none in the areas themselves, a lack which outsiders use to advantage by crossing over the boundaries into protected areas and logging illegally or converting forests to plantations and other agricultural uses. It is imperative that boundaries be clearly marked on site. Consideration is also needed on the possibilities for managing protected areas locally through decentralization or together with other countries.

Third, managing vast protected areas requires that governments provide budgets and personnel for the areas in their countries, but none of these could be considered adequate. Further, government administrative officials have insuf-

TABLE 1. Forest and Protected Areas

| | World | Nine Asian countries | China |
|---|---|---|---|
| A. Land area (1000 ha) | 13,063,900 | 1,605,053 | 932,742 |
| B. Forested area (1000 ha) | 3,869,455 | 412,570 | 163,480 |
| Forest cover (%) (2000) | 30 | 25.7 | 17.5 |
| C. Average annual forest loss, 1990–2000 (1000 ha) | 9,391 | −144 | −1,806 |
| Average annual forest loss rate, 1990–2000 (%) | 0.2 | −0.4 | −1.2 |
| D. Strict Nature Reserves (Ia) | | | |
| Number | 1,423 | 130 | 22 |
| Area (1000 ha) | 97,800 | 5,797 | 2,899 |
| Average area (1,000 ha per place) | 69 | 45 | 132 |
| E. Wilderness Area (Ib) | | | |
| Number | 654 | 15 | 15 |
| Area (1000 ha) | 93,901 | 45,729 | 45,729 |
| Average area (1,000 ha per place) | 144 | 3,049 | 3,049 |
| F. National Park (II) | | | |
| Number | 2,233 | 238 | 20 |
| Area (1000 ha) | 399,444 | 23,464 | 816 |
| Average area (1,000 ha per place) | 179 | 99 | 41 |
| G. Natural Monument (III) | | | |
| Number | 409 | 38 | 9 |
| Area (1000 ha) | 19,119 | 510 | 120 |
| Average area (1,000 ha per place) | 47 | 13 | 13 |
| H. Habitat/Species Management (IV) | | | |
| Number | 3,622 | 629 | 149 |
| Area (1000 ha) | 245,097 | 24,739 | 5,621 |
| Average area (1,000 ha per place) | 68 | 39 | 38 |
| I. Protected Landscape/Seascape (V) | | | |
| Number | 2,418 | 139 | 63 |
| Area (1000 ha) | 105,147 | 7,360 | 4,661 |
| Average area (1,000 ha per place) | 43 | 53 | 74 |
| J. Managed Resource Protected Area (VI) | | | |
| Number | 1,995 | 862 | 330 |
| Area (1000 ha) | 359,862 | 23,573 | 8,372 |
| Average area (1,000 ha per place) | 180 | 27 | 25 |
| K. Total | | | |
| Number | 12,754 | 2,051 | 608 |
| Area (1000 ha) | 1,320,369 | 131,169 | 68,218 |
| Average area (1,000 ha per place) | 104 | 64 | 112 |
| Proportion of area to land area (%) | 10 | 8 | 7 |
| Proportion of area to forested area (%) | 34 | 32 | 42 |

Note: Covers the protected areas in World Conservation Union (IUCN) categories I
    through VI that are 1,000 ha or larger. A few of the protected areas are marine
    areas. Forest cover equals forested area divided by land area. N.S. = not significant.
Sources: A-C: FAO, *State of the World's Forests*, 2001. D-K: WCMC and IUCN, *1997
    United Nations List of Protected Areas*, 1998.

| India | Indonesia | Japan | S. Korea | Malaysia | Philippines | Thailand | Vietnam |
|---|---|---|---|---|---|---|---|
| 297,319 | 181,157 | 37,652 | 9,873 | 32,855 | 29,817 | 51,089 | 32,549 |
| 64,113 | 104,986 | 24,081 | 6,248 | 19,292 | 5,789 | 14,762 | 9,819 |
| 21.6 | 58.0 | 64.0 | 63.3 | 58.7 | 19.4 | 28.9 | 30.2 |
| −38 | 1,312 | −3 | 5 | 237 | 89 | 112 | −52 |
| −0.1 | 1.2 | n.s. | 0.1 | 1.2 | 1.4 | 0.7 | −0.5 |
| 2 | 70 | 8 | 0 | 26 | 2 | 0 | 0 |
| 196 | 2,575 | 24 | 0 | 88 | 15 | 0 | 0 |
| 98 | 37 | 3 | — | 3 | 8 | — | — |
| 0 | 0 | 0 | 0 | 0 | 0 | 0 | 0 |
| 0 | 0 | 0 | 0 | 0 | 0 | 0 | 0 |
| — | — | — | — | — | — | — | — |
| 63 | 35 | 15 | 0 | 17 | 5 | 74 | 9 |
| 3,251 | 12,689 | 1,296 | 0 | 815 | 448 | 3,947 | 202 |
| 52 | 86 | 86 | — | 48 | 90 | 53 | 22 |
| 0 | 1 | 0 | 0 | 0 | 0 | 28 | 0 |
| 0 | 5 | 0 | 0 | 0 | 0 | 385 | 0 |
| — | 5 | — | — | — | — | 14 | — |
| 313 | 42 | 29 | 6 | 9 | 1 | 37 | 43 |
| 10,822 | 3,594 | 478 | 35 | 580 | 89 | 2,728 | 792 |
| 35 | 86 | 16 | 6 | 64 | 89 | 74 | 18 |
| 1 | 30 | 13 | 20 | 1 | 10 | 1 | 0 |
| 19 | 366 | 752 | 647 | 1 | 901 | 13 | 0 |
| 19 | 12 | 58 | 32 | 1 | 90 | 13 | n.s. |
| 0 | 531 | 0 | 0 | 1 | 0 | 0 | 0 |
| 0 | 15,180 | 0 | 0 | 21 | 0 | 0 | 0 |
| — | 29 | — | — | 21 | — | — | — |
| 379 | 709 | 65 | 26 | 54 | 18 | 140 | 52 |
| 14,287 | 34,407 | 2,550 | 682 | 1,505 | 1,453 | 7,073 | 994 |
| 38 | 49 | 39 | 26 | 28 | 81 | 51 | 19 |
| 5 | 19 | 7 | 7 | 5 | 5 | 14 | 3 |
| 22 | 33 | 11 | 11 | 8 | 25 | 48 | 10 |

ficient basic knowledge for protected area management and lack a firm grasp of world trends. It is vital that governments provide responsible personnel with the right training and, when involved parties alone cannot properly manage protected areas, further reinforce their management systems by involving outside parties such as domestic and foreign NGOs and researchers. It is also desirable to carry on dialog and build institutions for cooperation with government agencies in other countries, international NGOs, international agencies, and other parties that provide support for protected areas.

Harada Kazuhiro

# 8.    Illegal Wildlife Trade Flourishes

Species extinction nullifies biodiversity conservation efforts. The biggest causes of extinction are habitat loss and breakup, followed by the overexploitation of wildlife. According to a report by the World Conservation Union (IUCN) in 2000, capture and killing are the main causes of threats to 37% of the mammals, 34% of the birds, and all the reptiles that are endangered. The chief factor inducing overexploitation is the commercial value of wildlife on the international market. Thirteen percent each of the endangered mammals and birds mentioned above are affected by commercial trade,[1] and that is the reason for focusing on the international wildlife trade to conserve biodiversity.

From the time when modern capitalism developed until the present day when globalization is making inroads everywhere in the world, the biggest international market for wildlife "commodities" has been in the West. The primary commodities traded on that market are quite different from those of local markets where meat has always played a major role. Excluding wood and fisheries products, which are far larger than other categories, the predominant kinds of commodities include furs, reptile hides, live animals for pets and laboratory experiments, animal body parts for making ornaments (ivory, coral, tortoiseshell, shells, etc.), live plants for ornamental purposes, and animals for use in medicines.

Asia is rich in species (and therefore biological) diversity. For example, examining the 10 countries with the largest number of species by category reveals that three of the mammal-rich countries, four of the bird-rich countries, and four of the reptile-rich countries are Asian.[2] Consequently, Asia is one of the primary sources of wildlife commodities. This is evident from some of the trade data on species controlled by the Convention on International Trade in Endangered Species of Fauna and Flora (CITES) by virtue of being listed on one of its appendices (Table 1). Especially prominent are Indonesia (live primates, parrots, turtles, lizards, snakes, lizard hides, snake skins), China (live primates, turtles, wild orchids, feline furs), the Philippines (live primates), Thailand (wild orchids, snake skins), and Malaysia (lizard hides, snake skins).

But it would be a mistake to think that Asia is only a wildlife producer and supplier. While Indonesia does in fact tend to be so, some countries like Thailand and Malaysia are importers of certain items (Taiwan cannot join CITES and therefore does not appear in Table 1, but it imports many parrots[3] and other kinds of wildlife), and some countries are heavy consumers which only import. Representative of this category are Japan, South Korea, and Singapore. South Korea ranks first worldwide in net transactions (exports subtracted from imports) of feline furs, is the second-largest importer of wild orchids after Japan, and third-largest importer of live lizards, ranking above Japan.

Many parts of Asia, like Thailand, Singapore, and Hong Kong, are important transit or accumulation points for the international wildlife trade. Further, in countries like China with its wide economic gulf between urban and rural areas, there are considerable "exports" from rural areas to the cities, giving rise to the same kind of extensive capture and killing of wildlife that the international

TABLE 1. International Trade in CITES–Listed Species in 1997, Recorded Officially as Legal Transactions (Number)

| | Live primates | Live parrots | Live turtles | Live lizards |
|---|---|---|---|---|
| Total world imports | 25,733 | 235,336 | 76,079 | 948,497 |
| Net Asian transactions | –9,787 | 20,748 | 26,799 | 79,529 |
| Armenia | — | — | — | — |
| Azerbaijan | 4 | — | — | — |
| Bangladesh | — | 39 | — | — |
| Bhutan | — | — | — | — |
| Cambodia | — | — | — | — |
| China | –5,966 | 4,731 | –1,271 | 103 |
| Georgia | — | — | — | — |
| India | 9 | 7 | –1 | — |
| Indonesia | –3,995 | –18,334 | –1,784 | –10,296 |
| Japan | 3,556 | 9,413 | 30,670 | 39,255 |
| Kazakhstan | 1 | 4 | — | — |
| Korea, Dem People's Rep | — | — | 10 | 21 |
| Korea, Rep | 62 | 661 | 31 | 45,448 |
| Kyrgyzstan | 2 | — | — | — |
| Laos | — | –1 | — | — |
| Malaysia | 74 | 1,603 | –877 | 1,408 |
| Mongolia | — | — | — | — |
| Burma | — | — | — | — |
| Nepal | — | 574 | — | — |
| Pakistan | 14 | –702 | — | — |
| Philippines | –2,809 | 772 | 10 | 475 |
| Singapore | 17 | 9,277 | 60 | 99 |
| Sri Lanka | 0 | 364 | — | — |
| Tajikistan | — | — | — | — |
| Thailand | –6 | 3,449 | 1 | 1,809 |
| Turkmenistan | — | 1 | — | — |
| Uzbekistan | 31 | 12 | — | — |
| Vietnam | –819 | –246 | –394 | –81 |

Note: Data for all countries are net transactions (imports minus exports). Data are for Asia excluding the Middle East.

Source: World Resources Institute, *World Resources 2000–2001*, 2001.

market causes. An example is freshwater turtles. China is a major wildlife consumer and a fairly big importer (ranking second worldwide in net transactions of live snakes). This is cause for concern about the impacts on biodiversity in Asia and the world.

As this shows, Asia can be characterized as a major supplier, and an area for transit and accumulation, of "commodities" for the traditional international wildlife market in the West, and at the same time as a region that continues its

| Live snakes | Wild orchids | Cat skins | Crocodilian skins | Lizard skins | Snake skins |
|---|---|---|---|---|---|
| 258,715 | 343,801 | 21,864 | 850,198 | 1,637,973 | 1,457,767 |
| −19,677 | −52,312 | −8,013 | 448,884 | −397,934 | −794,741 |
| — | — | — | — | — | — |
| 2 | — | — | — | — | — |
| — | — | — | — | — | −4,301 |
| — | — | — | — | — | — |
| — | — | — | — | — | — |
| 46,222 | −100,210 | −17,999 | 105,946 | 13,236 | 867 |
| −1 | 1 | — | 16 | 15 | −510 |
| −76,862 | −102 | — | −260 | −511,000 | −441,902 |
| 4,772 | 128,911 | −354 | 82,166 | 318,159 | 120,999 |
| — | — | — | — | — | — |
| 72 | — | — | — | — | — |
| 279 | 100,300 | 9,550 | 70,332 | 20,140 | 36,416 |
| — | — | — | — | — | — |
| — | — | — | — | — | −2,657 |
| −8,708 | −2,837 | — | 2,702 | −237,993 | −432,512 |
| — | — | −40 | — | — | — |
| — | — | — | — | — | — |
| — | — | — | — | — | — |
| 1 | — | — | 1 | — | — |
| 15 | 197 | — | — | — | 0 |
| 2,376 | 5,043 | — | 5,327 | −53,097 | 10,082 |
| — | — | — | — | — | — |
| — | — | — | — | — | — |
| — | −207,579 | — | 146,613 | −7,367 | −427,079 |
| — | — | — | — | — | — |
| 2 | — | — | — | — | — |
| −7,416 | −1,951 | — | — | — | −55,718 |

growth as a extensive consumer of wildlife with a large volume of international trade carried on within the region.

In 1997 Japan conducted 41,819 transactions in CITES-listed species (Fig. 1), ranking second worldwide, and ranking first in terms of per capita transactions. Table 1 data on net transactions indicate that Japan is first worldwide in live turtles and wild orchids, second in live primates, and third in live parrots, lizard hides, and snake skins. Japan also serves as a nexus for the Asian wildlife trade,

FIG. 1 Imports of CITES-Listed Species by Japan, Recorded Officially as Legal Transactions

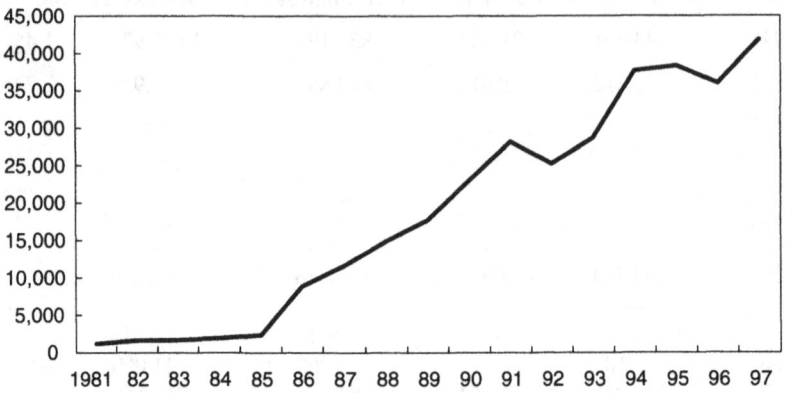

Source:  TRAFFIC Japan.

and it is conspicuous for its policy of actively promoting the international wildlife
trade. The trade in African ivory is a typical example in which many management efforts came to naught and the species was subject to destructive impacts
because a wildlife commodity with a long history was put on the modern-day
globalized and adaptable international market. Yet Japan wants the ban on the
ivory trade lifted.

Data presented here do not include illegal trade, that is, trade that violates
CITES. A 2002 report released by Britain's Department of the Environment,
Transport and the Regions states that the international wildlife trade is as high
as $20 billion a year, of which one-fourth is illegal. In a June 2002 major smug-
gling incident, authorities in Singapore seized six tons of ivory. In Japan there
were 1,695 cases in which imports were stopped at customs (11 of them were
penalized for violations of the customs act, such as the smuggling of 500 kg of
ivory in which a Japanese ivory dealer was involved), and in 2001 there were
1,658 cases (including 14 violations of the Customs Act).[4] Yet these figures likely
represent only a small portion of total smuggling.

Sakamoto Masayuki

# 9.    Coastal and Marine Environments in Crisis

Three factors underlie the crisis faced by coastal and marine environments in the Asia-Pacific Region: (1) marine pollution from land or sea pollution sources, (2) the threat to marine life from causes including overexploitation of marine resources, and upsetting marine ecosystem balance, and (3) direct damage to coastal and marine ecosystems by urban expansion, tourism development, and other causes.[1]

Marine pollution has a wide range of sources including household wastewater, industrial effluent, and oil spills. To help get a basic understanding of these sources they are grouped according to the relative weights of the pollutants flowing into the oceans, which in *State of the Environment in Asia 2002/2003*[2] are 40% land-based, 30% atmospheric, and 20% from sea transport and other sea-based activities. While they are all large and require remedial measures, this section will concentrate on land-based sources in the Asia-Pacific region.

Rivers in Cambodia, China, Malaysia, Thailand, and Vietnam carry at least 636,840 tons of nitrogen a year to the coastal region of the South China Sea. China's rivers account for 55% of that, Vietnam's for 21%, and Thailand's for 20%.[3] The amounts of pollutants flowing into the South China Sea from rivers (Table 1) and the biological oxygen demand (BOD) from household wastewater that is generated and removed in countries on the South China Sea (Table 2) indicate that these seven coastal countries alone generate about 6 million tons BOD. Four countries treat effluent, but only 11% of the BOD is removed. It is anticipated that if coastal populations in these countries continue growing at current rates, 6.6 million tons BOD will arise in 2005.[4] Industries in the same seven countries generate at least 430,000 tons BOD annually.[5]

One form of marine pollution via the atmosphere is the absorption of atmospheric $CO_2$ by the oceans. In recent years researchers have found that $CO_2$ absorbed by the oceans has increased seawater acidity and is hampering formation of carbonates in coral reefs. $CO_2$ absorption will probably decline, while the increase in its concentration in air and seawater will accelerate.[6]

The second threat to coastal and marine environments includes the overexploitation of marine resources. Long-term trends in marine resource use indicate that the percentage of marine resources used in excess of the maximum sustainable yield (MSY) in the tropical Pacific was under 10% until the 1980s, but rapidly increased in the 1990s and is already crossing the 20% mark (Fig. 1). Overexploitation of marine resources is encouraged mainly by developed countries' marine product imports, and Japan is an especially big importer. Part I Chapter 2 describes the example of Japan's tuna imports for sashimi.

Coastal ecosystems suffer damage from the compound impacts of all three factors, which shall be examined here in terms of mangrove damage. At least 40% of the world's mangrove forests are concentrated in South and Southeast Asia, which also have the highest mangrove species diversity. About 10% of the world's mangroves are in the South Pacific. Population growth and increasing demand for natural resources in these regions are putting heavy pressure on mangrove

TABLE 1. Pollutant Entry into the South China Sea via Rivers

(t/year)

| Countries and Rivers | BOD | Total N | Total P | Suspended solids | Oil |
|---|---|---|---|---|---|
| Cambodia | | | | | |
| Tonle Sap Lake and river water system | 6,022 | 1,084 | 303 | 13,250 | — |
| Mekong River (in Cambodia) | 4,964 | 894 | 255 | 10,950 | — |
| China | | | | | |
| Guangdong Province (5 rivers) | 566,385 | (340,050) | (3,768) | (58,531,000) | 9,698 |
| Guangxi Zhuang Autonomous Region (3 rivers) | 57,668 | (8,602) | (507) | — | 823 |
| Hainan Province (3 rivers) | 140 | — | — | — | 368 |
| Thailand | | | | | |
| Central, eastern, and southern rivers | 299,224 | 130,044 | 7,137 | 12,587 | — |
| South China Sea coast total | 1,015,936 | 636,840 | 58,202 | 58,642,827 | — |

Source: ESCAP and ADB, *State of the Environment in Asia and the Pacific 2000*, United Nations, 2000, p. 109, Table 5.6.

TABLE 2. BOD Generated by Residential Effluent and
Removed by Water Treatment

| | | (1,000 t/year) |
|---|---|---|
| | Generated | Removed |
| Cambodia | 36.2 | Untreated |
| China | 1,089.4 | <109 |
| Indonesia | 1,920.2 | 364 |
| Malaysia | 188.6 | 53 |
| Philippines | 431.3 | 149 |
| Thailand | 677.8 | 89 |
| Vietnam | 1,371.0 | Untreated |
| Total | 5,714.5 | 655 |

Source: China, L. S. and H. Kirkman, *Overview of Land-
Based Sources and Activities Affecting the Marine
Environment in the East Asian Seas*, Regional Seas
Report and Studies Series, UNEP/GPA Coordination
Office and EAS/RCU, 2000, p. 35, Table 11.

FIG. 1. Percentages of Tropical Marine Resources Used in Excess of Maximum Sustainable Yield

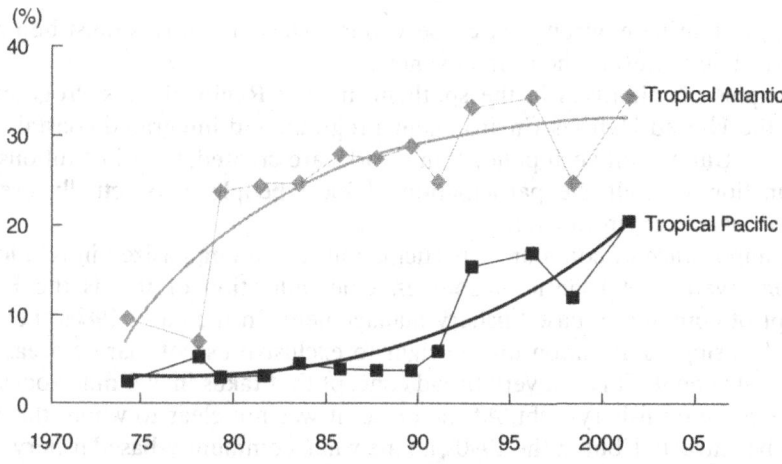

Source: FAO Fisheries Department, *The State of World Fisheries and Aquaculture 2000*,
FAO, 2000, p. 104, Figure 42.

ecosystems. Mangrove forests have been extensively destroyed for purposes
including industry, housing, and recreation. In particular, at least 60% of Asia's
mangroves have been lost through creation of aquaculture ponds (Fig. 2), which
is the fate of 3 million ha of mangroves in Southeast Asia alone.[7]

These developments indicate the need for quick action to conserve coastal and
marine environments, and already there are national and international initiatives

FIG. 2. Mangrove Forest Loss in the Asia-Pacific Region

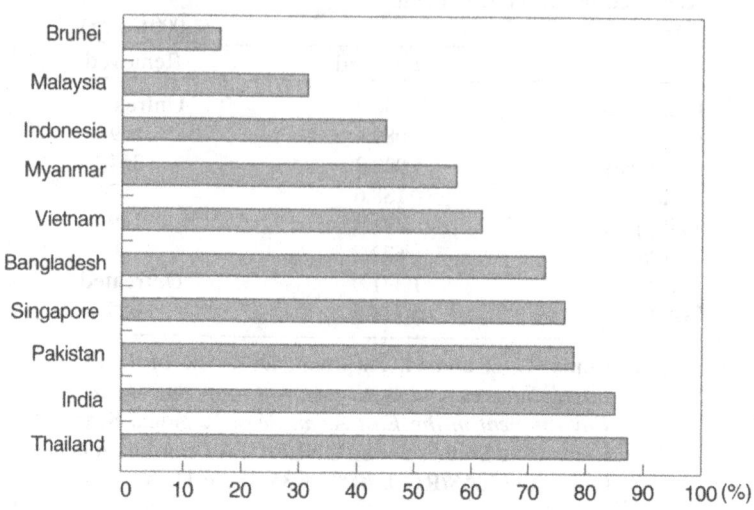

Source: ESCAP and ADB, *State of the Environment in Asia and the Pacific 2000*, United Nations, 2000, p. 103, Figure 5.2.

directed at marine environment conservation.[8] These initiatives must be implemented while bettering their effectiveness.

Examples of initiatives in the spotlight are the Regional Seas Programmes under the United Nations Environment Program, and integrated coastal management.[9] But even if such policy frameworks are created, their institutions will not function without the participation of local people who actually use the affected marine environments.

The importance of community participation is also emphasized in relation to the conservation of fisheries resources. One indication of this is the FAO's concept of community-based fishery management. In the early 1980s the FAO began focusing its attention on the right to exclusive use of marine areas as a policy instrument. This is a very broad concept that takes in 200-mile zones and Japan's common fishery right. At the outset it was not clear to whom this right would be entrusted, but in the 1990s, a time when community-based fishery management was formalized as a method of managing coastal fisheries in developing countries, it was determined that the right to exclusive use would be entrusted to cooperatives and other fishers' organizations. Authority and responsibility for conserving fisheries resources in the ocean near a certain area are therefore entrusted to fishers in that area, which is one way of providing for community participation in managing the natural resources of an area.[10]

The course of such community participation will doubtless be one primary factor influencing the success or failure of policy trends aimed at marine environment conservation.

Yokemoto Masafumi

# 10.  Deepening Water Crisis

In recent years many countries seem to be running out of water as their populations continue to grow. A report[1] released by Population Action International (PAI) states that in 2050, 18% of the world's population will experience water scarcity, while 24% will be water-stressed.

Of importance here are water shortage criteria. PAI employed the "water stress index" proposed by Malin Falkenmark.[2] Water is judged insufficient depending on the level of use to which annual per person water availability[3] corresponds. First, less than 500 m[3] per person annually is absolute water scarcity; second, 500 to 1,000 m[3] annually is chronic scarcity (the PAI report water scarcity combines these two levels), and third, 1,000 to about 1,700 m[3] annually is stress in the narrow sense. Water stress happens when difficulty in obtaining supplies becomes constant and widespread.

It is instructive to see the current state of water use in Asian countries, following the lead of existing research and using the latest data (Table 1). To the water stress indicator the table adds the basic water requirement (BWR) indicator,[4] whose criterion is 50 liters/person/day. When actual water household use drops below that mark, minimum necessary household water needs are not being met. In connection with the availability of statistical data, however, water use was calculated as annual household water withdrawals divided by total population.

The water stress indicator shows that Singapore has absolute water scarcity, which is probably because the country has hardly any water sources of its own.[5] South Korea also became water-stressed in 2000. Although India was not stressed in 2000, it is anticipated that in 2025 its stress indicator will be 1,385 m[3]/person/year, putting it below the stress line. On the other hand, this indicator suggests that seven countries including China will have no problems.

Turning our attention to the BWR, South Korea and Japan are seen to have very high values, which are likely because of extravagant water consumption due to the widespread use of flush toilets and washing machines. Although all countries are over the 50 liters/person/day mark, India, Indonesia, Singapore, Thailand, and Vietnam are below 100 liters/person/day.[6] People in these countries might not have sufficient water supplies for household use.

Of course since the BWR of a country is a leveled value for the entire country, it is necessary to take regional differences into account. For example, there are considerable differences in per capita water consumption according to main supply in India, Indonesia, Singapore, and Thailand (Fig. 1).

Generally, consumption by public supply is higher than the BWR in each country. Consumption is especially high in Bangkok and Delhi, and appears to be approaching extravagance. In China, meanwhile, Beijing and Tianjin have consumption levels that are lower than their BWR indicators, which is probably an indication of the tight water supply in northern China.

It is also highly significant that about 556,850,000 people have no public water supplies. In particular, China and India alone account for 79% of those people, and supplying them with water is a serious challenge for both countries.

TABLE 1. Water Use in 10 Asian Countries According to Two Indicators

| Country | | China | India | Indonesia |
|---|---|---|---|---|
| (1) Total resources (km³/year) | | 2,829.6 | 1,896.7 | 2,838.0 |
| | Total | 549.76 | 500.00 | 74.35 |
| (2) Withdrawals (km³/year) | Domestic | 60.47 (11%) | 25.00 (5%) | 4.46 (6%) |
| | Year of study | 2000 | 1990 | 1990 |
| (3) Total Population (thousands) | 2000 | 1,275,215 | 1,016,938 | 211,559 |
| | 2025 | 1,445,100 | 1,369,284 | 270,113 |
| (4) Utility supply (2000) | | 75% | 88% | 76% |
| Population not supplied in 2000 (thousands) | | 318,804 | 122,033 | 50,774 |
| Water stress indicator | 2000 | 2,219 | 1,865 | 13,415 |
| (m³/person/year) | 2025 | 1,958 | 1,385 | 10,507 |
| 2000 BWR indicator (L/day/person) | | 129.9 | 67.4 | 57.8 |

Notes: (a) Domestic consumption is percentage of total withdrawals. (b) Domestic withdrawals (excluding Japan) and Japan's domestic water percentage and population
not supplied were prepared by the author. (c) 2025 populations are medium
variant estimates. (d) The 2025 water stress indicator assumed that total resources
are constant. (e) China's (1) and (3) are mainland only.

Sources: (1) All from: FAO, *Review of World Water Resources by Country*, 2003, Annex 2
(Total resources: actual).

FIG. 1. Per Capita Water Consumption in Asian Cities

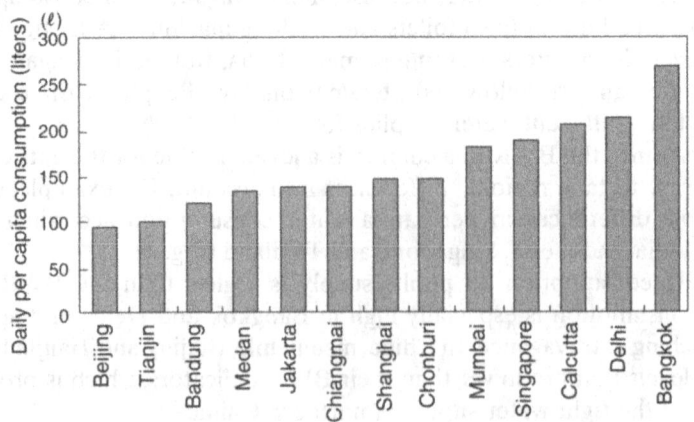

Source: Asia Development Bank, *Second Water Utilities Data Book*, 1997.

| Japan | S. Korea | Malaysia | Philippines | Singapore | Thailand | Vietnam |
|---|---|---|---|---|---|---|
| 430.0 | 69.7 | 580.0 | 479.0 | 0.6 | 409.9 | 891.2 |
| 87.70 | 23.67 | 12.73 | 55.42 | 0.19 | 33.13 | 54.33 |
| 16.40 (19%) | 6.15 (26%) | 1.40 (11%) | 4.43 (8%) | 0.09 (45%) | 1.66 (5%) | 2.17 (4%) |
| 1999 | 1994 | 1995 | 1995 | 1975 | 1990 | 1990 |
| 127,034 | 46,835 | 23,001 | 75,711 | 4,016 | 60,925 | 78,137 |
| 123,444 | 50,165 | 33,479 | 108,589 | 4,905 | 73,869 | 104,649 |
| 96.6% | 92% | — | 87% | 100% | 80% | 56% |
| 4,319 | 3,747 | — | 9,842 | 0 | 12,185 | 34,380 |
| 3,385 | 1,488 | 25,216 | 6,327 | 149 | 6,728 | 11,406 |
| 3,483 | 1,389 | 17,324 | 4,411 | 122 | 5,549 | 8,516 |
| 353.7 | 359.8 | 166.8 | 160.3 | 61.4 | 74.6 | 76.1 |

(2) China and Singapore: Gleick, Peter H., *The World's Water: The Biennial Report on Freshwater Resources 2002-2003*, 2002, Island Press, pp. 248–249; Japan: Ministry of Land, Infrastructure and Transport, *Japan's Water Resources, 2002*, pp. 51–52; others: World Resources Institute, *World Resources 2002-2004*, 2003, p. 274.
(3) All from: the "World Population Prospects: The 2002 Revision Population Database" Website, <http://esa.un.org/unpp/> (as of April 6, 2003).
(4) Japan: *Water Resources*, p. 52; others: World Health Organization and United Nations Children's Fund, *Global Water Supply and Sanitation Assessment 2000 Report*, 2000, pp. 47–48.

In that sense, the rapid increase in the use of bottled water should be taken into consideration. The urban poor are forced to rely on bottled water because their poverty prevents them from using existing water systems. Even people who have access to water supplies will depend on bottled water if tapwater is not fit for drinking. As such, the increasing use of bottled water suggests there are problems with existing water supply systems or water utilization systems as a whole.

Securing public water supplies is a serious matter of policy in many Asian countries, which will probably develop still more water resources to cope with future population growth, but they should not conduct the kind of ill-conceived development seen in Japan. Their water resource development must integrate water quality protection and demand-management policy to avoid damage to river ecosystems, and that makes it important to obtain statistical data that give an accurate picture of water utilization.

Noda Koji

## 11.  The Spreading Cadmium Scourge

Cadmium contamination has been found in only a few Asian countries and regions including Japan, China, Taiwan, and South Korea. This section will discuss China, which has recently reported the discovery of Itai-itai disease (cadmium poisoning); Taiwan, which has about 50,000 ha of farmland contaminated with cadmium and other heavy metals, and in 2000 established the Soil and Ground-water Remediation Act; South Korea, which in 2000 carried out a study of heavy metal contamination in the vicinities of closed metal mines; the amount of cadmium ingested in Asian countries through rice; the concentration of cadmium in rice; and standards for cadmium in rice.

### China

China has the largest metal resource reserves in East Asia. In particular, it has the world's highest production of zinc, whose sphalerite contains cadmium, 300 operating zinc and lead mines of various sizes, and at least 250 zinc smelters. Just as in Japan, cadmium pollution has been found in areas downstream from mines and smelters for nonferrous metals. A 1980 report writes that China has 19 such places totaling 11,000 ha in size. Studies on the exposure of local people to cadmium and impacts on their health have been conducted since the mid-1980s.

Dayu in Jiangxi Province has many mines including four large tungsten mines that earn this area the name "tungsten capital of the world." Four modern ore dressing plants have been in operation since 1960, and their cadmium-containing effluent has been released into rivers, contaminating about 5,500 ha of farmland that accounts for about half of China's total cadmium-contaminated farmland. Cadmium concentrations are reported to be 0.89 to 1.49 ppm in rice paddies (the natural level is 0.09 ppm) and 0.59 to 1.0 ppm in brown rice (natural level, 0.04 ppm). Both are over 10 times the concentrations found in nature.

Lead and zinc smelters have been operating in Wenzhou, Zhejiang Province since 1961, releasing about 100,000 tons of industrial wastewater into local rivers each year. Smelters discard untreated slag in uncovered piles on their grounds, and rain washes the material into nearby rivers. Concentrations of heavy metals including cadmium, lead, zinc, and copper in effluent exceed China's environmental standards, but polluted river water is nevertheless used for irrigation in downstream regions. In some places, high cadmium concentrations ranging between 1 and 5 ppm (average 3.7 ppm) in rice have been reported.

Exposure levels and body burdens of people living in these two cadmium-polluted areas were comparable with the high concentrations found in Japan's Jinzu River basin.

In September 2001 the Taiyuan Institute of Environmental Medicine in Shanxi Province announced the results of studies on cadmium health impacts conducted around a smelter that started operating in 1960 and around a lead and zinc mine operating since the 1950s (both locations unknown). Renal dysfunction was found near the smelter, and near the mine, researchers found four suspected cases of serious chronic cadmium poisoning, one of which was very likely Itai-itai disease.

On October 18, 2001 the Xinhua News Agency website reported that probable victims of Itai-itai disease had been discovered near a lead and antimony smelter operating since 1965 in Nanning City, Guangxi Zhuang Autonomous Region.

With cadmium concentrations in locally produced rice ranging between 1 and 6 ppm, victimized farmers filed a lawsuit against the smelter and won acquiescence to their demands for monetary compensation and soil remediation. Upstream from Nanning City in the vicinity of Guilin, a popular tourist destination, lies the large Dachang mine, which produces tin, lead, and zinc. A maximum cadmium concentration of 1.38 ppm has been detected in rice produced on farmland near the mine, making it quite possible that Dachang ore has been processed by the Nanning smelter.

Although human health damage by cadmium contamination has been reported in several places in China, it is likely that many more such places would be found by a detailed nationwide study.

## Taiwan

Cadmium-contaminated rice is nothing new to Taiwan. In 1988 rice in the villages of Kuanyin and Luchu were contaminated by the Kaoyin and Chili chemical plants in Taoyuan County. Subsequently an electroplating plant in the Hemei area of Changhua County caused another instance of cadmium contamination.

In September 2001 cadmium contamination of rice was discovered to have been caused by the Taiwan Dye Plant, a manufacturer of dyes, pigments, and plastic stabilizers in Huwei Village, Yunlin County. About 3 ha of farmland were seriously contaminated at over 10 ppm (59 ppm maximum), and the concentration in rice ranged between 0.58 and 2.13 ppm. Physical examinations of local people that December found abnormalities in the urine of 185 out of the 630 people tested, while 40 of them had urine protein irregularities, indicating that cadmium had caused renal dysfunction. In January 2002 some of the victims settled on the condition that the factory would pay compensation through the county government's pollution mediation committee under the Pollution Dispute Settlement Law, but farmers declining to settle are preparing a lawsuit.

Contaminated farmland can be remediated by washing the soil with dilute acid, but it cannot reduce the cadmium concentration below 5 ppm. Such remediation also washes away nutrients and acidifies the soil, resulting in barren farmland. More and more farmland has been contaminated with cadmium, arsenic, and copper, with a total of about 50,000 ha of land contaminated with heavy metals, an area exceeding the combined size of Taipei (27,200 ha) and Kaohsiung (15,400 ha). Sources of heavy metal contamination include pesticides, factories, and waste landfills.

## South Korea

In 2000 South Korea's Ministry of Environment (MOE) carried out detailed studies of soil contamination and its effects in the vicinities of 10 closed metal mines that are of concern. MOE collected and analyzed 998 samples from 613 sites near the mines, finding that 257 sites (41.9%) had concentrations exceeding concern levels for soil contamination, while 145 sites (23.7%) exceeded the threshold requiring remedial action. The breakdown of the latter 145 sites by type of land use from highest to lowest percentage is farmland (excluding rice paddies), mines, rice paddies, river beds, and mixed use.

At sites exceeding concern levels the numbers of sites affected by contaminants were 89 for cadmium (22.6%), 84 for copper (21.3%), 76 for arsenic (19.3%), 76 for mercury (19.3%), and 62 for lead (15.7%). The figures for exceeding levels requiring remedial action were 58 sites for mercury (31.4%), 47 for arsenic (25.4%), 32 for copper (17.3%), 24 for cadmium (13%), and 23 for lead (12.4%).

## Cadmium Contamination Throughout Asia

A team led by Watanabe Takao at the Miyagi University of Education in Japan conducted a dietary study in a number of Asian countries and regions, where they investigated the amount of cadmium ingested, cadmium concentrations in rice, and other indicators. High amounts of cadmium ingestion per day were found in Japan, South Korea, and Manila (Table 1). Investigating cadmium concentration in rice found that Japan, Taiwan, and Malaysia have high figures (Table 2). Soil contamination is reported in both Japan and Taiwan, and Malaysia has large numbers of mines and smelters for nonferrous metals that likely account for its high concentration.

Japan has the world's highest allowable level for cadmium in rice, 1 ppm, followed by Taiwan at 0.5 ppm, and the EU, South Korea, and China at 0.2 ppm.

TABLE 1. Daily Intake of Cadmium in Asia

| Country/region | (μg/day) Cadmium intake |
|---|---|
| Japan    1977–81 | 37.5 |
| 1991–97 | 25.5 |
| S. Korea | 21.2 |
| Kuala Lumpur | 7.3 |
| Tainan | 10.1 |
| China | 9.9 |
| Bangkok | 7.1 |
| Manila | 14.2 |

Source: Watanabe, T., "Proceedings of 20th Seminar of Itai-itai disease" (2001, in Japanese).

TABLE 2. Cadmium Concentration of Brown Rice in Various
Countries/Regions

|                  | (μg/kg)               |
|------------------|-----------------------|
| Country/region   | Cadmium concentration |
| Japan            | 55.70                 |
| US               | 7.43                  |
| Indonesia        | 21.77                 |
| Australia        | 2.67                  |
| S. Korea         | 15.70                 |
| Thailand         | 15.04                 |
| China            | 15.45                 |
| Taiwan           | 39.55                 |
| Philippines      | 20.14                 |
| Malaysia         | 27.47                 |

Source: Same as Table 1.

TABLE 3. Cadmium Standards for Brown Rice in Various
Countries/Regions

|                | (ppm)            |
|----------------|------------------|
| Country/region | Cadmium standard |
| Japan          | 1.0              |
| EU             | 0.2              |
| S. Korea       | 0.2              |
| Australia      | 0.1              |
| Thailand       | 0.1              |
| Taiwan         | 0.5              |
| China          | 0.2              |
| US             | No standard      |

Source: *Mainichi Shimbun*, May 22, 2002 (in Japanese).

Thailand and Australia have the lowest allowable level at 0.1 ppm (Table 3). The
Codex Alimentarius Commission, a joint commission of the UN Food and Agriculture Organization and the World Health Organization, has proposed an international standard of under 0.2 ppm as safe. Only Japan and Taiwan have yet to
satisfy this standard.

Hata Akio

## 12. Pollution Export and Rising Foreign Direct Investment

The phenomenon known as "pollution export" can be divided roughly into three types: (1) the export of dangerous and hazardous substances, (2) the transfer of pollution-intensive manufacturing processes and factories abroad, and (3) environmental damage through resource development. From the perspective of industrial siting, the mechanism by which it occurs is related to increases in the cost of environmental compliance that are induced by stricter environmental standards in industrialized countries, while in the developing countries that welcome pollution-intensive industries it is the wish to overcome poverty through development.

A representative example of type 1 is the transboundary movement of hazardous wastes. Although the Basel Convention bans such movement, many hazardous wastes such as spent automobile batteries and scrapped consumer appliances are exported for recycling and for the extraction of useful resources including lead, platinum, and palladium. Taiwanese factories retrieving the lead from spent automobile batteries exported by Japan are a cause of human lead poisoning, and in China dioxins are emitted by the open burning of discarded electric wire imported from Japan to recover the copper.

Type 2 is related to the declining competitiveness of goods due to rising production costs. Environmental taxes and other regulatory means impost additional costs for environmental measures on companies that produce goods with harmful environmental effects, thereby raising the cost of producing those goods. Consequently, consumer demand shifts from those higher-priced goods to other goods, which reduces the production of the higher-priced goods. From the perspective of an international market, industries producing goods with heavy environmental burdens lose export competitiveness and suffer a serious economic blow due to tougher environmental standards. If goods produced by pollution-intensive manufacturing processes with heavy environmental burdens are necessities domestically, then even if they are not export goods, companies will probably elect to make foreign investments and move their production sites abroad if the total costs of production abroad, transporting those goods back to their own countries, and duties are lower than the combined costs of production and environmental compliance at home. This thinking is behind the shift of pollution-intensive industries from countries in the North, whose environmental standards are strict, to countries of the South, whose standards are lax. However, a rebuttal (the Porter hypothesis) claims that tougher environmental standards will bring about technological innovation that will lead to lower costs and higher quality, which will in turn make companies more competitive internationally.

Examples of type 3 are pollution arising from development of underground resources through direct investment or ODA, and deforestation caused by the timber trade. When a manufacturing method presents environmental hazards but the goods themselves are clean and can be imported, businesses will invest and conduct development in other countries to create supplies there. And although companies might not be actively trying to avoid pollution in their own countries, importing those goods sometimes results in worsening the environmental impacts of their production processes in the producing countries. Heavy consumption of shrimp in Japan induces the widespread loss of mangrove forests in the shrimp-exporting Southeast Asian countries owing to aquaculture pond development. Producing eels for Japanese consumption has caused ground subsidence in China and Taiwan because of excess groundwater withdrawals.

Not everyone sees eye to eye on pollution export. Those who optimistically argue that globalization will make resource allocation more efficient and bring environmental improvements as well think that instances of pollution export are a small number of special cases induced by policy failures. More often than not, the real stories behind cases of pollution export in Asia are not clear (Table 1), and this is but a small sampling. The difficulties of solving pollution export problems are not only that we cannot easily discern the relationship between damage and its causes through international economic activities such as trade and direct investment, but also that even if a problem becomes apparent, there are no actors or rules for implementing remedial policies across national borders.

A number of empirical studies on trade and the environment verify that pollution-intensive industries are shifting from the industrialized to the developing countries, but results do not clearly show that the cause is environmental policy in the industrialized countries. Another view of pollution export sees it as "eco-dumping" that happens because developing countries do not enact environmental measures. But the critical point is that developing countries cannot cope well with environmental problems due to weakness stemming from poverty and other factors inherent in their social structures. One will misjudge the essential nature of pollution export by seeing it merely as a matter of industrial competitiveness.

There are hardly any data on pollution export, but some clues are afforded by examining Japanese direct investment in manufacturing industries in other Asian countries (Fig. 1). The number of investments began rising about 1986, skyrocketed in 1987, peaked in 1988, and has declined since then. The amounts of new investments have decreased somewhat, but are still big, and their cumulative amount is enormous (Table 2). In consideration of trends in Asian countries with their continuing economic growth, it is quite possible that more instances of pollution export will occur unless some kind of effective international measures are enacted.

After considering matters including international trade, in 1972 the OECD proposed the polluter pays principle, which holds that polluters themselves should shoulder the costs for environmental measures. But while that rule can perhaps be applied in affluent industrialized countries, it seems that developing countries cannot be realistically expected to implement this principle because they are trying to come up with the fiscal resources for development and overcome poverty. A recently developed concept is "indirect responsibility for emissions," which, having considered the life cycle of a product from production to disposal, and taking its environmental impact into account, holds that its importer

TABLE 1. Pollution Havens

| Year | Incident | Description |
|---|---|---|
| 1966 | Carbon disulfide poisoning, Wonjin Rayon Co., Ltd. (S. Korea) | In the second half of the 1980s a social issue was the $CS_2$ poisoning of workers at Wonjin Rayon, which was operating with used rayon spinning machines from the Shiga Factory of Japan's Toray Industries. In Japan $CS_2$ poisoning was an issue in the prewar years, and research for its prevention has been conducted primarily by the Japan Chemical Fibers Association. Wonjin went out of business in 1993, and sold its plant to a chemical factory in Dandong City, China's Liaoning Province. Plant amortized in seven years in Japan is still operating in China. |
| 1969 | Bhopal gas leak disaster (India) | On December 2, 1984 a pesticide factory of the US-owned Union Carbide company released a large amount of deadly poisonous methyl isocyanate gas, killing 2,500 and injuring 500,000. |
| 1970 | Mamut Copper Mine (Malaysia) | The Mamut Copper Mine was developed by the Overseas Mineral Resources Development Sabah Sdn Bhd, of which Japanese interests own 51%. Dumping wastes into rivers polluted downstream areas with heavy metals, preventing use of rivers for drinking water and adversely affecting agriculture. Japanese investors sold their stock in 1987, but ore is still exported to Japan. |
| 1975 | Hexavalent chrome contamination by Ulsan Inorganic Chemical Co., Ltd. (S. Korea) | Ulsan Inorganic Chemical Co., which produces chrome and other substances, is a joint Korean-Japanese venture, but the Japanese investor was Nippon Chemical Industrial Co, Ltd, which at that time had caused problems in Japan with chrome slag. Farmland in the area was devastated by Ulsan Inorganic, and in 1979 the drinking water for 3,500 workers in Ulsan industrial complex was found to be contaminated with hexavalent chrome. |
| 1976 | Philippine Sinter Corporation (Philippines) | The Philippine Sinter Corporation is a wholly owned subsidiary of Kawasaki Steel Corporation, and its job of sintering produces the most NOx and SOx of all steelmaking processes. Caused air pollution, worker asthma, and massive fish die offs by ocean pollution. |

| 1978 | Cadmium contamination by Korea Zinc Co., Ltd. (S. Korea) | Sited in the Onsan City industrial complex, Korea Zinc was a joint venture with Toho Zinc Co., Ltd., which caused sulfurous acid and cadmium pollution at Annaka, Japan. In 1979 Korea Zinc damaged farmed seaweed with water pollution, and in 1984 a toxic gas leak caused acute poisoning in nine elementary school students. |
| 1978 | River pollution by Semarang Diamond Chemical (SDC) (Indonesia) | Factory effluent from SDC (40% owned by Showa Chemical and 30% by Mitsubishi Corporation), which produces lime citrate, polluted the Tapak River and killed fish. Pollution spread through groundwater and also affected rice farming and drinking water wells. Local people suffered many skin diseases and tumors. |
| 1979 | Asian Rare Earth (ARE) waste (Malaysia) | Radioactive waste generated by ARE (35% owned by Mitsubishi Chemical), which produces rare earth metals, caused cancer, leukemia, and other illnesses among local people. In 1985 the factory closed down and lawsuits seeking compensation were filed against the company. ARE ceased operating in 1991. |
| 1983 | Philippine Associated Smelting & Refining Corporation (PASAR) copper smelter and refinery (Philippines) | The PASAR facility, which was built on the island of Leyte with funding from Japanese ODA and the Export-Import Bank of Japan, is owned 16%, 9.6%, and 6.4%, respectively, by Marubeni Corporation, Sumitomo Corporation, and Itochu Corporation. Marubeni received orders for the plant, while Mitsui Mining & Smelting (which caused itai-itai disease) and Furukawa Co., Ltd. (which caused Ashio copper mine pollution) did the designing and construction. The facility caused air pollution and heavy metal marine pollution in the area. |

Sources: Japan Federation of Bar Associations Committee on Pollution Measures and Environmental Conservation, ed., *Japanese Pollution Export and Environmental Damage*, and other sources.

TABLE 2.  Recent Direct Japanese Investment in Other Asian Countries/Regions

No. of contracts, 100 million yen

| Year | 1997 | | 1998 | | 1999 | | 2000 | | 2001 | |
|---|---|---|---|---|---|---|---|---|---|---|
| Country/region | Contracts | Amounts | Contracts | Amounts | Contracts | Amounts | Contracts | Amounts | Contracts | Amounts |
| Asia | 1,151 | 14,948 | 537 | 8,357 | 528 | 7,988 | 448 | 6,555 | 495 | 7,730 |
| ASEAN4 | 470 | 6,989 | 211 | 4,276 | 204 | 3,208 | 150 | 2,248 | 144 | 2,944 |
| Indonesia | 170 | 3,085 | 62 | 1,378 | 57 | 1,024 | 25 | 457 | 51 | 576 |
| Thailand | 154 | 2,291 | 72 | 1,755 | 72 | 910 | 61 | 1,029 | 51 | 1,102 |
| Malaysia | 82 | 971 | 32 | 658 | 44 | 586 | 23 | 256 | 18 | 320 |
| Philippines | 64 | 642 | 45 | 485 | 31 | 688 | 41 | 506 | 24 | 946 |
| NIEs | 331 | 4,186 | 183 | 2,259 | 212 | 3,567 | 177 | 2,964 | 145 | 2,672 |
| S. Korea | 53 | 543 | 47 | 387 | 62 | 1,093 | 52 | 899 | 47 | 680 |
| Taiwan | 67 | 552 | 27 | 287 | 26 | 318 | 51 | 563 | 31 | 399 |
| Hong Kong | 115 | 853 | 51 | 770 | 75 | 1,083 | 51 | 1,034 | 37 | 370 |
| Singapore | 96 | 2,238 | 58 | 815 | 49 | 1,073 | 23 | 468 | 30 | 1,223 |
| China | 258 | 2,438 | 112 | 1,363 | 76 | 838 | 102 | 1,099 | 187 | 1,802 |
| India | 28 | 532 | 17 | 329 | 12 | 232 | 10 | 185 | 6 | 181 |
| Vietnam | 45 | 381 | 12 | 65 | 17 | 110 | 5 | 24 | 9 | 97 |
| World Total | 2,489 | 66,229 | 1,597 | 52,169 | 1,713 | 74,390 | 1,684 | 53,690 | 1,753 | 39,548 |

Source: Japan Bank for International Cooperation, *Journal of JBIC Institute*, no. 12, September 2002, p. 38.

FIG. 1. Direct Japanese Investment in Other Asian Countries

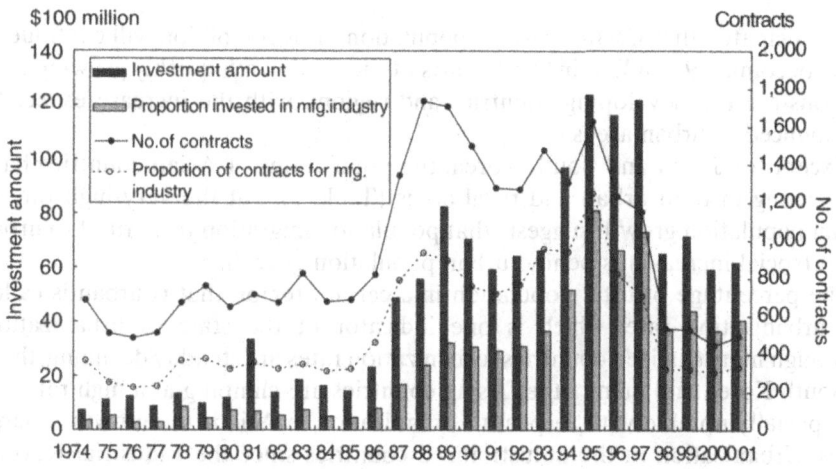

Sources: The Export-Import Bank of Japan, *Journal of Research Institute for International Investment and Development*; Japan Bank for International Cooperation, *Journal of Research Institute for Development and Finance*, various issues.

and beneficiaries such as the final consumer have indirect responsibility for the product's pollutant emissions. Eliminating pollution export requires adopting such a key concept, looking for ways that international trade policy and environmental policy can help each other, and monitoring the cross-border activities of transnational corporations and other entities.

Hayashi Tadashi

# 13. Urbanization Proceeds Rapidly

It is anticipated that Earth's current population of over 6 billion will continue to grow, becoming 9.3 billion in 2050.[1] Parts of the world with quickly growing populations are the developing countries and regions, with the increase especially pronounced in urban areas.

Except for Japan and South Korea, the populations of Asian countries have been rising in both urban and rural areas (Table 1), but the very high rate of urban population growth suggests that population migration from rural to urban areas (social increase) is behind urban population growth.

The percentage of total population in a certain region that is urban is called the urbanization rate, which is one indicator of the state of urbanization. Although industrialized countries' urbanization rates are steadily declining, those of South Korea and many other Asian countries are climbing at a high rate.

Especially notable with respect to urbanization in Asia is the growth of large cities. Urbanization in the industrialized countries since the Industrial Revolution created representative mega-cities such as New York, London, Paris, and Tokyo, but since the waning years of the 20th century dramatic urbanization has been occurring in developing countries, mainly those in Asia. It is anticipated that

TABLE 1. Population and Urbanization in Asia

| Year | | World | More developed regions | Asia | China | India |
|------|---|-------|------------------------|------|-------|-------|
| 1960 | Population | 3,021,908 | 915,841 | 1,702,320 | 657,492 | 442,344 |
| | Urban population | 1,016,665 | 562,521 | 353,895 | 105,249 | 79,413 |
| | Rural population | 2,005,243 | 353,320 | 1,348,425 | 552,243 | 362,931 |
| | Urbanization rate | 33.6% | 61.4% | 20.8% | 16.0% | 18.0% |
| 1980 | Population | 4,440,402 | 1,082,859 | 2,641,339 | 998,877 | 688,856 |
| | Urban population | 1,757,102 | 773,994 | 709,214 | 196,222 | 158,851 |
| | Rural population | 2,683,300 | 308,865 | 1,932,125 | 802,655 | 530,005 |
| | Urbanization rate | 39.6% | 71.5% | 26.9% | 19.6% | 23.1% |
| 2000 | Population | 6,055,049 | 1,187,980 | 3,682,550 | 1,277,558 | 1,013,662 |
| | Urban population | 2,845,049 | 902,993 | 1,351,806 | 409,965 | 288,283 |
| | Rural population | 3,210,000 | 284,987 | 2,330,744 | 867,593 | 725,379 |
| | Urbanization rate | 47.0% | 76.0% | 36.7% | 32.1% | 28.4% |
| 2020 | Population | 7,501,521 | 1,216,567 | 4,545,249 | 1,454,462 | 1,275,549 |
| | Urban population | 4,176,428 | 986,079 | 2,165,601 | 637,913 | 498,997 |
| | Rural population | 3,325,093 | 230,488 | 2,379,648 | 816,549 | 776,552 |
| | Urbanization rate | 55.7% | 81.1% | 47.6% | 43.9% | 39.1% |

Source: United Nations, *World Urbanization Prospects: The 1999 Revision*, 2001.

large cities will continue growing rapidly in Asia, and there is a tendency for population to concentrate in the larger cities (Table 2). Concentration around capital cities is especially pronounced, a trend that is apparent in Thailand and the Philippines, as well as in Japan and South Korea. These trends contrast with indicators in Europe, whose cities have already entered the postindustrial stage.

A characteristic of this large-scale urbanization in Asia is that people leaving impoverished urban areas to seek employment in the cities do not find a sufficiently industrialized environment to accommodate them. This phenomenon, called "overurbanization," happens when a city's population exceeds the size that is appropriate for its level of economic development. In other words, it is urbanization without industrialization, in which a city grows to huge size while its internal economic foundation is still underdeveloped. Asian cities whose central areas were formed by overurbanization lack modern, strong urban industries, instead forming low-quality residential areas with a poor employment base.

Such rapid urbanization breeds a variety of urban ills and increases environmental and social stresses. One problem is poor living environments that have not only inadequate housing but also highly unsanitary conditions. Only 30 to 40% of households in the cities of China and India have toilet facilities, and many city residents cannot get sufficient drinking water.

Thousands

| Indonesia | Japan | S. Korea | Malaysia | Philippines | Thailand | Vietnam |
|---|---|---|---|---|---|---|
| 96,195 | 94,096 | 25,004 | 8,140 | 27,561 | 26,392 | 34,743 |
| 14,031 | 58,812 | 6,929 | 2,165 | 8,350 | 3,302 | 5,107 |
| 82,164 | 35,284 | 18,075 | 5,975 | 19,211 | 23,090 | 29,636 |
| 14.6% | 62.5% | 27.7% | 26.6% | 30.3% | 12.5% | 14.7% |
| 150,958 | 116,808 | 38,124 | 13,764 | 48,316 | 46,718 | 53,711 |
| 33,514 | 88,990 | 21,678 | 5,787 | 18,110 | 7,961 | 10,338 |
| 117,444 | 27,818 | 16,446 | 7,977 | 30,206 | 38,757 | 43,373 |
| 22.2% | 76.2% | 56.9% | 42.0% | 37.5% | 17.0% | 19.2% |
| 212,108 | 126,714 | 46,844 | 22,244 | 75,967 | 61,399 | 79,832 |
| 86,833 | 99,788 | 38,354 | 12,772 | 44,530 | 13,252 | 15,749 |
| 125,275 | 26,926 | 8,490 | 9,472 | 31,437 | 48,147 | 64,083 |
| 40.9% | 78.8% | 81.9% | 57.4% | 58.6% | 21.6% | 19.7% |
| 262,291 | 123,893 | 51,894 | 29,254 | 102,404 | 70,975 | 102,532 |
| 152,636 | 102,333 | 46,310 | 20,082 | 71,579 | 23,082 | 27,992 |
| 109,655 | 21,560 | 5,584 | 9,172 | 30,825 | 47,893 | 74,540 |
| 58.2% | 82.6% | 89.2% | 68.6% | 69.9% | 32.5% | 27.3% |

TABLE 2.  Distribution of Urban Population of Asia and Europe by Size of Urban Settlement, 1975, 1995, and 2015

|  |  | 1975 | 1995 | 2015 |
|---|---|---|---|---|
| Asia | 10 million or more | | | |
| | Number of cities | 2 | 7 | 15 |
| | Population (thousands) | 31,214 | 98,086 | 240,750 |
| | Percentage distribution of urban population | 5.3 | 8.3 | 12.4 |
| | 5 to 10 million | | | |
| | Number of cities | 7 | 13 | 20 |
| | Population (thousands) | 51,101 | 100,557 | 137,431 |
| | Percentage distribution of urban population | 8.6 | 8.5 | 7.1 |
| | 1 to 5 million | | | |
| | Number of cities | 77 | 160 | 267 |
| | Population (thousands) | 142,410 | 285,593 | 510,024 |
| | Percentage distribution of urban population | 24 | 24.1 | 26.2 |
| Europe | 10 million or more | | | |
| | Number of cities | 0 | 0 | 0 |
| | Population (thousands) | 0 | 0 | 0 |
| | Percentage distribution of urban population | 0 | 0 | 0 |
| | 5 to 10 million | | | |
| | Number of cities | 5 | 5 | 5 |
| | Population (thousands) | 36,654 | 38,017 | 38,376 |
| | Percentage distribution of urban population | 8.1 | 7.1 | 6.8 |
| | 1 to 5 million | | | |
| | Number of cities | 42 | 58 | 64 |
| | Population (thousands) | 82,242 | 107,869 | 118,480 |
| | Percentage distribution of urban population | 18.1 | 20.1 | 20.9 |

Source: United Nations, *World Urbanization Prospects: The 1999 Revision*, 2001.

Meanwhile, since the latter half of the 1980s Asian countries have pursued policies for actively accepting foreign capital, thereby inducing rapid industrialization including the appearance of large industrial estates on the outskirts of cities.
At the same time, the growth of the urban middle class is causing an explosive
increase in motor vehicle traffic and expansion of the consumer society. The
combination of these factors is behind the heavy pollution in Asia's cities
(Table 3).

TABLE 3. Air Pollution in Major Asian Cities

|  | Population (thousands) | Total suspended particulates ($\mu g/m^3$) | $SO_2$ ($\mu g/m^3$) | $NO_2$ ($\mu g/m^3$) |
|---|---|---|---|---|
| Bangkok | 7,281 | 223 | 11 | 23 |
| Beijing | 10,839 | 377 | 90 | 122 |
| Mumbai | 18,066 | 240 | 33 | 39 |
| Calcutta | 12,918 | 375 | 49 | 34 |
| Delhi | 11,695 | 415 | 24 | 41 |
| Jakarta | 11,018 | 271 | — | — |
| Kuala Lumpur | 1,378 | 85 | 24 | — |
| Metro Manila | 10,870 | 200 | 33 | — |
| Osaka | 11,013 | 43 | 19 | 63 |
| Seoul | 9,888 | 84 | 44 | 60 |
| Shanghai | 12,887 | 246 | 53 | 73 |
| Tokyo | 26,444 | 49 | 18 | 68 |

Notes: 1. Pollution values are annual means.
    2. WHO annual guidelines for air quality standards are $50\,\mu g/m^3$ for $SO_2$ and $NO_2$.
Source: The World Bank, *World Development Indicators*, 2002.

TABLE 4. Agenda 21: Number of Formally Committed
Municipalities

|  | 1996 | 2001 |
|---|---|---|
| World | 1,812 | 6,416 |
| Europe | 1,576 | 5,291 |
| Asia | 87 | 461 |
| China | 14 | 25 |
| India | 20 | 14 |
| Indonesia | 6 | 8 |
| Japan | 26 | 110 |
| S. Korea | 9 | 172 |
| Malaysia | — | 9 |
| Philippines | 3 | 28 |
| Thailand | 6 | 21 |
| Vietnam | 2 | 20 |

Source: Prepared from the World Resource Institute's
    Website <http://www.wri.org>.

It is anticipated that environmental problems arising with urbanization will further worsen, and it will be important for city governments to address them. One indicator will be Local Agenda 21s, which are local governments' action plans for achieving sustainable societies (Table 4). The plans of European local governments make up most of the world total, while Japan and South Korea account for most such plans in Asia. Although the substance of the plans is of course most important, the active formulation and implementation of action plans by Asian cities in an effort to create sustainable societies is important not only to their urban environments, but also to global environmental conservation.

Asazuma Yutaka

# 14. Motor Vehicles and Air Pollution

Transportation demand rises in tandem with economic growth in many of Asia's cities, which is aggravating air pollution, noise, and other environmental problems stemming from transportation. One characteristic of Asia is that the concentration of population in large cities is more extreme than in Western countries. But hardly any large cities have developed in a planned manner, and they have not succeeded in providing cheap and low-pollution mass transit services that can serve as satisfactory transportation for their residents. As a consequence, streets are clogged with automobiles, motorcycles, and motorized three-wheelers, as well as buses and other vehicles with unacceptably high emissions. The result is traffic congestion and deteriorating urban environments. This is why Asian—especially Chinese—cities account for over half of the world's cities with the worst air pollution.

Annual average values for air pollution in major Asian cities (Table 1) show that while Japan, South Korea, and Malaysia generally satisfy the WHO guideline of $90 \mu g/m^3$ for total suspended particulates (TSP), this level is exceeded by China, India, Indonesia, the Philippines, and Thailand. Cities with three times the guideline are not unusual. Beijing's level is $377 \mu g/m^3$. The WHO guideline for $SO_2$ and $NO_2$ is $50 \mu g/m^3$. A few Japanese and South Korean cities slightly exceed this level, while a few of China's cities are far higher.

However, in some cities the causes of air pollution are not limited to motor vehicles, for a significant portion comes from coal-fired power plants, space heating equipment, and other fixed sources. According to a report by Japan's former Environment Agency, in 1994, 46% of the anthropogenic suspended particulate matter (SPM, particulates smaller than $10 \mu m$ in size) in the Kanto region and 59% in the Kansai region was estimated to be from motor vehicles. The contribution of vehicular traffic to air pollution in five Asian cities (Table 2) reveals that while there are differences, motor vehicles are responsible for between 46 and 82% of NOx and 24 to 88% of SPM.

The sizes of the vehicle fleets (passenger cars and commercial vehicles combined) in five Asian countries/regions and the world (Table 3) reveals that in 2002 Japan was far ahead in Asia with nearly 74 million vehicles, followed in a distant second place by China with somewhat over 20 million and third place by South Korea with close to 14 million. But in recent years growth has been remarkable in countries other than Japan. If the 1990 fleet is given the value 100, Japan is about the same as the world as a whole at 128, while the fleets of China and South Korea are growing rapidly at 430 and 411. India and Taiwan are also increasing quickly at 299 and 195. In terms of the number of vehicles per 1,000 people, Japan has 580 followed by Taiwan and South Korea at less than half, with 255 and 292. China and India are far lower. These figures suggest that as income levels in Asian countries rise, their fleet sizes will enlarge rapidly.

Motor vehicles emit comparatively few pollutants if traffic flows smoothly, but when traffic congestion prevents this, vehicles emit more pollutants per km traveled, which aggravates air pollution. Generally more roads are quickly built to

TABLE 1.  Air Pollution in Major Asian Cities

| Country | City | Population (thousands) 2000 | TSP ($\mu$g/m$^3$) 1990–95 | SO$_2$ ($\mu$g/m$^3$) 1990–98 | NO$_2$ ($\mu$g/m$^3$) 1990–98 |
|---|---|---|---|---|---|
| China | Anshan | 1,453 | 305 | 115 | 88 |
| | Beijing | 10,839 | 377 | 90 | 122 |
| | Changchun | 3,093 | 381 | 21 | 64 |
| | Chengdu | 3,294 | 366 | 77 | 74 |
| | Chongqing | 5,312 | 320 | 340 | 70 |
| | Dalian | 2,628 | 185 | 61 | 100 |
| | Guangzhou | 3,893 | 295 | 57 | 136 |
| | Guiyang | 2,533 | 330 | 424 | 53 |
| | Harbin | 2,928 | 359 | 23 | 30 |
| | Jinan | 2,568 | 472 | 132 | 45 |
| | Kunming | 1,701 | 253 | 19 | 33 |
| | Lanzhou | 1,730 | 732 | 102 | 104 |
| | Liupanshui | 2,023 | 408 | 102 | — |
| | Nanchang | 1,722 | 279 | 69 | 29 |
| | Pingxiang | 1,502 | 276 | 75 | — |
| | Qingdao | 2,316 | — | 190 | 64 |
| | Shanghai | 12,887 | 246 | 53 | 73 |
| | Shenyang | 4,828 | 374 | 99 | 73 |
| | Taiyuan | 2,415 | 568 | 211 | 55 |
| | Tianjin | 9,156 | 306 | 82 | 50 |
| | Urumqi | 1,643 | 515 | 60 | 70 |
| | Wuhan | 5,169 | 211 | 40 | 43 |
| | Zhengzhou | 2,070 | 474 | 63 | 95 |
| | Zibo | 2,675 | 453 | 198 | 43 |
| India | Ahmedabad | 4,160 | 299 | 30 | 21 |
| | Bangalore | 5,561 | 123 | — | — |
| | Calcutta | 12,918 | 375 | 49 | 34 |
| | Chennai | 6,002 | 130 | 15 | 17 |
| | Delhi | 11,695 | 415 | 24 | 41 |
| | Hyderbad | 6,842 | 152 | 12 | 17 |
| | Kanpur | 2,450 | 459 | 15 | 14 |
| | Lucknow | 2,568 | 463 | 26 | 25 |
| | Mumbai | 18,066 | 240 | 33 | 39 |
| | Nagpur | 2,062 | 185 | 6 | 13 |
| | Pune | 3,489 | 208 | — | — |
| Indonesia | Jakarta | 11,018 | 271 | — | — |
| Japan | Osaka | 11,013 | 43 | 19 | 63 |
| | Tokyo | 26,444 | 49 | 18 | 68 |
| | Yokohama | 3,178 | — | 100 | 13 |
| S. Korea | Pusan | 3,830 | 94 | 60 | 51 |
| | Seoul | 9,888 | 84 | 44 | 60 |
| | Taegu | 2,675 | 72 | 81 | 62 |

TABLE 1. *Continued*

| Country | City | Population (thousands) 2000 | TSP (µg/m³) 1990–95 | SO₂ (µg/m³) 1990–98 | NO₂ (µg/m³) 1990–98 |
|---------|------|------|------|------|------|
| Malaysia | Kuala Lumpur | 1,378 | 85 | 24 | — |
| Philippines | Manila | 10,870 | 200 | 33 | — |
| Singapore | Singapore | 3,567 | — | 20 | 30 |
| Thailand | Bangkok | 7,281 | 223 | 11 | 23 |

Note: Air pollution data are mostly from 1995. Due to the diversity of data sources, different data years, differences in pollution even within the same city, and other factors, this table is not suitable for a strict comparison between cities.

Source: The World Bank, *2002 World Development Indicators*.
Data in *World Development Indicators* are based on a wide variety of sources compiled by the WHO Healthy Cities Program's Air Management Information System and the World Resources Institute.

TABLE 2. Contribution of Motor Vehicles to Air Pollution in Five Asian Cities

| | Year | CO | HC | NOₓ | SO₂ | SPM |
|---|---|---|---|---|---|---|
| | | | | | | (%) |
| Beijing | 1989 | 39 | 75 | 46 | — | — |
| Mumbai | 1992 | — | — | 52 | 5 | 24 |
| Cochin (India) | 1993 | 70 | 95 | 77 | — | — |
| Colombo | 1992 | 100 | 100 | 82 | 94 | 88 |
| Delhi | 1987 | 90 | 90 | 59 | 13 | 37 |

Source: The World Resources Institute, The United Nations Environment Program, The United Nations Development Program and The World Bank (eds.), *World Resources 1996–97*, 1996.

alleviate congestion when an increase in road traffic is anticipated. As seen from the fleet sizes and road length in 12 Asian countries/regions (Table 4), road density is very high in Japan and the city state Singapore, with 3.1 km and 5.0 km per square km, respectively. Yet, owing to the large fleets in these two countries, their road length per vehicle is only 16 m and 5 m. This indicates that even if more roads are built, the growing vehicle fleet will make it impossible to maintain smoothly flowing traffic conditions unless road use is appropriately controlled. Especially in areas with high population concentrations such as large Asian cities, it is physically impossible to create enough road capacity to accommodate unlimited traffic demand. Accordingly, merely building more roads is not the answer because traffic demand management is also essential.

TABLE 3. Motor Vehicle Fleet Sizes of Five Asian Countries/Regions and the World

| Year | | China | India | Japan | S. Korea | Taiwan | World |
|---|---|---|---|---|---|---|---|
| 1970 | Fleet | 487,557 | 1,041,600 | 17,581,843 | 128,298 | 98,500 | 244,760,200 |
| | 1990 = 100 | 10 | 26 | 30 | 4 | 3 | 42 |
| | Vehicles per 1,000 population | 1 | 2 | 170 | 4 | 7 | 67 |
| 1975 | Fleet | 946,833 | 1,215,500 | 28,090,558 | 193,927 | 264,302 | 323,833,000 |
| 1980 | Fleet | 1,680,960 | 1,666,843 | 37,856,174 | 527,729 | 680,968 | 413,174,604 |
| | 1990 = 100 | 35 | 42 | 66 | 16 | 23 | 71 |
| | Vehicles per 1,000 population | 2 | 3 | 323 | 14 | 39 | 93 |
| 1985 | Fleet | 2,887,126 | 2,536,952 | 46,157,261 | 1,113,430 | 1,344,969 | 483,787,816 |
| 1990 | Fleet | 4,776,382 | 3,972,000 | 57,697,669 | 3,394,803 | 2,937,694 | 579,103,904 |
| | 1990 = 100 | 100 | 100 | 100 | 100 | 100 | 100 |
| | Vehicles per 1,000 population | 4 | 5 | 467 | 79 | 139 | 110 |
| 1995 | Fleet | 10,400,029 | 6,058,000 | 66,853,500 | 8,468,901 | 4,487,195 | 665,844,845 |
| | 1990 = 100 | 218 | 153 | 116 | 249 | 153 | 115 |
| | Vehicles per 1,000 population | 9 | 6 | 534 | 189 | 211 | 116 |
| 2000 | Fleet | 16,089,100 | 7,430,000 | 72,649,099 | 11,164,319 | 5,599,517 | 752,983,667 |
| | 1990 = 100 | 337 | 187 | 126 | 329 | 191 | 130 |
| | Vehicles per 1,000 population | — | 7 | 573 | 236 | — | — |
| 2001 | Fleet | 18,020,400 | 11,880,000 | 73,407,762 | 12,914,613 | 5,731,900 | 779,560,490 |
| | 1990 = 100 | 377 | 299 | 127 | 380 | 195 | 135 |
| | Vehicles per 1,000 population | — | 12 | 580 | 270 | 259 | — |
| 2002 | Fleet | 20,531,700 | 11,880,000 | 73,989,350 | 13,949,441 | 5,731,900 | 814,887,248 |
| | 1990 = 100 | 430 | 299 | 128 | 411 | 195 | 141 |
| | Vehicles per 1,000 population | — | 11 | 580 | 292 | 255 | — |

Note: Fleet sizes are year-end totals for passenger and commercial vehicles. In China, passenger vehicles include buses.
Source: Nikkan Jidosha Shimbun and Automobile Business Association of Japan, eds., *Automotive Years Hand Book*, *various years*.
Nikkan Jidosha Shimbun and Automobile Business Association of Japan, eds., *Automotive Yearbook 2004*, 2004.
This publication's data are from various countries' automobile manufacturers' associations and other sources.

Table 4 also shows motorcycle fleet size. In Japan, South Korea, and Singapore, which have high income levels, motorcycle ownership is only about 30% of passenger car ownership, while in Hong Kong it is under 10%. In other Asian countries, however, motorcycles outnumber passenger cars. India and Thailand have over five times as many motorcycles as cars, and in Vietnam the ratio is almost 40 to one. This heavy bias toward motorcycles is a characteristic of Asia. Even seen on a worldwide basis, the six countries/regions with the highest numbers of motorcycles are China, India, Thailand, Japan, Indonesia, and Taiwan. These six alone have about 60% of the world's motorcycles.

Southeast and South Asian cities are also characterized by their three-wheelers. Some examples of these are Thailand's *tuktuk*, India's *autoriksha*, Nepal's *tempo*, and Bangladesh's *mishuk*. Fleet sizes of various vehicle types in five South Asian countries (Table 5) indicate that the total of two- and three-wheeled vehicles in each country comes to between 51 and 73%. Three-wheeled vehicles alone account for 14% of Bangladesh's fleet.

These vehicles not only are indispensable as transportation for the public, but also endow cities with a rich individuality. However, they are a troublesome presence in terms of air pollution and noise. Most two- and three-wheeled vehicles have two-stroke engines, which have smaller NOx emissions than four-stroke engines, but higher emissions of particulates, hydrocarbons, and CO, as well as being noisy. A study in Bangladesh's capital city of Dhaka found that particulate emissions per unit distance traveled by three-wheelers with two-stroke engines were eight times those of passenger cars and 10 times those of three-wheelers with four-stroke engines.

Controlling exhaust emissions is currently achieved mainly by controls on individual new vehicles. These countries are beginning to set exhaust standards for hydrocarbons and CO based upon those of Europe. Some of the tuktuks in Bangkok have been converted to LPG fuel. In 1999 Nepal's capital of Kathmandu banned the diesel minibuses called *bikram tempo* and replaced them with electric three-wheeled minibuses called *safa tempo*. Generally, however, Asian cities are behind in regulating particulate emissions. In addition to regulating new vehicle exhaust, it needs to regulate that of existing vehicles.

Just as important as tougher controls on individual vehicles is instituting areawide control. Although areawide control is not necessarily implemented widely, there are a few places in Asia that are advanced by world standards. Since 1975 Singapore has implemented an "Area Licensing Scheme" that imposes a fee on vehicles entering the city center and has been successful in mitigating road congestion. In 1998 Singapore adopted an electronic system for collecting the fee and switched to "Electronic Road Pricing," which makes it possible to set the fee in a flexible manner according to time and place. In 1996 South Korea's capital city of Seoul instituted road pricing for vehicles that use a tunnel leading to the city center, and in Tokyo, despite uncertainty about whether road pricing is feasible, there are plans to start it as soon as possible in 2003 or thereafter.[1]

Singapore has not relied on areawide control of vehicle traffic alone, for in 1990 it launched a vehicle quota system in which the government controls the

TABLE 4. Two- and Four-Wheeled Vehicle Fleets and Roads in 12 Asian Countries/Regions

|  |  | Year | China | Hong Kong | India | Indonesia | Japan |
|---|---|---|---|---|---|---|---|
| Fleet size | Passenger | 2000 | 5,805,600 | 350,362 | 4,820,000 | 2,900,000 | 52,437,375 |
|  | Commercial | 2000 | 6,012,300 | 139,719 | 2,610,000 | 2,305,000 | 20,211,724 |
|  | Total | 2000 | 11,817,900 | 490,081 | 7,430,000 | 5,205,000 | 72,649,099 |
| Total road length (km) |  | 1999–2000 | 1,402,698 | 1,831 | 3,319,644 | 342,700 | 1,161,894 |
| Road density (km/km$^2$) |  | 1999–2000 | 0.15 | 1.70 | 1.01 | 0.19 | 3.07 |
| Road length per vehicle (m/vehicle) |  |  | 119 | 4 | 447 | 66 | 16 |
| Two-wheeler fleet |  | 1998–2001 | 37,720,000 | 33,079 | 24,691,876 | 13,563,017 | 13,719,898 |

Note: Fleet size does not include two-wheelers. Does not agree with Table 3 because of different data sources. Data for China show especially great disagreement.

Source: Fleet sizes: Society of Motor Manufactures and Traders Limited, *Motor Industry of Great Britain 2001 World Automotive Statistics*, 2001. Total road length and road density: International Road Federation, *World Road Statistics 2002*, 2002. Two-wheeler fleet: Honda Motor Co., Ltd., *World Overview of Two-Wheeled Vehicles 2001*, 2002 (in Japanese).

TABLE 5. Distribution of Vehicles by Type in Five South Asian Countries

|  |  | Bangladesh 1999 |
|---|---|---|
| Cars | Vehicles | 92,000 |
|  | % of whole | 18 |
| Taxis | Vehicles | 2,300 |
|  | % of whole | 0 |
| Light-duty gasoline | Vehicles | 52,000 |
|  | % of whole | 10 |
| Heavy-duty diesel | Vehicles | 55,000 |
|  | % of whole | 11 |
| Two-stroke three-wheelers | Vehicles | 68,000 |
|  | % of whole | 13 |
| Four-stroke three-wheelers | Vehicles | 7,600 |
|  | % of whole | 1 |
| Two-stroke two-wheelers | Vehicles | 200,000 |
|  | % of whole | 38 |
| Four-stroke two-wheelers | Vehicles | 35,000 |
|  | % of whole | 7 |
| Total | Vehicles | 523,000 |
|  | % of whole | 100 |

Note: Table totals do not match sums of individual categories because vehicles such as tractors are not included. Most three-wheelers in Nepal are diesel-powered. Ninety percent of Bangladesh's three-wheelers are assumed to have two-stroke engines. It is assumed that 85% of all two-wheelers have two-stroke engines.

Source: Kojima M., C. Brandon and J. Shah, *Improving Urban Air Quality in South Asia by Reducing Emmissions form Two-Stroke Engine Vehicles*, Working Paper 21911, The World Bank, 2000. Data were obtained from governments or manufacturers' associations.

| S. Korea | Malaysia | Philippines | Singapore | Taiwan | Thailand | Vietnam |
|---|---|---|---|---|---|---|
| 7,837,251 | 4,212,567 | 773,835 | 413,545 | 4,716,217 | 2,044,565 | 142,000 |
| 3,327,068 | 1,029,633 | 1,587,501 | 147,325 | 883,300 | 4,075,537 | 83,600 |
| 11,164,319 | 5,242,200 | 2,361,336 | 560,870 | 5,599,517 | 6,120,102 | 225,600 |
| 86,990 | 65,877 | 201,994 | 3,066 | 35,931 | 64,600 | 93,300 |
| 0.88 | 0.20 | 0.60 | 4.96 | 1.00 | 0.13 | 0.33 |
| 8 | 13 | 86 | 5 | 6 | 11 | 414 |
| 2,274,766 | 5,609,351 | 1,216,000 | 133,358 | 11,733,202 | 14,500,000 | 5,500,000 |

| India 1997 | Nepal 1999 | Pakistan 1999 | Sri Lanka 1997 |
|---|---|---|---|
| 3,500,000 | 49,000 | 670,000 | 122,000 |
| 9 | 21 | 17 | 13 |
| 420,000 | — | 68,000 | 6,000 |
| 1 | — | 2 | 1 |
| 740,000 | 2,600 | 310,000 | 14,000 |
| 2 | 1 | 8 | 1 |
| 5,200,000 | 46,000 | 750,000 | 235,000 |
| 14 | 20 | 19 | 25 |
| 1,180,000 | — | 91,000 | 59,000 |
| 3 | — | 2 | 6 |
| 210,000 | 5,900 | — | — |
| 1 | 3 | — | — |
| 21,800,000 | 110,000 | 1,700,000 | 424,000 |
| 59 | 47 | 43 | 45 |
| 3,900,000 | 19,000 | 250,000 | 75,000 |
| 10 | 8 | 6 | 8 |
| 37,200,000 | 232,000 | 4,000,000 | 936,000 |
| 100 | 100 | 100 | 100 |

number of new vehicles registered, and auctions "certificates of entitlement," or the right to buy a car. Since the 1980s Shanghai has also controlled ownership of private automobiles by auctioning license plates. Although the main purpose of traffic and ownership control is to mitigate traffic congestion, the close connection between traffic congestion level and pollutant emissions means that such means of control are also effective in cutting air pollution. Still better-planned and more effective urban transport systems will be needed also with an eye to combatting global warming and preventing traffic accidents.[2]

Koyama Shinya

# 15. Burgeoning Wastes and New Management Policies

Accurate data on wastes in Asia are not easily obtainable. Estimates in the revised version of a 1995 ESCAP study, released in 2000[1] and discussed in the first volume in this series,[2] indicate that total municipal solid waste (MSW) generated in the Asia-Pacific region was 700,000 tons per day in 1992, which increased to about 1.3 million tons in 2000. It is anticipated to be 1.9 million tons in 2010.

It is evident from ESCAP's estimates of MSW composition in certain Asian cities (Fig. 1) that kitchen waste and other organic wastes account for the major part of MSW in many cities, while in Tokyo and Singapore other kinds of waste account for the largest share. However, one cannot go very far in interpreting these results because ESCAP has not clearly described its criteria for measuring composition.

Industrial waste is more varied than MSW and often contains hazardous substances, requiring more rigorous control. Nevertheless, data are even harder to

FIG. 1. Comparison of Municipal Solid Waste Composition in Selected Asian Cities

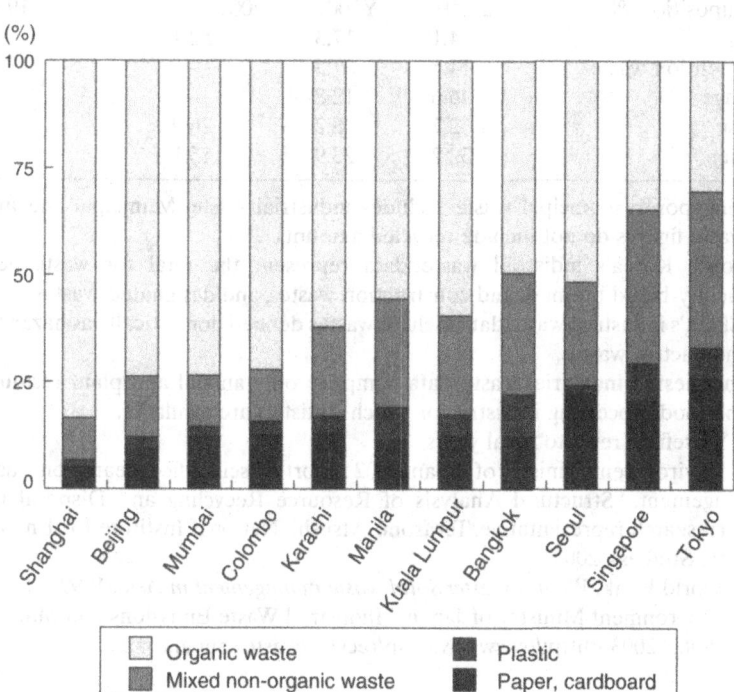

Source: Based on ESCAP, *State of the Environment in Asia and the Pacific*, 2000, Figure 8.3.

TABLE 1. Wastes Generated in Selected Asian Countries/Regions

|   |   | Japan | S. Korea | China | | Taiwan |
|---|---|---|---|---|---|---|
| A | Municipal solid waste | Nationally | Nationally | Nationally | Hong Kong | Nationally |
|   |   | FY2000 | FY2001 | 2000 | 1997 | 1999 |
| A1 | Total, 1,000 t/year | 52,360 | 17,702 | 138,824 | 3,168 | 7,889 |
| A2 | Per capita, kg/day | 1.13 | 1.01 | 1.81 | 1.335 | 0.96 |
| A3 | % by weight | FY2000 | FY2001 | Shanghai, 1998 | | 1999 |
|   | Organic | 34 | 32.8 | 67.3 | 24.9 | — |
|   | Paper, cardboard | 33 | 23.8 | 8.8 | 29.1 | 35.8 |
|   | Plastic | 13 | — | 13.5 | 19.6 | 19.9 |
|   | Glass | 5 | 2.8 | 5.2 | 5.0 | 4.99 |
|   | Metal | 3 | 1.0 | 0.7 | 5.3 | 3.8 |
|   | Other | 12 | 40.6 | 4.5 | 12.7 | 65.67 |
| A4 | Recycling rate, % | 16.7 | 43.1 | | 6.5 | |
| A5 | Incineration rate, % | 77.4 | 13.6 | | | 53.83 |
| A6 | Landfill rate, % | 5.9 | 43.3 | | | 39.02 |
| B | Industrial waste | Nationally | Nationally | Nationally | | Nationally |
|   |   | FY2000 | FY2001 | 2001 | | 1999 |
| B1 | Total, 1,000 t/year | 406,000 | 68,541 | 888,400 | | 2,350 |
| B2 | Composition, % | FY2000 | FY2001 | 2001 | | 1999 |
|   | Slag | 4.1 | 17.3 | 50.4 | | 28.4 |
|   | Demolition waste | 14.5 | 37.8 | — | | 0.5 |
|   | Sludge | 46.6 | 12.8 | — | | 17.7 |
|   | Dust | 2.7 | 8.2 | 26.9 | | 11.9 |
|   | Other | 32.2 | 23.9 | 15.1 | | 41.4 |

Notes: Singapore's municipal waste includes industrial waste. Municipal and industrial waste figures do not include recycled amounts.

South Korea's indistrial waste data represent the total for waste generator facility-based business and construction wastes, and designated wastes.

China's industrial waste data include wastes defined domestically as hazardous and radioactive wastes.

Indonesia's industrial waste data comprise only animal and plant residues from the food processing industry, for which statistics are available.

FY prefixes refer to fiscal years.

Sources: Environment Ministry of Japan, 2002 report on scientific research on waste management, "Structural Analysis of Resource Recycling and Disposal in Aisa" (research representative, Terasono Atsushi, National Institute for Environmental Studies), 2003.

World Bank, *What a Waste: Solid waste management in Asia*, 1999.

Environment Ministry of Japan, "Industrial Waste Emissions and Management, 1998," 2003 <http://www.env.go.jp/recycle/waste/sanpai.html>.

| Philippines | Malaysia | Singapore | Indonesia | Thailand | India |
|---|---|---|---|---|---|
| Manila | Kuala Lumpur | Nationally | Jakarta | Bangkok | 23-city total |
| 2000 | 1997 | 2002 | 1996/97 | 1996 | 1995 |
| 10,670 | 720 | 1,436 | 10,070,000 m³ | 2,955 | 11,000 |
| 0.71 | 1.42 | 0.94 | | 1.450 | 0.456 |
| 1999 | 1993 | 2002 | Bandung 1994/95 | | 1995 |
| 49 | 32.5 | 39.1 | 63.6 | | 41.8 |
| 19 | 28.4 | 25.0 | 10.4 | | 5.7 |
| 17 | 17.7 | 20.0 | 9.8 | | 3.9 |
| — | 2.2 | 1.8 | 1.5 | | 2.1 |
| 6 | 3.3 | 2.5 | 1.0 | | 1.9 |
| 9 | 12.2 | 11.5 | 13.6 | | 44.6 |
| | 6.3 | | | | |
| | | 92.2 | 2 | | |
| 69 | | | 70 | | |
| Nationally | Nationally | Nationally | Nationally | Nationally | Nationally |
| 2000 | 1999 | 2002 | 1998 | | 1999 |
| | 379 | 1,189 | 156,000 | 1,200 | 147,050 |
| | 1996 | | 1998 | | 1999 |
| | 33.6 | | — | | 12.8 |
| | — | | | | — |
| | 44.2 | | — | | 6.3 |
| | — | | — | | 39.4 |
| | 22.2 | | 100 | | 41.5 |

Environment Ministry of Japan, "Municipal Solid Waste Emissions and Management, 1998," 2003 <http://www.env.go.jp/recycle/waste/ippai.html>.
S. Korea Ministry of Environment, MoE Website <http://www.me.go.kr/>.
China Environment Yearbook Company, *China Environment Yearbook 2002.*
Environment Protection Agency, *Monitoring of Solid Waste in Hong Kong 1997.*
Taiwan Executive Yuan, Environmental Protection Administration, *Environment White Paper*, 2001 edition.
Malaysia, Ministry of Science, Technology and the Environment, *Environmental quality report 1999.*
Singapore, National Environment Agency Website <http://www.nea.gov.sg/>.
Indonesia, Central Bureau of Statistics, *Environmental Statistics of Indonesia*, 1999.
Thailand, Pollution Control Department, *State of Thailand's Pollution in Year 1996*, 1998.
India, Ministry of Environment & Forests, *The State of Environment—India: 2001* <http://envfor.nic.in/soer/2001/soer.html>.

FIG. 2. Comparison of Waste Intensity of Industrial Production in Selected Asian Countries/Regions

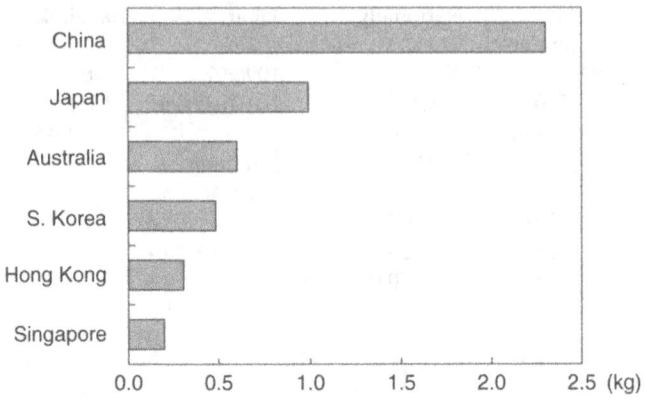

Waste intensity is waste generated per US$ industrial production

Source:  Based on ESCAP, *State of the Environment in Asia and the Pacific*, 2000, Figure 8.4.

come by than those for MSW. ESCAP has therefore estimated the waste intensity, or amount of industrial waste generated per unit production in Asia (Fig. 2). Under this criterion, Japan generates 1 kg of waste for each dollar of production. The smaller this figure, the lighter the environmental burden. This value reflects differences in industrial structures, technological levels, and other characteristics. Currently available data have been gathered and presented here (Table 1).

Japan, South Korea, Taiwan, China, and other countries/regions have introduced laws to facilitate recycling as a way of dealing with the huge volumes of MSW. Japan's Container and Packaging Waste Recycling Law of 1997 and Home Electrical Appliance Recycling Law of 2001 created a system to encourage recycling of several categories and a framework for shouldering the costs.

Since 1992 when South Korea's Act Relating to Promotion of Resources Saving and Reutilization took effect, a deposit refund system and a quantity-based waste collection fee system have been gradually implemented. In 2003 recycling was required for televisions and other scrapped consumer appliances, and old products are collected free of charge when purchasing new ones. Taiwan introduced a recycling system in 1997 based on payment of costs by manufacturers and distributors, and it is encouraging the recycling of packaging, containers, scrapped consumer appliances, and other waste.

China's rate of MSW increase is so fast as to be comparable to its economic growth rate, and it aims to make advances in waste management and recycling during the 10th Five-Year Plan, which began in 2001. For that purpose it is developing a policy to introduce measures such as charging for MSW management and separate collection. China is also considering legislation for recycling scrapped consumer appliances.

Kojima Michikazu, Yamashita Hidetoshi

# 16.  The Internationalization of Recycling

Trade in waste for recycling is brisk around the world. The waste trade is also discussed in the second volume of this series,[1] and in Part I Chapter 2 (Trade and the Environment) of the present volume.

Two of the recent trends in Asia are that advances in obligatory recycling have quickly increased exports from Japan to other Asian countries, and that China's imports are growing, as noted in Part I Chapter 2.

Examining the waste trade alone when analyzing the internationalization of recycling reveals only a small part of the whole picture. Questions that need to be addressed include: In what countries are the products that become wastes produced and consumed? What resources are consumed in producing those products? And what countries produced those Resources? Assessments must trace the material flow of the wastes back to the point of resource extraction. Also, the question of what kind of products the wastes are used in after being recycled shows that the material flow must be traced in both directions.

Therefore the material flow must be analyzed as a whole by combining the waste (reclaimed resources) being studied, substitute primary resources (pulp for recovered paper, iron ore for steel scrap, etc.), and the final products produced using both (paper products made from recovered paper and pulp).

The author created a figure showing the integrated material flow of paper resources in Asia based on production statistics according to the UN Food and Agriculture Organization and *Pulp and Paper International*, and Asian countries' trade statistics (Fig. 1). This one figure combines information on Asian countries' paper products, pulp and post-consumer production (i.e., recovery) and consumption amounts, trade volume, and trading partners. Each element in the figure is a square whose size is proportional to each country's production or consumption amount.

Take for example Indonesia, found just left and below center. Production of paper products exceeds consumption, showing that Indonesia's production of such products exceeds its domestic demand, and that they are widely exported. It is the same for pulp, which Indonesia exports to South Korea and China. As this shows, Indonesia supplies other parts of Asia with paper resources in the form of finished products and feedstock.

Comparing Indonesia's pulp and recovered paper consumption reveals that recovered paper makes up over 50% of the feedstock in its paper production. But most of the recovered paper consumed in Indonesia is imported from Western countries and Singapore. Thus the paper resources used in making paper products in Indonesia consist mainly of a domestically produced component and recovered paper from the West. In this way, examining the elements for a country in Fig. 1 makes it possible to determine the state of paper recycling in that country and the connections with other countries in its material flow.

Using Fig. 1 to get a general idea of the material flow in Asia reveals that Indonesia and Thailand supply paper products and pulp, while South Korea imports pulp, and Taiwan and Japan import paper products. From those importing countries the material flow leads to China, Malaysia, Singapore, and Hong Kong. Further,

FIG. 1. Material Flow of Paper Resources in Asia, 2000

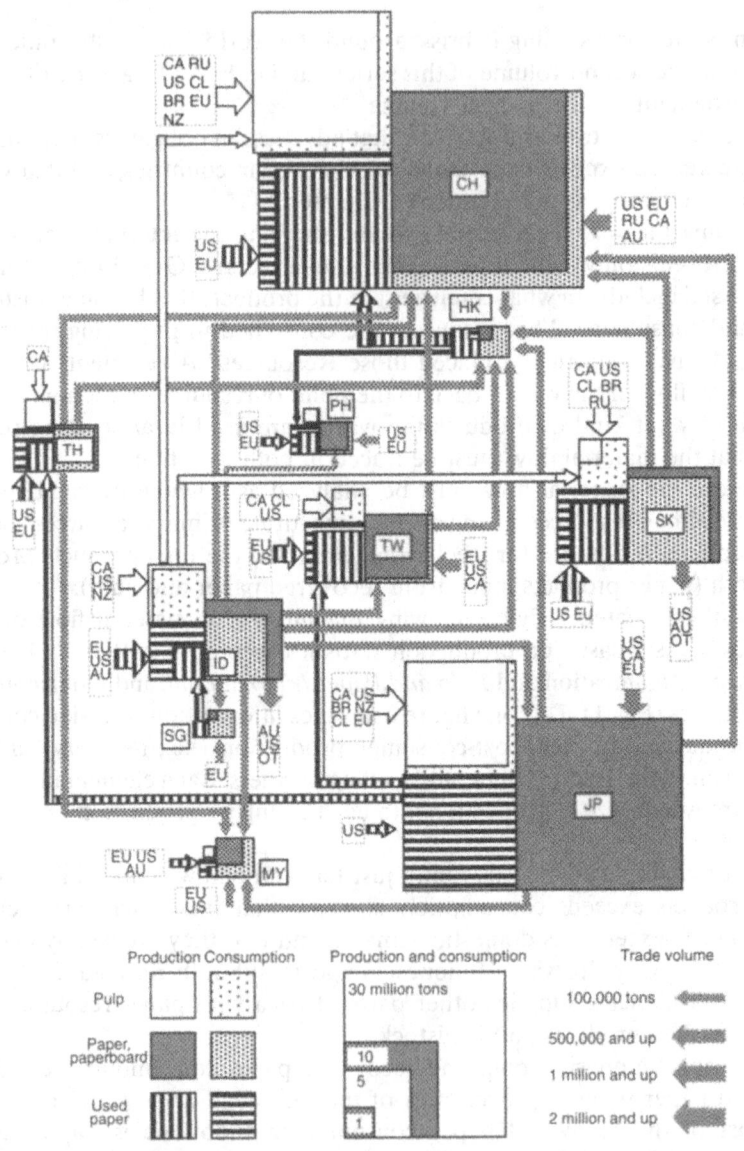

Note:  AU, Australia; CA, Canada; RU, Russia; US, United States; CL, Chile; CH, China; BR, Brazil; EU, European Union; NZ, New Zealand; HK, Hong Kong; TH, Thailand; PH, Philippines; TW, Taiwan; SG, Singapore; ID, Indonesia; OT, Others; SK, South Korea; MY, Malaysia; JP, Japan

"Other" exports by Indonesia are mainly Iran, Saudi Arabia, and other Middle Eastern countries.

"Other" exports by South Korea include Bangladesh and Vietnam.

Partners with higher trade volumes are above when written vertically and to the left when written horizontally.

TABLE 1. Material Flow of Paper Products in Asia (2000)

(10⁴ tons) — rendered as $(10^4 \text{ tons})$

| | (a) | (b) | (c) | (d) | (e) | (f) | (g) | (h) | (i) | (j) | (k) | (l) | (m) | (n) | (o) | $x$ |
|---|---|---|---|---|---|---|---|---|---|---|---|---|---|---|---|---|
| (a) | 3,005 | 12 | 1 | 7 | 4 | 2 | 1 | 1 | 6 | 1 | 2 | 0 | 1 | 1 | 46 | 3,090 |
| (b) | 0 | 19 | 0 | 0 | 0 | 0 | 0 | 0 | 0 | 0 | 0 | 0 | 0 | 0 | 0 | 19 |
| (c) | 63 | 10 | 404 | 14 | 8 | 29 | 5 | 7 | 18 | 7 | 12 | 1 | 15 | 14 | 90 | 698 |
| (d) | 36 | 8 | 2 | 3,038 | 5 | 16 | 2 | 6 | 11 | 7 | 5 | 0 | 2 | 15 | 26 | 3,179 |
| (e) | 90 | 18 | 1 | 6 | 696 | 9 | 3 | 6 | 4 | 2 | 12 | 0 | 2 | 40 | 87 | 976 |
| (f) | 2 | 1 | 1 | 0 | 0 | 52 | 0 | 6 | 0 | 1 | 0 | 0 | 0 | 0 | 16 | 79 |
| (g) | 4 | 0 | 0 | 0 | 0 | 1 | 74 | 2 | 3 | 1 | 0 | 0 | 0 | 0 | 2 | 87 |
| (h) | 0 | 0 | 0 | 0 | 0 | 0 | 0 | 9 | 0 | 0 | 0 | 0 | 0 | 0 | 0 | 9 |
| (i) | 42 | 16 | 1 | 4 | 0 | 6 | 1 | 0 | 330 | 2 | 0 | 0 | 0 | 1 | 46 | 449 |
| (j) | 24 | 4 | 1 | 1 | 0 | 12 | 3 | 5 | 2 | 150 | 3 | 0 | 4 | 4 | 18 | 232 |
| (k) | 11 | 0 | 1 | 0 | 0 | 2 | 1 | 2 | 3 | 1 | 239 | 1 | 14 | 5 | 4 | 284 |
| (l) | 13 | 0 | 2 | 26 | 5 | 6 | 2 | 2 | 10 | 1 | 12 | 580 | 874 | 304 | 254 | 2,092 |
| (m) | 65 | 10 | 8 | 27 | 12 | 20 | 10 | 15 | 22 | 10 | 75 | 146 | 5,766 | 920 | 358 | 8,463 |
| (n) | 92 | 5 | 4 | 76 | 15 | 14 | 15 | 6 | 18 | 5 | 9 | 145 | 249 | 7,822 | 181 | 8,655 |
| (o) | 42 | 0 | 1 | 0 | 2 | 11 | 3 | 4 | 11 | 6 | 10 | 39 | 695 | 242 | 2,979 | 4,045 |
| $x^*$ | 3,488 | 104 | 427 | 3,201 | 747 | 180 | 121 | 70 | 438 | 192 | 379 | 912 | 7,623 | 9,369 | 5,107 | |

Note: (a) through (o) stand for these countries/regions: (a) China, (b) Hong Kong, (c) Indonesia, (d) Japan, (e) South Korea, (f) Malaysia, (g) Philippines, (h) Singapore, (i) Taiwan, (j) Thailand, (k) Australia, (l) Canada, (m) EU, (n) United States, (o) other regions. $x$ is the production and $x^*$ is the consumption of each country/region. Production and collection statistics are based on the FAO database <http://apps.fao.org/page/collections?subset=forestry>, and supplemented by the PPI *Annual Review*. In the calculation of trade statistics, exports from A to B are incorporated into A's export statistics and B's export statistics, but ordinarily the two do not match. This analysis uses their averages as trade volume. Consumption was calculated from production/collection and trade. When apparent trade is negative owing to the effect of intermediate trade, data were corrected by converting the transit trade amount into direct trade from the initial exporter to the final importer to avoid having negative values in the table.

FIG. 2. Pulp Self-Sufficiency Rates in Asia, 2000

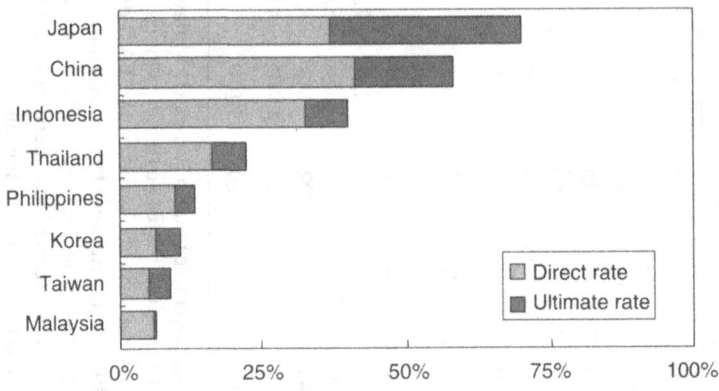

Indonesia supplies paper products to Singapore, as does China to Hong Kong, while at the same time recovered paper flows between them in the opposite direction, thereby creating closed cycles within the two pairs. Analyzing the material flow not only of wastes, but also with the inclusion of products and resources, makes it possible to get a more accurate picture of the international specialization.

In comparing the paper resource self-sufficiency rates of selected countries (Fig. 2), analyses require data on the material flow of paper products, pulp, and recovered paper. In the data for paper products (Table 1), the direct self-sufficiency rate is the percentage of domestically produced pulp in paper products consumed domestically. Because the material flows of feedstock and products are connected in this analysis, the feedstock mix at the time products are made by the exporter is used for items imported as products.

Types of feedstock other than domestic pulp are domestic recovered paper, foreign pulp, and foreign recovered paper. Tracing the origin of domestic recovered paper back to its original paper products will likely reveal that it contains domestic pulp and domestic recovered paper in the same proportions as paper products consumed domestically. For example, if the feedstock mix is 50% domestic pulp, 40% domestic recovered paper, and 10% other, going back one step in the material flow should find that half the 40% domestic recovered paper, or 20% of the total feedstock, is the indirect primary self-sufficiency rate. Tracing this indirect self-sufficiency rate ad infinitum yields the ultimate self-sufficiency rate.

Although Japan ranks below China in the direct self-sufficiency rate, its ultimate self-sufficiency rate is 70% as opposed to China's 58%. The same reversal occurs between Taiwan and Malaysia. Large changes in the difference between the ultimate and direct self-sufficiency rates depending on the country/region are due to their different recovered paper self-sufficiency rates.

In the region overall, only Japan and China have self-sufficiency rates topping 50%. And although not considered in this analysis, both countries are also pulp importers, indicating that as a whole Asia is highly dependent on non-Asian paper resources. Such information is obtainable only through an integrated assessment of product and resource material flows.[2]

Yamashita Hidetoshi

# 17. The Challenges of Controlling Greenhouse Gas Emissions

$CO_2$ emissions in nine East Asian countries/regions account for about one-fourth of the world's total, and are the largest after Europe's. Japan is the only one of these countries listed among Annex I countries (industrialized countries plus former Eastern European countries) of the UN Framework Convention on Climate Change, but large emitters include China with its huge population and South Korea, which is an OECD member. Taiwan has the highest per capita emissions at 11.5 t ($CO_2$ equivalent), while Vietnam has only 0.5 t, underscoring the large differences within Asia. $CO_2$ emissions due to fuel use are mainly from coal in China, but mostly from oil in other countries (Table 1).

By sector, a characteristic of East Asia is that industry has a large share of emissions. Although the industrial sectors of Western countries make up 30 to 40% of total emissions, the figures are 75% in China and 60% in Taiwan.

The US Energy Information Agency predicts that total world $CO_2$ emissions will exceed 30 billion t in 2020, and that developing countries' emissions will top those of industrialized countries (Fig. 1). Emission increases in Asia (excluding the Middle East) will be high in part because of China and India, coming to about 40% of the world total.

Fig. 1. Outlook of $CO_2$ emissions in 2020

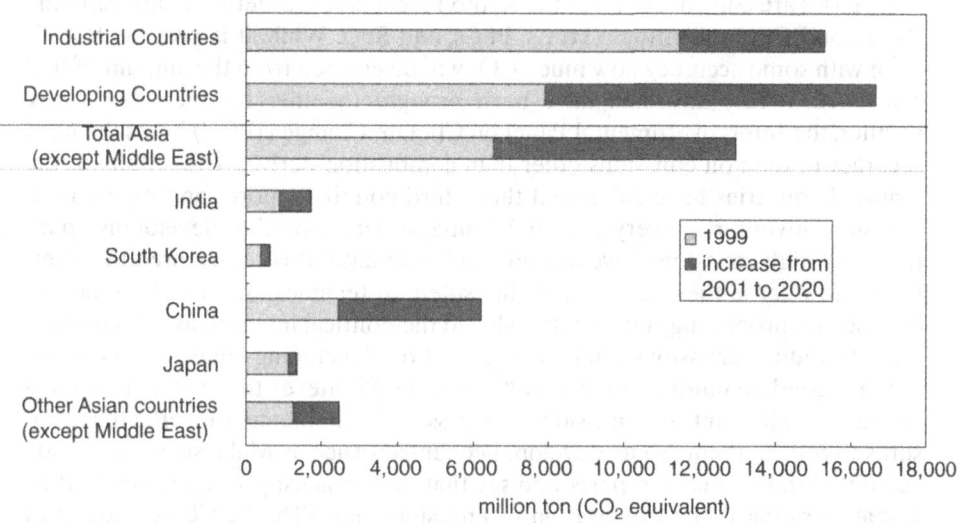

Source: Energy Information Administration (EIA), International Energy Outlook 2002, DOE/EIA-0484(2002), Washington, DC, March 2002, p.189.
Notes: 1. Outlook of emissions from 2001 to 2020 is based on EIA's reference case.
     2. $CO_2$ emission is energy-derived.

TABLE 1. Greenhouse Gas Emissions of East Asian Countries

|  |  | China | Indonesia | Japan |
|---|---|---|---|---|
| 1. Fossil fuel | total emission | 2,842 | 252 | 1,150 |
| 2000 | coal | 2,157 | 27 | 321 |
| (million ton—$CO_2$ equivalent) | natural gas | 59 | 71 | 152 |
|  | oil | 626 | 154 | 677 |
| 2. Emission rate of world (2000) |  | 12.0% | 1.1% | 4.9% |
| 3. Emission rate by sectors | Industry | 75.1% | 46.1% | 50.1% |
| 1998 | Household | 8.6% | 23.5% | 13.2% |
|  | Business | 10.9% | 3.4% | 11.6% |
|  | Transportation | 5.3% | 27.0% | 25.1% |
| 4. Emissions per capita (2000) |  | 2.23 | 1.20 | 9.07 |
| (ton—$CO_2$ equivalent) |  |  |  |  |

Source: 1&2; Energy Information Administration (EIA), *International Energy Annual 2000*, February 2002 <http://www.eia.doe.gov/emeu/iea/carbon.html>.
3; EIA, *Country Analysis Briefs: East Asia and South Asia*, March 2002, <http://www.eia.doe.gov/emeu/cabs/ cabsfe.html>.
4; EIA, *Per Capita (Person) Total Carbon Dioxide Emissions form the Consumption of Petroleum, Natural Gas, and Coal, and the Flaring of Natural Gas, All Countries, 1980–2000* (Million ($10^6$) Metric Tons Carbon Equivalent per Person), 2002, <http://www.eia.doe.gov/emeu/international/total.html#IntlCarbon>.

Six gases are controlled under the Kyoto Protocol: $CO_2$, methane, nitrous oxide ($N_2O$), and CFC substitutes (HFCs, PFCs, and $SF_6$). While it is possible to estimate with some accuracy how much $CO_2$ will be emitted from the amount of fuel consumed, hardly any data have been brought together for the other gases. Further, the Intergovernmental Panel on Climate Change (IPCC) notes the poor accuracy of data on emissions other than combustion-derived $CO_2$. Many of the Annex I countries have submitted their third country reports, and report their emissions inventories every year, but China, Brazil, and other developing countries with high emissions have not released their data. It is conjectured that their failure to release these data is not due solely to technical and fiscal limitations imposed on processing statistics, but also to the political intent to avoid the obligation to reduce emissions, which is imposed on developing countries beginning in the second commitment period (2013–2017). Whatever the case, determining the state of all countries' emissions is necessary for implementing effective measures to reduce them. Some developing countries such as Malaysia have already submitted their country reports, and say that their highest priority in combatting global warming is the preparation of emissions data. The FCCC secretariat at the UN encourages developing countries to calculate their emissions data inaccordance with IPCC guidelines (1995, revised in 1996). It provides financial assistance for some of the costs, and helps in other ways such as training specialists.

| S. Korea | Malaysia | Philippines | Taiwan | Thailand | Vietnam | 9 countries and areas |
|---|---|---|---|---|---|---|
| 423 | 110 | 72 | 255 | 166 | 41 | 5,309 |
| 146 | 8 | 19 | 119 | 31 | 13 | 2,843 |
| 40 | 40 | 0 | 13 | 34 | 3 | 411 |
| 237 | 61 | 52 | 123 | 101 | 25 | 2,056 |
| 1.8% | 0.5% | 0.3% | 1.1% | 0.7% | 0.2% | 22.5% |
| 49.8% | 51.9% | 55.0% | 60.4% | 41.3% | 38.1% | |
| 9.1% | 7.3% | 16.1% | 10.8% | 10.4% | 19.6% | |
| 17.1% | 10.6% | 6.7% | 6.5% | 12.1% | 8.8% | |
| 24.0% | 30.2% | 22.2% | 22.2% | 36.2% | 33.5% | |
| 8.95 | 4.71 | 0.93 | 11.47 | 2.66 | 0.53 | |

In convention negotiations parties are discussing the requirement for cutting emissions in the second commitment period. At COP 8 in Delhi in 2002 Brazil proposed that the parties consider setting reduction targets in the second commitment period and beyond based on historical emissions data. The "Brazil proposal" holds that setting targets should be equitable and employ the polluter pays principle instead of making decisions according to political compromises. It uses a simplified climate model, and suggests that emission requirements be decided scientifically. In this sense, it is a meaningful proposal.

Prior to this proposal, 13 research groups from industrialized countries around the world complied with a Brazilian government request by dividing the world into four regions (OECD countries as of 1990, former Eastern European countries, African and South American countries, and Asian countries), and calculated the responsibility for temperature rise up to 2000 based on the emissions of $CO_2$, methane, and $N_2O$ from 1890 to 2000. This interregional comparison of global warming responsibility found that about 40% of the responsibility lies in OECD countries, about 15% in former Eastern European countries, about 20% in African and South American countries, and 25% in Asian countries. Further, the results of the 13 research groups were very similar (Fig. 2). There are plans to assign numerical values to not only responsibility for past emissions, but also future emissions. Additionally, because improving the accuracy of country data will affect climate model results, it is urgent that the capability of research institutions in developing countries to build models be improved, especially for the purpose of preparing data on those countries.

Uezono Masatake

FIG. 2. Comparison of Reference Case Calculations: Relative Contribution to Temperature Change in 2000 with Emissions of $CO_2$, $CH_4$, and $N_2O$ Attributed Between 1890 and 2000 as Provided in the EDGAR Database

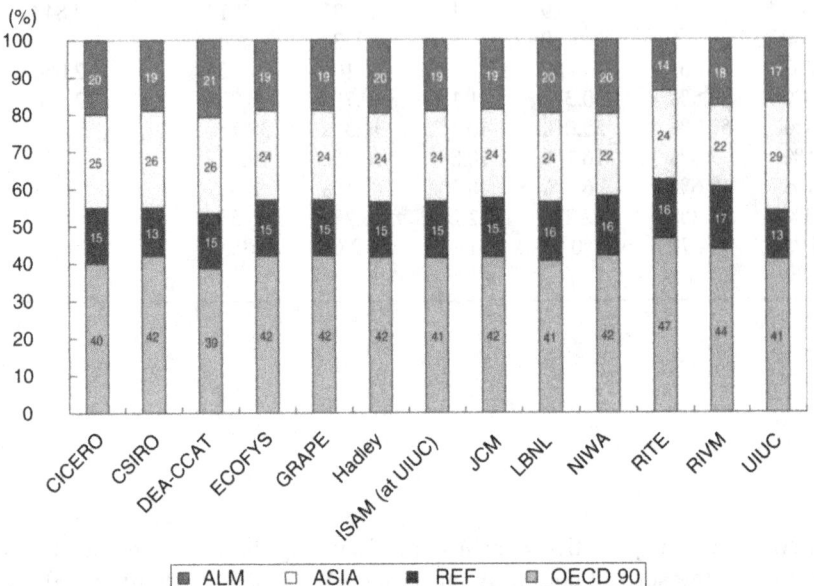

Notes: 1. OECD 90 (members of the OECD in 1990); REF (Eastern Europe and the former Soviet Union), ALM (Africa and Latin America and the Middle East); ASIA (Asia).

2. (a) Centre for International Climate and Environmental Research (CICERO), Norway;
   (b) Commonwealth Scientific and Industrial Research Organization (CSIRO), Australia;
   (c) Climate Change Advisory Team of the Danish Energy Agency (DEA-CCAT), Denmark;
   (d) ECOFYS Energy and Environment, Germany;
   (e) Hadley Centre, Meteorological Office, United Kingdom;
   (f) Institute of Applied Energy (GRAPE), Japan;
   (g) Universite catholique de Louvain using the Java Climate Model (JCM), Switzerland/Belgium;
   (h) Lawrence Berkeley National Laboratory (LBNL), United States of America;
   (i) National Institute of Water and Atmospheric Research (NIWA), New Zealand;
   (j) Research Institute of Innovative Technology for the Earth (RITE), Japan;
   (k) National Institute of Public Health and the Environment (RIVM), the Netherlands;
   (l) University of Illinois at Urbana Champaign using the Integrated Science Assessment Model (ISAM), United States of America;
   (m) University of Illinois at Urbana Champaign (UIUC), Climate Research Group (CRG), United States of America.

Source: UNFCCC, Scientific and Methodological Assessment of Contributions to Climate Change, FCCC/SBSTA/2002/INF.14, p.7.

# 18.  Nuclear Power Development and KEDO

The state of nuclear power in Asia over the last few years can be divided into two contrasting situations: stagnation in Japan and South Korea, and rapid expansion in India. In Japan, the September 1999 JCO criticality accident at Tokaimura, the MOX fuel data fabrication incident, and other events forced electric utilities to downsize their construction target of 16 to 20 new power plants by 2010[1] to 13 plants. As a result, the annual rate of 1.5 new plants in the 1970s and 1980s, which was a time of expansion for the industry, fell back to one per year, which generated concerns about inducing a decline in the nuclear power industry's supply capacity, and its repercussions for the economy. At the end of 2001 the big nuclear plant manufacturers Hitachi, Toshiba, and Mitsubishi Heavy Industries had a combined capacity of six plants a year. For this reason they are actively working on exports, construction of nuclear fuel cycle facilities, and the development of small reactors (300 MW or smaller). Yet, due to factors including retail deregulation in the electric power market and the slump in power demand, it is anticipated that the nuclear power industry will have a substantial decrease in sales. Additionally, the private sector, especially mining/manufacturing and power producers, are continuing to make cuts in research expenditures and personnel. It is also likely that the Japan Atomic Energy Research Institute and the Japan Nuclear Cycle Development Institute will merge.

In a bid to ensure that new plant sitings proceed smoothly, in January 2001 Japan's government created the Nuclear Power Promotion Law, which raises the subsidy rate for public works from 50 to 55%, and incorporated the 70% local burden rate into the criteria for calculating the local allocation tax. But as nuclear plants age (attaining about 30 years of service), breakdowns and accidents are expected to increase, as with Hamaoka I. It will likely remain highly difficult to surmount the hurdles to new plant sitings. There are 10 nuclear plants that started operating prior to 1975, including Tsuruga I and Fukushima I, which exceeded 30 years of operation as of August 2002. In late August 2002 a whistleblower revealed that the Tokyo Electric Power Company had concealed and fabricated inspection data at Fukushima I.

South Korea is actively pursuing nuclear plant construction in a bid to meet its anticipated electricity demand increase of 3.3% per year, but opposition by local communities is presenting hurdles to securing new sites. South Korea's government has already removed eight locations from its list of new site candidates, leaving it with only one new site as of August 2002 and forcing it to designate areas near existing plants as new construction sites. The First Draft Basic Plan for Electric Power Supply and Demand (2002–2015) anticipates that 28 nuclear plants will be operating by 2015, and that nuclear power facilities will represent 34.6% of total generating capacity, but difficulty in securing new sites means that eight plants would have to be concentrated at a single site. Additionally, the government has plans to extend the operating lifetimes of the Kori I and Wolsong I plants from 30 to 40 years. The former went on line in 1978 and was slated for decommissioning in 2008, while the latter started operating in 1983 and was to

TABLE 1. Nuclear Power in Asia (Number of Reactors as of August 2002)

|  | Japan | S. Korea | China | DPR of Korea | Taiwan | India | Pakistan |
|---|---|---|---|---|---|---|---|
| Operating | 53 | 18 | 5 | — | 6 | 14 | 2 |
| Under construction | 4 | 2 | 6 | 2 | 2 | 2 | — |
| Planned | 8 | 8 | 2 | — | — | 10 | — |

be decommissioned in 2013. Structural reform carried out in April 2001 divided the former electric power public corporation into six companies. It appears that the Korean Hydroelectric and Nuclear Power Company, which operates small-scale hydropower and all nuclear plants, will have to ask that nuclear plants be left in service longer so as to bolster its balance sheet.

Meanwhile, opposition by local communities continues to stymie progress in selecting a disposal site for low-level radioactive waste. The government and electric utilities want a site like that of Japan's Rokkasho-mura: a permanent disposal area for low-level radioactive waste, and an intermediate storage facility for high-level radioactive waste. When publicly seeking a site in 2001 the government increased monetary assistance from 200 billion won to 300 billion won and extended the application period. Nevertheless, not a single municipality responded. As of August 2002, the candidate sites are two islands south of the Korean Peninsula.

Taiwan's political parties have agreed that nuclear plants in operation will be decommissioned after 32 years, and that no new plants will be ordered other than the fourth (Liumen), making this island's nuclear phaseout effort the most advanced in Asia. In particular, the fact that in May 2000 President Chen Shuibian himself addressed the issue of canceling construction of the fourth plant, and the television broadcast of hearings on that issue, contributed significantly to the shaping of public opinion for phasing out nuclear power. Jinmen Island was chosen as the top candidate for Taiwan's radioactive waste final disposal site in place of the intermediate storage site on Lanyu Island (opened in 1982, closed in 1996), where community opposition stopped the further entry of low-level radioactive waste.

In China there is less active discussion on nuclear power plant construction owing to several factors, among them development of thermal power which aims to actively use domestic resources, and efforts to proceed smoothly with hydropower from the Three Gorges Dam. Even if all nuclear plants under construction were completed by 2002 and all 11 were in operation, it would account for only slightly more than 2% of total generating capacity. Still, China has not given up on sustainable nuclear power development, and the government intends to put more effort into domesticating and standardizing nuclear power. China is also working on natural gas pipelines from Central Asian countries and from Irkutsk in Russia, as well as other projects.

India faces a constant shortage of electricity (usually 8%), and maintains an electricity supply policy centered on nuclear power. Recently its nuclear power policy is shifting away from the CANDU reactors that have been its mainstay, and toward the introduction of VVER-1000 reactors with Russian assistance. However, in 2001 stiff local opposition induced authorities to hold India's first ever public hearings on nuclear power concerning the Koodankulam plant, which was to be built with 85% Russian assistance.

Although Indonesia's nuclear plant construction plans fell victim to financing difficulties, recently the Communist Party of Vietnam's Ninth Party Congress developed plans for a study on the construction of a nuclear plant of 3,000 to 4,000 MW capacity starting in 2010 or thereafter, and bringing it online in 2017.

In August 2002 the main construction phase of two light-water reactors in North Korea began under the Korean Peninsula Energy Development Organization (KEDO), and in preparation for completion of the first reactor in 2008, a new initiative is underway by Japan, South Korea, the US, and North Korea concerning the nuclear power compensation system that is essential for operating nuclear plants. Unless nuclear plants are state-run, governments exempt component and equipment makers from the enormous compensation payments in a nuclear accident, and make the plant operator assume all compensation liability. KEDO created a "Law on Compensation for Nuclear Damage" that includes such terms. There are also regional and international compensation agreements for cross-border contamination by nuclear accidents such as the Paris Convention, the Vienna Convention (and protocol to amend the convention), and the Convention on Supplementary Compensation for Nuclear Damage. In particular, the Law on Compensation for Nuclear Damage was created for the purposes of not only providing restitution for victims, but also to protect the nuclear power industry from the enormous damage of a nuclear accident, and to alleviate the burden on the insurance industry.

Companies supplying parts and equipment for the North Korean LWRs demand to North Korea and to the KEDO secretariat that they be exempted from operator liability, i.e., exclusive liability is channelled to the operator of the nuclear installation. In December 2000 the turbine and generator suppliers were changed to a consortium of the Japanese companies Hitachi and Toshiba from the US company General Electric, which feared liability for compensation. This is also a vital concern to Japan and South Korea, which must consider the issue of compensation in case of cross-border contamination. KEDO parties, especially Japan, South Korea, and the US, are hurrying to ratify the Vienna Convention Protocol and the Convention on Supplementary Compensation for Nuclear Damage, which they have until now avoided joining, and they are urging North Korea to create a domestic nuclear power compensation system and to ratify the two compensation conventions. As of August 31, 2002 only two countries had ratified those two conventions, but if Japan, South Korea, and the US ratify them, they will enter into force.

As of August 2002 the North Korean government, although not averse to creating a compensation system or becoming a party to the international compensation conventions, disagreed on who the compensating party should be. Specifically, North Korea feels that the state should be liable for compensation, but other KEDO parties express concerns about the North Korean government's ability to pay compensation. The only way to resolve this disagreement is to create an arrangement under which North Korea establishes a nuclear power insurance pool and enters into a reinsurance agreement with a foreign pool. Fortunately, it appears that North Korea is getting advice from foreign insurance companies.

Due to construction delays caused by North Korea's long-range missile tests and other events, it is anticipated that construction costs for the two LWRs will skyrocket from the 1998 sum of $4.6 billion to $9 billion. The $4.6 billion amount was to have been covered mainly by South Korea ($3.2 billion) and Japan ($1 billion), but it has not been decided who will shoulder the remaining 8% in addition to the EU (€87.5 million annually for five years).

Chang Jung-Ouk

# 19. International Contact among Environmental NGOs Grows More Vigorous

When environmental NGOs, citizens, and citizens' groups work on the solution to a problem, no matter what it may be, they cannot avoid trying to influence government policy, and especially administrative agencies, and their efforts perforce embody certain political stands and activities. Hence NGO activities will change considerably depending on the extent to which governments acknowledge the political views expressed by citizens and NGOs. The activities of an environmental NGO are thus necessarily affected significantly by its country's national polity.

Nearly all Asian countries restricted the activities of citizens and their organizations, but primarily in East and Southeast Asia during recent years the political freedom of citizens and NGOs has expanded with economic development, which has in turn energized their activities. Progress is also seen in NGO-related legislation. An example is the strong state control of NGOs in China. In October 1989 after the Tiananmen Square incident the government created an "Ordinance for the Registration and Control of Social Organizations," which tightened control of NGOs. But restrictions on NGOs were loosened under the policy of reform and opening the country, and the private organization control system launched in 1998 recognized "social organizations"[1] and "nongovernmental and noncommercial enterprises."[2] There are also many NGOs that register as companies, or create subordinate organizations within existing social organizations. Additionally, as part of the government's 1998 organizational reform, "encouraging the development of nongovernmental organizations" was included among the duties and authority the State Environmental Protection Administration.

In the 1980s South Korea emerged from a long military dictatorship that had continued since the Korean War and entered an era of democratization. A representative event was the democratization struggle of June 1987. While the democracy movement had opposed the dictatorship up to that time, it could not grow as a citizens' movement because it was illegal. The end of military rule and the advent of democracy made legal citizens' movements possible, and from the mid-1980s onward South Korea's citizens' movements experienced a renaissance.

With the fall of the Marcos government in 1987 and the establishment of the Aquino government in the Philippines, the 1987 constitutional amendment included provisions that applauded the role of the private nonprofit sector and encouraged it, leading to the sector's rapid development. Taiwan also exercised strict control over the activities of private organizations, but NGOs increased and their activities flourished in tandem with advances in democratization beginning in the mid-1980s.

## Status of Asian Environmental NGOs

There are few resources or studies that tell how many environmental NGOs Asian countries have, or give an accurate picture of them. It is therefore nearly impossible to find accurate figures or details. Even the definition of NGO differs from one country to another. This situation holds in Japan as well. The limited number of studies, reports, and descriptions available give the following general picture of environmental NGOs in major Asian countries.

*China.* According to a report by Wang Canfa, who is affiliated with the Center for Legal Assistance to Pollution Victims at the China University of Political Science & Law, 2000 statistics indicated that China had 136,000 social organizations several thousand of which call themselves NGOs. A questionnaire survey conducted on Beijing NGOs by the Tsinghua University NGO Center in 1999 and 2000 found that in the distribution by activity area, 16 of 299 or 15.4% (multiple responses) claimed to be nature protection NGOs. Some specific examples of environmental NGOs are the China Environmental Protection Fund (founded in 1993), said to be a representative association-type NGO, while grassroots NGOs are Friends of Nature (1994), founded mainly by faculty members at Beijing University, and the Beijing Global Village Environmental Center (1996). The Center for Legal Assistance to Pollution Victims mentioned above was founded in 1998 by legal specialists and academics. Mr. Wang says that as of November 2001 the Center was the only Chinese NGO that directly extends help to pollution victims.

*India.* India has traditional charity activities with a religious context, and flourishing NGO activities have earned India a reputation as an NGO superpower. Especially since the 1980s the number of NGOs has soared, so that nearly 80% of all NGOs have been created between then and now. Environmental problems are actively addressed by the government and vigorous NGO efforts. Data from a 1998 study on 3,289 NGOs indicate that 2,301 of them (70.4%) claimed to be working in the "environment and pollution" field, which was second after "health and nutrition" (multiple responses). A representative environmental NGO is the Centre for Science and Environment (CSE), founded in 1980. With a full-time staff of about 75 people, and the purpose of research and advocacy on environmental issues, CSE is also very active on climate change.

*Indonesia.* Indonesia's NGOs started gaining momentum in the 1970s, and environmental NGOs initiated activities in the 1980s. Environmental NGOs numbered only 12 in 1978, but the count swelled to about 750 in the mid-1990s. The 2000 "NGO Directory" lists about 400 NGOs, of which 160 (40%) are active in the environmental arena. The Indonesia Forum for the Environment (WALHI), an environmental NGO network, was founded in 1980. In 1982, shortly after its inception, WALHI worked with the Ministry of Natural Environment in drafting the Law Concerning Environmental Management, but in the 1990s its relationship with the government soured over issues such as forest protection and opposition to dam construction. WALHI has a full-time staff of 20, and 260 participating organizations.

*South Korea.* The "Inventory of Korean Private Environmental Organizations 2000" cites a 1999 survey that found 4,023 civil society organizations, a figure that rises to about 20,000 when including local chapters. Of these, environmental NGOs numbered 287, or 7.1%. A characteristic of South Korea's environmental NGOs is that they are larger than those of Japan. The Korean Federation for Environmental Movement, which was founded in 1993 as a federation of eight organizations involved in anti-pollution campaigns and other initiatives throughout the country, has a membership of 70,000 and 70 full-time staff members. Green Korea, a federation of three organizations formed in 1994, has 10,000 members and a full-time staff of 20. South Korean environmental NGOs and citizens turned out in large numbers at the August 2002 Johannesburg Summit and made the strength of South Korea's environmental movement felt by their vigorous presence there.

*Taiwan.* In 1999 there were 15,328 social organizations registered with the government, of which 4,470 were "social services and philanthropic organizations," but it is not clear how many of these are NGOs addressing environmental issues. A representative environmental NGO is the Taiwan Environmental Protection Union (TEPU, which was formed as a network of environmental movements in November 1987, just after martial law had been lifted that July, and in the context of the 1980s campaigns against pollution, large-scale development, and nuclear power. TEPU apparently as 11 chapters and at least 1,000 members throughout Taiwan.

## International Contact among Asian Environmental NGOs

International contact among Asian environmental NGOs began to flourish beginning with the 1992 Earth Summit, and a number of Asia-level networks have formed. Such international contact among environmental NGOs, and their networks for the everyday exchange of information and experiences are increasingly vital as global environmental problems become more serious. Yet many NGOs have difficulty carrying on continuous activities for lack of funding.

*Climate Action Network.* Founded in 1989, CAN is an international network of environmental NGOs addressing climate change, currently comprising 334 NGOs from 81 countries (Fig. 1). There are 60 Asian NGOs, the largest number for any region after western Europe. CAN is not organized as a network on the East Asia level. Formerly only Japanese NGOs were participating, but recently they were joined by the Korean Federation for Environmental Movement and the Taiwan Environmental Protection Union.

*Atmosphere Action Network East Asia.* AANEA is a network of East Asian environmental NGOs created at convention in Seoul in August 1995 with participating organizations from seven regions: South Korea, China, Taiwan, Hong Kong, Russia, Mongolia, and Japan (Table 1). AANEA's purposes include building an information network, sharing experiences, performing joint research and studies, setting up monitoring systems, and upgrading the capabilities of citizens and environmental NGOs to address problems such as regional air pollution, acid

FIG. 1.  CAN Member Organizations in Asian Countries/Regions (2002)    FIG. 2.  CAN Member Organizations by Region (2002)

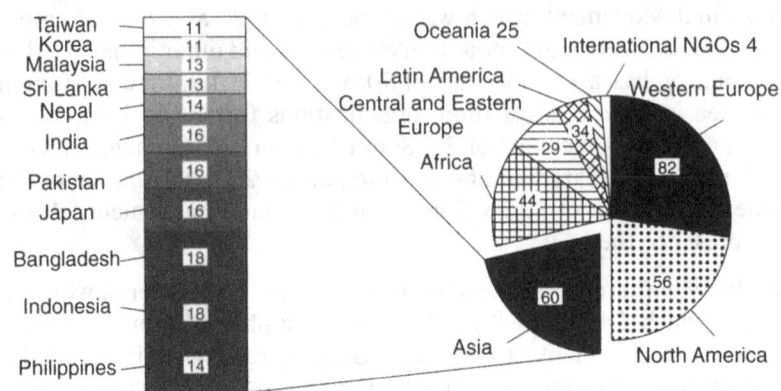

Total: 60 Organizations                    Total: 334 Organizations

Source: Climate Action Network International NGO Directory 2002 <http://www. climnet.org/>

TABLE 1.  AANEA's Members list

| Korea | Semin Foundation of Korea |
|---|---|
| | Citizen's Movement for Environmental Justice (CMEJ) |
| | Korean Federation for Environmental Movement (KFEM) |
| | Green Korea United |
| China | Friend of Nature |
| | Peking University Center for Environment Science |
| Hong Kong | The Conservancy Association (CA) |
| Taiwan | Taiwan Environmental Protection Union (TEPU) |
| | Climate Action Network Taiwan |
| Japan | Citizens' Alliance for Saving the Atmosphere and the Earth (CASA) |
| | Peoples' Forum 2001 Japan/Global Warming Research Group |
| | The Japan Air Pollution Victims Association |
| | Japan Acid Rain Monitoring Network |
| | Japan Tropical Forest Action Network (JATAN) |
| | Citizens' Air Pollution Survey (CAPS) |
| Mongolia | The Mongolian Association for Conservation of Nature and Environment (MACNE) |
| | Development and Environment NGO (D&E) |
| Russia | Geographical Society, Public Initiative for Ecology (PIE) |
| | Khabarovsk Branch of The Wild Life Foundation |

rain, ozone layer depletion, and global warming. So far AANEA has held six conventions in various countries to share information and experiences.

*Sustainable and Peaceful Energy Network Asia.* Sparked by the December 1997 Third Conference of the Parties to the United Nations Framework Convention on Climate Change in Kyoto, SPENA was founded in 1998 when environmental NGOs and researchers tackling energy issues gathered in South Korea from countries including Indonesia, Thailand, the Philippines, Malaysia, South Korea, Japan, Taiwan, and India. SPENA's purpose is to promote the formulation of policy for sustainable and peaceful energy in East Asia. It has held workshops for environmental NGOs and researchers in Bangkok (1999), Tokyo (2000), Jakarta (2001), and the Johannesburg Summit (2002).

*NGO International Convention for Environmental Rehabilitation.* In November 2001 an NGO international convention called "May the 21st Century Be the Age of Environmental Rehabilitation" was held in Kitakyushu City, Japan with the participation of environmental NGOs and researchers from countries/regions including India, Japan, South Korea, the Philippines, Taiwan, and Thailand. The convention's purpose was billed as: "Members of Japanese, other Asian, and European NGOs who are working to regenerate the environment even while living with pollution and environmental damage will describe their efforts, learn from each other, and develop proposals." In addition to the two-day meeting, participants from outside Japan visited the sites of itai-itai disease and Minamata disease.

Hayakawa Mitsutoshi, Kawasaka Kyoko

## 20. Trends in Environmental ODA, and the Need for Post-Project Evaluation

At the 1989 Summit of the Arch and the 1992 UN Conference on Environment and Development, Japan pledged to increase the amount of its environmental official development assistance (ODA) as part of its international contribution. Environmental ODA is defined as assistance for improvement of living conditions, disaster prevention, forest conservation and afforestation, pollution prevention, environmental conservation (wildlife protection, etc.), and addressing global environmental problems such as ozone layer depletion. ODA amounts have been given on the basis of these classifications.

But this is just the definition decided on by Japan. Other donor countries show their ODA amounts using other definitions, or do not create such ODA categories at all. Some of the underlying reasons are: there are few agencies having systems able to ascertain the funding amounts set aside for environmental conservation; it is difficult to determine the percentage of funds allotted to the environmental component when conservation plays a secondary role in the project purpose; except for a few donor countries, governments are reluctant to pledge quantitative targets for funds allotted to the environment; and, the difficulty of defining the environment leads to judgmental errors and confusion in classification. In fact, Japanese aid projects have also included an increasing number of projects that have no conservation components at all despite being classified as tap water/sewerage projects, and that intend to promote cleaner production being classified under the production sector instead of the environmental sector.

Hence the classifications and definitions used until now have lost much of their relevancy.

For this reason the OECD's Development Assistance Committee (DAC) released its own proposal for the classifications and definitions pertaining to environmental ODA, and has used it to classify donor countries' aid. At this time, statistics prepared according to this definition are the only way to quantitatively compare donor countries' environmental ODA.

Under the DAC-proposed classification, environmental aid is defined as not only ODA classified in the environmental sector, but also aid which aims "to improve the physical or biological environment of the recipient country, area, or target group concerned," and "specific action to integrate environmental concerns with a range of development objectives through institution building and/or capacity development." It further defines environmental aid as "environmental" only if the environment is a "principal objective" (type 2) or a "significant objective" (type 1), thereby excluding projects in which the environment is incidental or those whose environmental benefits are mere secondary effects. A feature of this classification is that it recognizes that aid projects in sectors other than the environment can elicit positive effects for environmental conservation.

Under this definition of environmental ODA, from 1995 to 2000 a total of 2,470 environmental ODA projects worth $28.1 billion was provided to in East and

Southeast Asia. The breakdown was 1,107 type 2 projects at about $8 billion and 1,363 type 1 projects at about $20.1 billion. In terms of the number of projects, these come to 14.6% of all environmental ODA provided to developing countries, even fewer projects than sub-Sahel Africa and South and Central America (Table 1). But these projects account for 46.7% of the total in monetary terms, the largest proportion of all the regions (Table 2). Further, the monetary amount of environmental ODA provided to East and Southeast Asia was 45.3% of total ODA, which gives the impression that ODA is becoming "green," but in terms of number, environmental ODA projects are only 17% of the total, belying the "green" image suggested by the monetary-base percentage.

Japan's environmental ODA significantly influences this characteristic of environmental ODA in the East/Southeast Asia region, which has the most such aid

TABLE 1. Number of Environmental ODA Receipts by Region, 1995–2000

|  | Receipts | % of total |
|---|---|---|
| Europe, Russia | 1,001 | 5.93 |
| Africa | 4,542 | 26.90 |
| Central and South America | 3,112 | 18.43 |
| Middle East, North Africa | 984 | 5.83 |
| South Asia | 1,536 | 9.10 |
| East and Southeast Asia | 2,470 | 14.63 |
| Oceania | 292 | 1.73 |
| Developing countries in general | 2,948 | 17.46 |
| Total | 16,885 | |

Source: Prepared by author using OECD, *International Development Statistics 2003*.

TABLE 2. Environmental ODA Receipt Amounts by Region, 1995–2000

|  | US$1,000 | |
|---|---|---|
|  | Amount | % of total |
| Europe, Russia | 2,186,599 | 3.64 |
| Africa | 5,628,239 | 9.37 |
| Central and South America | 6,560,811 | 10.92 |
| Middle East, North Africa | 4,653,249 | 7.75 |
| South Asia | 9,885,452 | 16.45 |
| East and Southeast Asia | 28,083,846 | 46.75 |
| Oceania | 259,527 | 0.43 |
| Developing countries in general | 2,818,538 | 4.69 |
| Total | 60,076,261 | |

Source: Same as Table 1.

on a monetary base, but little on a project-number base. One major characteristic of Japanese ODA is that the monetary amount per project is higher than that of other donor countries. As a result, the contribution made by Japanese aid tends to be large monetarily but small in the number of projects. ODA received in East and Southeast Asia from 1995 to 2000 was 64% Japanese in monetary terms but only 40% by number of projects. This difference is even wider for environmental ODA, with the figures being 89% on a monetary base but under 20% by number of projects.

Data on environmental ODA in East and Southeast Asia on a monetary base according to recipient country from 1995 to 2000 indicate that China was the biggest recipient, followed in descending order by Indonesia, Thailand, the Philippines, and Vietnam (Table 3). Aid to Malaysia increased from 1999, while that to Thailand and Indonesia declined gradually. Malaysia's increase was largely due to its switch from private financing to ODA because of the unavailability of private financing caused by the economic crisis, and due to Japan's assistance for overhauling a large thermal power plant to reduce greenhouse gas emissions. Transitional countries such as China, Vietnam, Mongolia, Cambodia, and Laos received more ODA especially starting in 1998. Therefore Mongolia, Cambodia, and Laos, which received less than 1% of the region's environmental ODA in monetary terms, got about 4% according to the number of projects (Table 4).

A fact of importance here is that even though a country might receive much environmental ODA in terms of either projects or money, it will not necessarily reduce that country's environmental burden or result in environmental improvements.

The reasons are, first, that an ODA project whose primary purpose is environmental conservation has clear limits on the environmental effectiveness of the technologies provided. For example, building a sewage treatment plant may

TABLE 3. Environmental ODA Receipt Amounts in East and Southeast Asia, 1995–2000

|  |  | US$1,000 |
| --- | --- | --- |
|  | Amount | % of total |
| Cambodia | 119,586 | 0.43 |
| China | 8,148,881 | 29.02 |
| Indonesia | 6,043,093 | 21.52 |
| Laos | 111,778 | 0.40 |
| Malaysia | 1,676,867 | 5.97 |
| Mongolia | 182,005 | 0.65 |
| Philippines | 4,255,293 | 15.15 |
| Thailand | 4,460,944 | 15.88 |
| Vietnam | 2,794,970 | 9.95 |
| Others | 290,429 | 1.03 |

Source: Same as Table 1.

TABLE 4. Number of Environmental ODA Receipts in East and Southeast Asia, 1995–2000

|  | Receipts | % of total |
| --- | --- | --- |
| Cambodia | 116 | 4.70 |
| China | 602 | 24.37 |
| Indonesia | 379 | 15.34 |
| Laos | 104 | 4.21 |
| Malaysia | 72 | 2.91 |
| Mongolia | 62 | 2.51 |
| Philippines | 305 | 12.35 |
| Thailand | 166 | 6.72 |
| Vietnam | 384 | 15.55 |
| Others | 280 | 11.34 |

Source: Same as Table 1.

improve the regional environment by treating effluent, but it may worsen the environment near the facility with the discharge of treated water. Such concerns underlay the temporary suspension of the construction of a treatment plant on the outskirts of Bangkok that was to treat both household and factory effluent.[1]

Second, there are political, administrative, and social restrictions which prevent the assumed beneficiaries from having the incentives and capabilities for effective implementation of environmental ODA. Many developing countries put economic growth first and do not assign high priority to environmental conservation. If a donor follows the request-driven principle as Japan has done in the past, then there will be few requests and grants for environmental improvement. Therefore if donor countries want to provide more environmental ODA, it will have to be donor-driven. Yet donor-driven projects are not always consistent with the priorities and political will of the recipients and beneficiaries. It is often the case that projects will be cancelled, or will not be very effective due to a lack of political and social acceptance. This tendency is especially pronounced with projects that aim to support changes in recipient countries' environmental and sectoral policies through environmental ODA.

Further, even if a recipient country has an incentive for implementation, a project will not necessarily be very effective in conserving the environment if that country has not built the institutions and management capabilities to take advantage of the environmental technologies provided.[2]

Finally, it is possible that ODA in which environmental conservation is a significant objective will give little regard to conservation if the project pursues its principal objective. For example, ODA for building expressways or elevated railways aims directly to improve transportation infrastructure, but it is expected that the project can afford the secondary benefits of diminishing traffic congestion, and in turn mitigating health damage. For this reason such projects are often classified as environmental ODA. But if roads are built without measuring environment capacity and developing a plan for road structure and land use that are commensurate with that capacity, a project could end up degrading the environment.[3] For this reason ODA whose significant objective is environmental should include carefully performed post-implementation evaluation from a perspective of not only its principal objective, but also environmental conservation. And as knowledge and lessons from such evaluations are built, the classification of environmental ODA should be given even closer scrutiny.

Mori Akihisa

# 21. Environmental Conventions and Agreements

## Ratification and Implementation of Multilateral Environmental Agreements

Definite progress is being realized in the ratification and implementation of multilateral environmental agreements (MEAs) by major Asian countries (Table 1).

One environmental agreement that has achieved impressive progress is the Ramsar Convention. Over the last three-odd years Japan's designated sites have increased from 10 to 13, expanding total designated area by about 550 ha. China's designated wetlands increased three-fold, and its total designated area over four-fold. Thailand has exponentially expanded its Ramsar sites from one to 10 and from 494 ha to 370,600 ha. India more than tripled its number of sites from six to 19. In Asia as a whole, 76 new wetlands were designated between May 1999 and November 2002, which approximately doubled the number of sites and raised total designated wetland area to 8.2 million ha, about 8% of worldwide designated wetland area. Underlying this progress are factors including sensitivity to Ramsar obligations in the domestic environmental policies and plans of signatories; the creation of wetland centers; getting more Ramsar sites designated as a means of conserving and managing wetlands; and progress in regional cooperation for joint training, capacity-building, and information exchange.

Excellent progress is also being achieved in implementing the Montreal Protocol on Substances That Deplete the Ozone Layer. Four amendments since the protocol took effect in 1989 have brought more substances under controls that have themselves been further tightened.

Ratification of these four amendments by major countries indicates their positive efforts to protect the ozone layer (Table 2). Japan, Malaysia, and India have already ratified all the amendments. Even China, which is behind in ratifying them, notified all its domestic automakers in 2000 that they would have to replace the CFC-12 used in automobile air conditioners with HFCs by the end of 2001. In this sense the protocol has benefited significantly from arrangements by which providing funds through the Multilateral Fund and the Global Environment Facility to developing countries' ozone-depleting substance (ODS) reduction programs definitely cuts production of regulated substances and helps them meet their obligations. In Asia, considerable funding goes to China and India in particular. In 2002 the funding for China and India approved by the Multilateral Fund's Executive Committee from sources including the World Bank, UNEP, UNDP, and bilateral funding was about US$62.5 million and US$14.5 million, respectively.

Attention is also focused on progress in Kyoto Protocol ratification. The 2001 agreement on the future implementation rules (Marrakesh Accords) added new momentum toward bringing the protocol into force, and good progress in ratification was occasioned by the Johannesburg Summit in August and September of 2002. The protocol was ratified by China, which accounts for 12.6% of total worldwide emissions, Japan (5.1%), India (4.8%), and South Korea (1.7%). In

TABLE 1. Ratification Status of Multilateral Environmental Agreements by Asian Countries

|  | Ramsar Convention | CITES | Basel Convention | UN Framework Convention on Climate Change |
|---|---|---|---|---|
| Signature or adoption | Feb 2, 1971 | Mar 3, 1973 | Mar 22, 1989 | May 9, 1992 |
| Entry into force | Dec 21, 1975 | Jul 1, 1975 | May 5, 1992 | Mar 21, 1994 |
|  | Entry into force Ramsar sites (surface area) | Ratification* | Ratification* | Ratification* |
| Japan | Oct 17, 1980 13 sites (84,089 ha) | Aug 6, 1980 (Ac) | Sep 17, 1993 (A) | May 28, 1993 (Ap) |
| S. Korea | Jul 28, 1997 2 sites (960 ha) | Jul 9, 1993 (A) | Feb 28, 1994 (A) | Dec 14, 1993 (R) |
| China | Jul 31, 1992 21 sites (2,547,763 ha) | Jan 8, 1981 (A) | Dec 17, 1991 (R) | Jan 5, 1993 (R) |
| Philippines | Nov 8, 1994 4 sites (68,404 ha) | Aug 18, 1981 (R) | Oct 21, 1993 (R) | Aug 2, 1994 (R) |
| Vietnam | Jan 20, 1989 1 site (12,000 ha) | Jan 20, 1994 (A) | Mar 13, 1995 (A) | Nov 16, 1994 (R) |
| Malaysia | Mar 10, 1995 4 site (48,745 ha) | Oct 20, 1977 (A) | Oct 8, 1993 (A) | Jul 13, 1993 (R) |
| Indonesia | Aug 8, 1992 2 sites (242,700 ha) | Dec 28, 1978 (A) | Sep 20, 1993 (A) | Aug 23, 1994 (R) |
| Thailand | Sep 13, 1998 10 sites (370,600 ha) | Jan 21, 1983 (R) | Nov 2, 1997 (R) | Dec 28, 1994 (R) |
| India | Feb 1, 1982 19 sites (648,507 ha) | Jul 20, 1976 (R) | Jun 24, 1992 (R) | Nov 1, 1993 (R) |

Note: Ratification as of June 2004. "Ratification (R)," "acceptance (Ac)," "approval (Ap)," and "accession (A)" all mean the act by which a State's consent to be bound by a treaty may be expressed (Vienna Convention on the Law of Treaties, Article 2(b)), and have the same effect under international law. "Accession" is often used by countries which express their consent even though they did not take part in treaty negotiations or sign, but "acceptance" and "approval" may be used in the same sense as "accession." The lower items for countries under the Ramsar Convention are the numbers and total areas of wetlands designated by these countries as wetlands of international importance and registered with the secretariat.

| Kyoto Protocol | Biodiversity Convention | Convention to Combat Desertification | Rotterdam Convention |
|---|---|---|---|
| Dec 11, 1997 | Jun 5, 1992 | Jun 17, 1994 | Sep 10, 1998 |
| Not yet in force | Dec 29, 1994 | Dec 26, 1996 | Feb 24, 2004 |
| Ratification* | Ratification* | Ratification* | Ratification* |
| Jun 4, 2002 (Ac) | May 28, 1993 (Ac) | Sep 11, 1998 (A) | Jun 15, 2004 (Ac) |
| Nov 8, 2002 (R) | Oct 3, 1994 (R) | Aug 17, 1999 (R) | Aug 11, 2003 (R) |
| Aug 30, 2002 (Ap) | Jan 5, 1993 (R) | Feb 18, 1997 (R) | |
| Nov 20, 2003 (R) | Oct 8, 1993 (R) | Feb 10, 2000 (R) | |
| Sep 25, 2002 (R) | Nov 16, 1994 (R) | Aug 25, 1998 (A) | |
| Sep 4, 2002 (R) | Jun 24, 1994 (R) | Jun 25, 1997 (R) | Sep 4, 2002 (A) |
| | Aug 23, 1994 (R) | Aug 31, 1998 (A) | |
| Aug 28, 2002 (R) | Jan 29, 2004 (R) | Mar 7, 2001 (A) | Feb 19, 2002 (A) |
| Aug 26, 2002 (A) | Feb 18, 1994 (R) | Dec 17, 1996 (R) | |

TABLE 2. Montreal Protocol Amendments Ratification Status

| | London Amendment<br>Jun 27–29, 1990<br>Aug 10, 1992 | Copenhagen Amendment<br>Nov 23–25, 1992<br>Jun 14, 1994 | Montreal Amendment<br>Sep 15–17, 1997<br>Nov 10, 1999 | Beijing Amendment<br>Nov 29–Dec 3, 1999<br>Feb 25, 2002 |
|---|---|---|---|---|
| Adopted / Entry into force | | | | |
| Japan | Sep 4, 1991 (Ac) | Dec 20, 1994 (Ac) | Aug 30, 2002 (Ac) | Aug 30, 2002 (Ac) |
| S. Korea | Dec 10, 1992 (A) | Dec 2, 1994 (Ac) | Aug 19, 1998 (Ac) | Jan 9, 2004 (Ac) |
| China | Jun 14, 1991 (A) | Apr 22, 2003 (A) | — | — |
| Philippines | Aug 9, 1993 (R) | Jun 15, 2001 (R) | — | — |
| Vietnam | Jan 26, 1994 (A) | Jan 26, 1994 (A) | — | — |
| Malaysia | Jun 16, 1993 (A) | Aug 5, 1993 (A) | Oct 26, 2001 (R) | Oct 26, 2001 (R) |
| Indonesia | Jun 26, 1992 (A) | Dec 10, 1998 (A) | — | — |
| Thailand | Jun 25, 1992 (R) | Dec 1, 1995 (R) | Jun 23, 2003 (R) | — |
| India | Jun 19, 1992 (A) | Mar 3, 2003 (A) | Mar 3, 2003 (A) | Mar 3, 2003 (A) |

view of the large proportion of total world emissions that Asian countries are responsible for, and of the predicted emission increases brought about by economic development, as well as from the perspective of universal commitment addressing global warming, these countries' ratification of the Kyoto Protocol has no small significance.

MEA ratification and implementation are making progress overall, but implementation still has problems. First, developing countries especially have difficulty implementing MEAs owing to funding and personnel shortages. For example, to implement the Biodiversity Convention and CITES, developing countries impose stiff penalties for illegal logging, poaching, and other illegal acts, and train personnel for enforcement, but they cannot perform adequate regulation and monitoring due to insufficient funds and personnel.

Second, in consideration of the effects that convention implementation has on other environmental issues, implementation at the national level sometimes requires international coordination. An example is the switch to HFCs and other greenhouse gases in an attempt to facilitate implementation of the Montreal Protocol. HFCs and other gases used as substitutes for ODSs are powerful greenhouse gases regulated by the Kyoto Protocol. Although the switch to substances with low greenhouse effects is already technically possible, switching to such greenhouse gases is used as a way to protect the ozone layer. Further, the Multilateral Fund recognizes ODS reduction programs as eligible for funding if they help the switch to HFCs for refrigerants in refrigerators, car air conditioners, and other applications. Even recent Multilateral Fund Executive Committee meetings have continued to approve such projects. On a bilateral basis, funds provided by Japan and several other countries have been used for projects that facilitate the switch to such greenhouse gases. Although the ozone layer protection regime and the climate change regime are separate regimes with different purposes, appropriate coordination should be exercised not only in each country when implementing domestic policies, but also between the two conventions to prevent mutual interference between the regimes.

Third, developing Asian countries continue receiving hazardous waste, some of which is exported to them illegally. Typical examples are the illegal export from Japan to the Philippines of medical waste that was subsequently reimported by the Japanese government in place of the exporter (1999–2000), and the illegal export to China of waste industrial plastic by a German company that later reimported the containers (2000). In 2000 China banned the import of electroscrap such as televisions, microwave ovens, and computers, and in 2001 it ratified the Basel Convention amendment banning the export of hazardous waste by OECD countries (industrialized countries) to non-OECD countries (developing countries). Ratification of the Basel amendment in Asia is limited to five countries including China and Malaysia, and ratification by waste-exporting industrialized countries is also overdue.

Therefore it is increasingly important to build and strengthen domestic institutions for ensuring waste reduction and environmentally sound management, and for monitoring illegal transboundary movement. It is also necessary to reinforce arrangements to internationally provide funding and technical assistance to countries which lack such capabilities and financial resources.

## Regional Environmental Agreements in Asia

Southeast Asian countries are carrying out vigorous regional initiatives centered on ASEAN, while in other parts of Asia, international cooperation in the environmental arena is based primarily on bilateral agreements, mainly because of political reasons. In addition to its preexisting agreements, in 2002 ASEAN countries signed an Agreement on Transboundary Haze Pollution.

A regional environmental agreement recently in the spotlight is the Convention for the Conservation of Southern Bluefin Tuna, concluded by Japan, Australia, and New Zealand in 1993. Based on the awareness that this tuna's stock may become depleted unless restrictions are imposed on its catch, this convention has endeavored to facilitate stock recovery by setting quotas for each party. In one case involving the convention, Australia and New Zealand submitted a dispute to the arbitral tribunal in July 1999 under Annex VII of the United Nations Convention on the Law of the Sea (UNCLOS), claiming that Japan had violated the UNCLOS by carrying out an experimental fishing program without the consent of Australia and New Zealand. They later filed suit with the International Tribunal for the Law of the Sea (ITLOS), seeking provisional measures for the immediate cessation of Japan's program. While waiting for the arbitral tribunal to hand down its judgment, ITLOS decided on provisional measures including: parties should ensure that their annual catches do not exceed the quotas they last agreed upon, and that in calculating annual catches for 1999 and 2000 the catch taken in 1999 under the experimental fishing program must be taken into account; and Japan should refrain from its experimental fishing program for southern bluefin tuna. In August 2000 the arbitral tribunal decided that it was without jurisdiction to rule on the merits of this case.

While the ITLOS provisional order displayed a judgment of interest with regard to some issues such as the legal status of the precautionary principle, it also suggested the limitations of the Convention for the Conservation of Southern Bluefin Tuna in effectively conserving the species. Catches of the tuna in the convention's applicable region by Taiwan, Indonesia, South Korea, and other non-parties have increased, totaling about 4,500 tons in 1997, an amount equalling about 40% of the total catch allowed by the convention. Consequently, the convention's purpose might not be fulfilled even if its three parties enact and implement measures for tuna conservation. An especially serious problem is the flag-of-convenience vessels that register themselves with a non-party country or a country with lax fishing regulations in order to circumvent fishing regulations. A convention mechanism will be needed to, for all practical purposes, either encourage non-parties operating in the affected region to become parties, or to have them comply with the convention's obligations.

## Bilateral Environmental Agreements

Under the Ganges Water Sharing Treaty signed by India and Bangladesh in 1996, joint monitoring is conducted and a joint scientific committee comprising experts from both countries is trying to elucidate reasons for water shortages. At a 1999 meeting of the Indo-Bangladesh Joint Rivers Commission, the two countries agreed to undertake a scientific study with a view to assessing the water availability of the Teesta River, and the following year they decided to sign a short-term agreement on sharing its water.

In the area of nature protection, the Philippines and Malaysia in 1999 signed an agreement for sea turtle protection, called the Agreement Establishing the Turtle Islands Heritage Protected Area, to jointly manage sea turtle islands in the Sulu Sea. In 2000, India and Bangladesh entered into a protocol for protecting tigers in their border zone.

## Broadening Environmental Cooperation in East Asia

There have been no regional treaties or conventions in East Asia owing to political factors, but in recent years a wide variety of cooperative relationships have emerged.

Since 1993 meetings of Senior Officials on Environmental Cooperation in North-East Asia have been held as this region's first venue for intergovernmental negotiations meant to build and promote cooperative relationships to address environmental issues in Northeast Asia. Participating in these meetings are China, Japan, Mongolia, South Korea, Russia, and, since the second meeting in 1994, North Korea. At the first meeting in 1993, participants launched the North-East Asian Subregional Programme of Environmental Cooperation (NEASPEC), and confirmed that priority areas would be a) energy and air pollution, b) ecosystem management, in particular deforestation and desertification, and c) capacity-building. The third meeting in 1996 made decisions on NEASPEC's applicable geographic coverage, its purposes, the functions and responsibilities of Senior Officials Meetings, coordination of cooperation, management, financial mechanisms, and other matters. The fourth meeting in 1998 reached agreement on a trust fund, and the sixth meeting in 2000 adopted a Vision Statement on environmental cooperation in Northeast Asia.

In 1994 Japan, China, South Korea, and Russia adopted the North-West Pacific Action Plan (NOWPAP) for the Sea of Japan and Yellow Sea, as one of the UNEP Regional Seas Programmes meant for conservation of international marine areas with a high degree of closure. Currently NOWPAP has seven projects: building a data base on marine environments in the target region, studies of different countries' laws and regulations on marine environmental conservation, developing programs to monitor the target region, preparation for and responses to marine pollution accidents (oil spills), designating regional activity centers that serve as focal points for activities in various fields, increasing public awareness about marine and coastal environments, and assessment and control of land-based pollution. In 1999 the locations of the regional activity centers

(RAC), which take responsibility for project implementation, were decided. The sixth meeting in 2000 decided to establish a regional coordinating unit with two secretariats in Toyama, Japan and Pusan, South Korea.

Against the backdrop of the threat of acid rain damage stemming from growing SOx and NOx emissions, and employing the achievements of experts' meetings that started in 1993, the Acid Deposition Monitoring Network in East Asia (EANET) conducted preparatory phase activities for about two years starting in 1998. The 10 countries that participated in preparatory phase activities—China, Indonesia, Japan, Malaysia, Mongolia, the Philippines, South Korea, Russia, Thailand, and Vietnam—have been running the formal network since January 2001. EANET's purposes are (1) develop a common understanding of the acid rain problem in East Asia, (2) provide information for decision-making at the national and regional levels to prevent the damaging environmental impacts of acid rain, and (3) promote cooperation on acid rain among participating countries. UNEP will eventually take over the secretariat, until which time Japan's Environment Ministry serves as the provisional secretariat. Japan's Acid Deposition and Oxidant Research Center has been designated as EANET's network center.

Takamura Yukari

# 22. Environmental Policy Instruments Diversify

In many Asian countries, environmental rights are recognized by the constitution or judicial precedents, and there is an increasing trend toward enacting basic laws on the environment, and toward developing and implementing environmental policy in a systematic and planned manner.

Generally, Asian countries have ministry-level government agencies serving as environmental administrative organizations. Indonesia formerly had a two-part arrangement comprising a Ministry of Environment for policy development and an Environmental Impact Management Agency (BAPEDAL) for policy implementation, but BAPEDAL was integrated into the Ministry of Environment to reinforce environmental administration. By contrast, Singapore wanted to allow its Ministry of the Environment to concentrate on a long-term policy agenda, so in 2002 it created the National Environment Agency by spinning off the policy implementation division responsible for tasks such as monitoring compliance with environmental regulations and running waste management facilities.

In addition to the traditional regulatory instruments, governments consider planning instruments, economic instruments, public education instruments, and others to be important as environmental policy instruments. China is ahead in the use of economic instruments, with lighter commodity taxes on low-emission vehicles, the collection of pollutant fees related to water and waste, and, in some geographical areas, trading in water rights and $SO_2$ emission rights. There are also calls for the incorporation of an emissions trading scheme modeled on the US Clean Air Act into the recently developed plan for reducing $SO_2$ emissions.

Surcharges and environmental taxes are being introduced in more countries. Thailand plans to levy a container and packaging tax, while in South Korea a global warming mitigation bill whose main features included a surcharge on the industrial use of fossil fuels was introduced in the National Assembly, but a strong industry backlash has prevented its passage.

South Korea is actively encouraging voluntary initiatives, which it is backing in ways that include development of the Strategies for Environmental Industry Development, an eco-labeling certification system based on the Act Relating to Environmental Technology Support and Development, the release of environmental reporting guidelines, and the introduction of the Environment-Friendly Company Designation System. South Korea uses low-interest financing and tax incentives to encourage the participation of businesses in a voluntary program for $CO_2$ emission reductions.

In the Philippines companies are asked to disclose environmental information about their compliance with environmental laws and regulations, and administrative authorities use an Industrial Ecowatch System (IES) to rank companies on a six-step scale.

In the area of achievements related to policy instruments, an especially notable event was China's creation of its Law on Environmental Impact Assessment in 2002. Formerly environmental assessments were performed only for large-scale construction projects in accordance with administrative rules, but after this law

TABLE 1. Environmental Laws in Asia

|  | China | India | Indonesia | Japan |
|---|---|---|---|---|
| A. Managing administrative agencies | State Environmental Protection Administration | Ministry of Environment and Forests | State Ministry of Environment | Ministry of the Environment |
| B. Basic environmental laws | Environmental Protection Law | The Environmental (Protection) Act | Law Concerning Environmental Management | Basic Environment Law |
| C. Environmental impact assessment laws | Law on the Environmental Impact Assessment | Environmental Impact Assessment Notification | Government Regulation No. 51 of 1993 on Environmental Impact Analysis | Environmental Impact Assessment Law |

| S. Korea | Malaysia | Philippines | Thailand | Taiwan | Vietnam |
|---|---|---|---|---|---|
| Environment Ministry | Ministry of Science, Technology and Environment | Department of Environment and Natural Resources, Environment Management Bureau | Ministry of Natural Resources and Environment National Environmental Board | Environmental Protection Administration | Ministry of Natural Resource and Environment |
| Basic Environmental Policy Act | Environmental Quality Act | Environmental Policy (Presidential Decree No. 1151) Environment Code (Presidential Decree No. 1152) | Enhancement and Conservation of National Environmental Quality Act | Basic Environmental Conservation Law (draft) | Law on Environmental Protection |
| Environmental Impact Assessment Act | Environmental Quality (Prescribed Activities) (Environmental Impact Assessment) Order | Presidential Decree No. 1586, Establishing an Environmental Impact Statement System Including Other Environmental Management (Related Research and Other Purposes) | Notice concerning Prescription of Types and Capacity of Projects or Activities of Government Agencies, State Enterprises or Private Sector Requiring Creation of Reports on Environmental Impacts (1996) | Environmental Impact Assessment Act | Law on Environmental Protection and Decree 175/CP |

TABLE 1. *Continued*

|  | China | India | Indonesia | Japan |
|---|---|---|---|---|
| D. Water laws | Law on the Prevention and Control of Water Pollution | The Water (Prevention and Control of Pollution) Act | Government Regulation No. 82 of 2001 on Water Quality Management | Water Pollution Control Law |
|  | Tentative Regulation on the Collection of Levies on Pollutant Emissions |  |  | Law Relating to the Prevention of Marine Pollution and Maritime Disaster Soil Contamination Countermea- sures Law |
| E. Air laws | Law on the Prevention and Control of Atmospheric Pollution | The Air (Prevention and Control of Pollution) Act |  | Air Pollution Control Law Law Concerning the Promotion of the Measures to Cope with Global Warming |
| F. Noise/vibration laws | Law on the Prevention and Control of Pollution from Environmental Noise | Noise Pollution (Regulation and Control) Rules | Decree of the State Minister for the Environment No.KEP- 48/ MENLH/11/ 1996 Concerning Noise Level Standards Decree of the State Minister for the Environment No.KEP-49/ MENLH/11/ 1996 Concerning Vibration Level Standards | Noise Regulation Law Vibration Regulation Law |

| S. Korea | Malaysia | Philippines | Thailand | Taiwan | Vietnam |
|---|---|---|---|---|---|
| Water Quality Preservation Act | Environmental Quality (Sewage and Industrial Effluents) Regulations | Clean Water Act (RA No. 9275, 2004) | Factory Act | Water Pollution Control Act | Law on Environmental Protection |
| Sewer System Act | Merchant Shipping (Oil Pollution) Act | | Groundwater Act | Drinking Water Management Act | |
| Soil Environment Conservation Act | | | Public Health Act | Soil and Groundwater Remediation Act | |
| Air Quality Preservation Act Underground Living Space Air Quality Control Act | Environmental Quality (Clean Air) Regulations | Clean Air Act | Factory Act | Air Pollution Control Act | Law on Environmental Protection |
| Noise and Vibration Control Act | Environmental Quality (Motor Vehicle Noise) Regulations | | | Noise Control Act | Law on Environmental Protection |

TABLE 1. *Continued*

|  | China | India | Indonesia | Japan |
|---|---|---|---|---|
| G. Recycling- and waste-related laws | Law on the Prevention and Control of Environmental Pollution by Solid Waste | The Hazardous Waste (Management and Handling) Rules The Bio-Medical Waste (Management and Handling) Rules The Re-cycled Plastics Manufacture and Usage Rules The Batteries (Management and Handling) Rules The Municipal Solid Wastes (Management and Handling) Rules | Government Regulation No. 19 of 1994 regarding Hazardous and Toxic Waste Management | Basic Law for Establishing a Recycling-based Society Wastes Disposal and Public Cleansing Law Law for the Promotion of Utilization of Recycled Resources Container and Packaging Waste Recycle Law Law for Recycling of Specified Kinds of Home Appliances Demolition Waste Recycling Law Food Recycling Law Automobile Recycling Law |
| H. Chemical regulations |  | The Manufacture, Storage and Import of Hazardous Chemical Rules |  | PRTR Law  Law Concerning the Examination and Manufacture etc. of Chemical Substances Law Concerning Special Measures against Dioxins |

| S. Korea | Malaysia | Philippines | Thailand | Taiwan | Vietnam |
|---|---|---|---|---|---|
| Waste Management Act<br><br>Act Relating to the Promotion of Resource Saving and Reutilization<br>Promotion of Waste Treatment Facilities and Local Communities Act | Environmental Quality (Scheduled Wastes) Regulations | Act to Control Toxic Substances, Hazardous and Nuclear Wastes (RA No.6969, 1990)<br>Ecological Solid Waste Management Act | Factory Act<br><br>Public Health Act | Waste Disposal Act<br><br>Resource Recycling Act | |
| Toxic Chemicals Control Act | | | Hazardous Substance Act | Toxic Chemical Substances Control Act | |

TABLE 1. *Continued*

| | China | India | Indonesia | Japan |
|---|---|---|---|---|
| I. Nature and resource protection laws | Law on the Protection of Wildlife | The Indian Wildlife (Protection) Act | Act No. 5 of 1990 on Conservation of Natural Resources and Ecosystems, dated 10 August 1990 | Natural Parks Law |
| | Grassland Law | Forest (Conservation) Act | Decree of the Minister for Forestry and Plantations No.260/ KEP II/ 1995 on Guidelines for the Prevention and Control of Forest Fires | Nature Conservation Law |
| | Forestry Law | The Biological Diversity Act | | Law for the Conservation of Endangered Species of Wild Fauna and Flora |
| | Law on Desert Prevention and Transformation | | | Wildlife Preservation and Game Act |
| | Marine Environment Protection Law Law on Coal Resources | | | Law for Promoting the Restora-tion of Nature Law on Ensuring Biological Diversity by Regulating the Use of Genetically Modified Organisms |
| | Fishery Law | | | Cultural Properties Law |

| S. Korea | Malaysia | Philippines | Thailand | Taiwan | Vietnam |
|---|---|---|---|---|---|
| Nature Environment Preservation Act | National Parks Act | RA 7586, National Integrated Protected Areas System Act of 1992 | Wildlife Conservation (and Protection) Act | | Law on Forest Protection and Development |
| Wetlands Conservation Act | National Forestry Act | Order on Prospecting Biological and Genetic Resources (EO No. 247, 1995) | Forestry Act | | Law on Minerals |
| Act Relating to Protection of Birds, Mammals and Hunting | Protection of Wild Life Act | Mining Code (RA No. 7942, 1995) | Fishery Act | | Law on Petroleum |
| Natural Park Act | Mineral Development Act | Act to Manage and Protect Caves and Cave Resources and for Other Purposes | Mineral Act | | Law on Fishery |
| | Fisheries Act | Animal Welfare Act of 1998 | | | Law on Aqua Resource Protection |
| | Petroleum Development Act | Fisheries Code | | | |

TABLE 1. *Continued*

|  | China | India | Indonesia | Japan |
|---|---|---|---|---|
| J. Pollution dispute resolution, etc. |  | The Public Liability Insurance Act | Regulation on Establishing Environmental Dispute Resolution Agencies | Environmental Pollution Disputes Settlement Act Pollution-Related Health Damage Compensation Law |
| K. Environmental criminal law, etc. | Criminal Law |  |  | Pollution Crimes Law |
| L. Facilitation of voluntary initiatives | Law on the Promotion of Clean Production |  |  | Green Purchasing Law Environmental Education Law |

took effect in September 2003, it required assessments of more projects including government development plans. South Korea amended its Basic Environmental Policy Law, which requires giving environmental impact assessment indicators when formulating policy. This is a new development in South Korea related to strategic environmental assessments.

In relation to environmental risk management and the regulation of chemical substances, China is working on a law for the environmental control of new chemical substances, which would require advance notice for the manufacture and import of new chemicals. Taiwan and Japan have created new laws for contaminated soil remediation. In Taiwan's case, a dispute have arisen over certain provisions on the shouldering of costs because the system requires business operators linked to contaminated sites to also pay costs.

To deal with waste, the Philippines passed the Ecological Solid Waste Management Act in 2000. This comprehensive law includes provisions on encouraging the three Rs (reduction, reuse, recycling), preferential treatment for pioneering businesses, the required establishment of recycling centers by local governments, and the closure of waste dumps in headwater areas. China promulgated a "policy on technologies for preventing contamination by hazardous waste," whose purpose is the reduction, recycling, and neutralization of such waste, and in 2002 it passed the Law on Promotion of Clean Production, which requires manufacturers and sellers to collect products and packaging that must be legally recycled. Other planned laws and regulations include an ordinance on the recovery, use, and management of recycled resources. South Korea amended

| S. Korea | Malaysia | Philippines | Thailand | Taiwan | Vietnam |
|---|---|---|---|---|---|
| Environmental Dispute Settlement Act | | | | Public Nuisance Dispute Mediation Act | |
| Acts Relating to Punishment for Environmental Crime | | | | | Criminal Code |
| Act Relating to Environmental Technology Support and Development | | | | | |

its Act Relating to Promotion of Resources Saving and Reutilization in order to increase extended producer responsibility. Further, it set reuse targets for certain products, and adopted a system that imposes surcharges on manufacturers that do not meet the targets. India promulgated "Batteries (Management and Handling) Rules" in 2001 and "Recycled Plastics Manufacture and Usage Rules" in 2000. Delhi State created even stricter regulations on plastic bags. Singapore amended its 1999 Environmental Public Health Act, thereby raising the penalties for illegal dumping, adding a provision to confiscate vehicles used for illegal dumping, and making other changes.

Notable air pollution-related events are the passage of the Clean Air Act in the Philippines, and China's 2000 amendment of its Law on the Prevention and Control of Atmospheric Pollution, which beefed up the control of motor vehicle emissions (Beijing's regulations are tougher than those of the national government). To deal with water pollution, in 2001 South Korea created a river region designation system, and for individual rivers passed special laws that include measures to address non-point sources.

In the area of nature protection and global environmental conservation, events of significance are the passage of India's Biological Diversity Act (2002) and China's Law on Desert Prevention and Transformation (2001), as well as progress in the Philippines on writing a sustainable forest management law that would prohibit commercial logging for 25 years.

A serious problem still shared by Asian countries is that their environmental laws lack effectiveness. Reasons for this include (1) a government might pass a law, but not its enforcement order, (2) a law's requirements are uncertain, (3) conflicts with other laws, and (4) low awareness among administrative authorities. An example of remedial action is that China is integrating the motor vehicle emission standards developed by multiple administrative agencies, and running a campaign that imposes stiff penalties on violators. South Korea has an early notification system giving businesses a five- to 10-year notice before regulations take effect to help them comply.

An instructive law related to alternative dispute resolution is South Korea's 1997 amended Environmental Dispute Settlement Act, which allows environmental NPOs to dispute instances of serious environmental damage.

Okubo Noriko

# Notes

*Introduction*

1. United Nations Development Program. *Human Development Report 1999: Globalization with a Human Face*, UNDP, 1999.
2. See pp. 194–197 of *The State of the Environment in Asia 2002/2003*.
3. Sen, Amartya. *Development as Freedom*, Alfred A. Knopf, New York, 1999.
4. Inoue, Makoto. "From Government to Governance: Forest Policy with Local Participation," *Ronza*, August 2003 (in Japanese).
5. Teranishi, Shun'ichi. *The Political Economy of Global Environmental Problems*, pp. 136–140, Toyo Keizai Inc., 1992 (in Japanese).
6. References used

   Sato, Hiroshi, ed. *Development Aid and Bangladesh*, Institute of Developing Economies, 1998 (in Japanese).

   Japan International Cooperation Agency. *Report on Available Environmental Information by Country: Bangladesh*, 1999 (in Japanese).

   International Federation of Red Cross and Red Crescent Societies. *World Disasters Report 1999*, 2000.

   Oya, Masahiko, ed. *Applied Geography for Disaster Prevention and Environmental Conservation*, Kokin Shoin, 1994 (in Japanese).

   UNICEF. *State of the World's Children Report*, 2000.

Part I

*Chapter 1*

1. *Asahi Shimbun*. Editorial, March 19, 2003 (in Japanese).

   Toyota, Toshiyuki. "Eisenhower's Great Regret," *Gunshuku Mondai Shiryo*, April 2003, pp. 2–7 (in Japanese).
2. "Closing and Returning Military Bases, and Restoring the Environment," a feature section in *Research on Environmental Disruption*, vol. 32, no. 4, April 2003. See the statement by Miyamoto Ken'ichi on pp. 14–15 of the round-table discussion, "The Military and the Environment." (In Japanese)
3. *Special Issue of Sekai: NO WAR!* Iwanami Shoten, June 2003 (in Japanese).
4. Stockholm International Peace Research Institute. <http://projects.sipri.org/index/mex_major_spenders.pdf>
5. Defense Research Center. *Security of Japan and the World as Seen in Military Data*, Soshisha, 2002, p. 28 (in Japanese).

6. Worldwatch Institute. *State of the World* 1991.

7. Ibid.

8. Estimated from: Office of the Deputy Undersecretary of Defense. *Department of Defense Base Structure Report*, 1999.

9. See: <http://www.city.nago.okinawa.jp/nago_koho/hiroba0209/futenma.html#01> (in Japanese).

10. IUCN. CGR2.CNV004xCNV005 Rev1, Conservation of Dugong (*Dugong dugon*), Okinawa Woodpecker (*Sapheopipo noguchii*), and Okinawa Rail (*Gallirallus okinawae*) in and around the Okinawa Island.

11. Harashina, Sachihiko. "Introduction: Environmental Considerations for Achieving Sustainable Development," *Research on Environmental Disruption*, vol. 3, no. 4, April 2001, pp. 10–15 (in Japanese).

12. Ryu, Je-hon and Harashina Sachihiko. "Orientation of NGOs' Role in the EIA System: Case Study of the East Shore Development Plan at Awase Tideland in Okinawa City," *Research on Environmental Disruption*, vol. 32, no. 2, October 2002, pp. 58–64 (in Japanese).

13. Umebayashi, Hiromichi. *US Forces in Japan*, Iwanami Shoten, 2002 (in Japanese). pp. 152–172 have a detailed explanation about the introduction of nuclear weapons into Japan.

14. From a March 2003 interview with the US Marines Okinawa Environmental Branch by one of the authors. The Environmental Branch showed itself ready and willing to cooperate with local environmental NGOs and Japanese research institutions.

15. Ui, Jun. "US Military Bases and Environmental Problems: Okinawa's Current Situation," *Gunshuku Mondai Shiryo*, May 2003, pp. 18–25 (in Japanese).

16. *Kanagawa-ken Hoken'i Shimbun*, December 15, 1999.
Goto, Masahiko. "Heavy Metal Contamination at Yokosuka Naval Base Berth 12, and Home Port Plan for Nuclear-powered Aircraft Carriers," *Proceedings of the First International Workshop on the Military and the Environment*, March 2003, pp. 18–27.
June 2002 interview with the Kanagawa Prefecture Health Insurance Doctors Association (all in Japanese).

17. For an overview of this topic in the Philippines, see:
Yokemoto, Masafumi. "Pollution by US Military Bases in the Philippines: The Case of Subic Bay Naval Base," Tokyo Keizai University Academic Research Center Working Paper Series, 2002-E-01, September 20, 2002 (in Japanese).
Oshima, Ken'ichi. "Pollution Harm on Clark Air Base in the Philippines," *Ritsumeikan Journal of International Relations and Area Studies*, no. 21, March 2003, pp. 65–77 (in Japanese).
*Asahi Shimbun*. September 23, 2002 (in Japanese).
*The Daily Yomiuri*. March 15, 2003.

18. Details are available in the final report issued by a special Senate committee on base pollution damage: *Committee Report No. 237*, Eleventh Congress of the Republic of the Philippines, Second Regular Session, May 16, 2000, pp. 12–15.

19. Military Base Closure: U.S. Financial Obligations in the Philippines.

20. *Committee Report No. 237*, pp. 35–47.

21. *Philippines Daily Inquirer*, December 3, 2002.

22. Komatsu, Ken'ichi and Shimobara Mamoru. "Former Workers at US Naval Base in the Philippines Afflicted with Asbestos Ailments," *Shukan Kinyobi*, July 5, 1996, p. 31 (in Japanese).

23. Campaign Headquarters for Eliminating Crimes by US Forces in South Korea. *Report on Crimes by US Forces in South Korea 2000–2002*, 2002 (in Korean).

24. There was a 1973 US-Japan "Co-operation Concerning Environmental Matters," but it was not released to the general public until January 23, 2003. This agreement permitted local government officials access to bases, but was kept secret for 30 years.

25. Fukuchi, Hiroaki. *Military Bases and Environmental Damage*, Dojidaisha, 1996 (in Japanese).

26. Ibid., p. 72.

27. For a list of lawsuits filed up to 1995, see: Japan Federation of Bar Associations, ed. *Japan's Security and the Base Issue: The Right to Live in Peace*, Akashi Shoten, pp. 353–357 (in Japanese).

28. Campaign Headquarters for Eliminating Crimes by US Forces in South Korea. *Ryukyu Shinpo*, February 22, 2002.

29. Depleted Uranium Research Group, ed. *Radioactive Ordnance and Depleted Uranium: Nuclear Battlefields and Uranium-Contaminated Areas*, Gijutsu to Ningen, 2003, pp. 96–105 (in Japanese).

30. Quy, Vo. "Military Activities During the Vietnam War and Environmental Problems," *Anthology of Reports from the First International Workshop on Military Activities and the Environment*, March 2003, pp. 55–65.

31. Ota, Masahide. *Testimony of Okinawa's Governor in the Proxy Signature Lawsuit: For a Peaceful Island without Military Bases*, Niraisha, 1996, p. 105.
    Interviews conducted with Okinawa Prefecture officials (both in Japanese).

32. Quy, Vo. "Environmental Research on the Effects of Massive Defoliation on Forest Ecosystems of Southern Vietnam During the War," "The Effect of Massive Defoliation on the Forest Ecology of Southern Vietnam During the War and Environmental Restoration Program," *Anthology of Reports from the First International Workshop on Military Activities and the Environment*, March 2003, pp. 85–89, 122–126.

33. United Nations Environment Programme. *Desk Study on the Environment in Iraq*, 2003, p. 68. <http://postconflict.unep.ch/publications/Iraq_DS.pdf>

34. Ibid., p. 80.

35. See the following UNEP documents on the use of DU munitions. All are downloadable from the UNEP Post-Conflict Assessment Unit: <http://postconflict.unep.ch/> Since 1999 UNEP has been issuing reports on the environmental impacts of major conflicts and wars.
    *Desk Study: The Potential Effects on Human Health and the Environment Arising from Possible Use of Depleted Uranium during the 1999 Kosovo Conflict*, 1999.
    *Final Report: Depleted Uranium in Serbia and Montenegro—Post-Conflict Environmental Assessment in the Federal Republic of Yugoslavia*, 2002.
    *Final Report: Depleted Uranium in Kosovo—Post-Conflict Environmental Assessment*, 2001.
    *Depleted Uranium in Bosnia and Herzegovina: Post-Conflict Environmental Assessment*, 2003.

36. Morizumi, Takashi. *Children of Iraq and the Gulf War: What Have DU Munitions Wrought?* Kobunken, 2002 (in Japanese).
    Egawa, Shoko and Morizumi, Takashi. *Report from Iraq*, Shogakkan, 2003 (in Japanese).
    Morizumi, Takashi. *A Different Nuclear War: Children of the Gulf War*, 2002.

37. Umebayashi, Hiromichi. "Base Closings and Environmental Remediation in the US," *Research on Environmental Disruption*, vol. 32., no. 4, pp. 4–9, April 2003 (in Japanese).
38. *Ryukyu Shimpo*, April 12, 2002 (in Japanese).
39. *People's Daily Online* (Japanese-language version), August 18, 2003 and subsequent reports.

*Chapter 2*
1. "Trends in Marine Product Consumption and Trade," pp. 144–146.
2. Ibid., pp. 142–143.
3. "Tuna" in the broad sense also includes bonito and marlin. Order Perciformes, family Scombridae, genus *Thunnus* include seven species: the bluefin tuna, southern bluefin tuna, bigeye tuna, yellowfin tuna, albacore, Atlantic tuna, and longtail tuna. The first four species are especially important as sashimi.
4. Total for seven tuna species. See: http://www.fao.org.
5. Miyake, Makoto, "Current State of Atlantic Salmon Stocks and Future Outlook," in: Yamamoto, Tadashi and Shindo, Shigeaki, eds., *World Fisheries, Vol. 1, Worldwide Fishery Trends*, Overseas Fishery Cooperation Foundation, 1998 (in Japanese).
6. About two-thirds of the world catch of bonito, tuna, and marlin are canned. See: *Bonito and Tuna Yearbook 1997*, Suisan Shinchousha, 1997, p. 206 (in Japanese).
7. On fresh tuna, see: Yokemoto, Masafumi, "Managing Tuna Resources by Regulating Japan's Import," *The Journal of Tokyo Keizai University*, no. 234, 2003 (in Japanese).
8. Fuke, Yosuke, "Flying Tuna and 'Modernization,'" in Murai, Yoshinori ed., *A Close Look at Japan's ODA (Second Edition)*, Commons, 1997, pp. 50–53 (in Japanese).
9. Hiroyoshi, Katsuji, "Progress in the Reorganization of the Seafood Wholesale Market, with Focus on Consuming Region Wholesale Markets," *Japanese Journal of Fisheries Economics*, vol. 26, no. 4, 1981, p. 12 (in Japanese).
10. *Bonito and Tuna Yearbook 2001*, Suisan Shinchousha, 2001, p. 409 (in Japanese).
11. Nakai, Akira, "Tuna Consumption and Distribution," in: Ono, Seiichiro, ed., *Tuna: From Production to Distribution*, Seizando Shoten, 1998, p. 288 (in Japanese).
12. Kim, Ki Soo, et al., "Structural Change in Korean Tuna Fishery," *Japanese Journal of Fisheries Economics*, vol. 43, no. 3, 1999, p. 6 (in Japanese).
13. Tokyo Fisheries Promotion Foundation, ed., *Structure of International Trade in Marine Products: The Marine Products Industries in South Korea, India, Canada, and the US, and the Activities of Japanese Trading Companies*, Tokyo Fisheries Promotion Foundation, 1983, pp. 26–28 (in Japanese).
14. Kim, et al., pp. 9–10.
15. Miyahara, Masanori, "International Management of Tuna Resources and Flag-of-convenience (FOC) Fishing Vessels," *Journal of International Fisheries*, vol 5, no. 1, 2002, p. 21, and Tsai, Kunchou and Miyazawa, Haruhiko, "The Development of Tuna Fishing in Taiwan," *Japanese Journal of Fisheries Economics*, vol. 43, no. 1, 1998, pp. 69–71 (both in Japanese).
16. Miyahara, p. 21. Including marlin but excluding albacore. As of 2002 this agreement had been extended every year. In the late 1990s each ship was assigned a quota.
17. Miyahara, and *Bonito and Tuna Yearbook 1997*, p. 236.
18. Document supplied by Miyahara Masanori, an official at Japan's Fisheries Agency, at a meeting of the Japan Fisheries Association on December 25, 2002. Below, "Miyahara document" (in Japanese).
19. Worldwatch Institute, *State of the World 2001*, 2001.

20. Miyahara document.
21. Safina, Carl, *Song for the Blue Ocean*, Henry Holt, 1998, pp. 13–15, 64–66.
22. Ibid., pp. 110–112.
23. In addition to import limits, a statistical certification system was instituted to disallow international trade in Atlantic tuna products without a statistical document issued by the flag country. Miyahara, pp. 20–21.
24. Ross, Michael L., *Timber Booms and Industrial Breakdown in Southeast Asia*, Cambridge University Press, 2001.
25. Tachibana, Satoshi, Kato Takashi, Yamamoto Nobuyuki, and Furuido Hiromichi, "Effects of Royalty System on Timber Production: A Case Study in Sabah," "*Asia Keizai*, vol. 37, no. 1, pp. 22–39, 1996 (in Japanese).
26. FAO, *State of the World's Forests 2001*, 2001.
27. Ibid.
28. Araya, Akihiko, *Indonesia's Plywood Industry*, J-FIC, 1998 (in Japanese).
29. Forest Trends, *Market for Forest Conservation*, September 2001.
30. Huang, Xia, Hu Hongying, Fujie Koichi, and Kubota Hiroshi, "Series—China's Paper/Pulp Industry and Effluent Measures: First Report on Paper/Pulp Industry," *Resource and Environmental Measures*, vol. 32, no. 13, 1996. Otsuka, Kenji, "Implementation of Controls on Industrial Pollution Sources in China: Effectiveness and Conditions of Regulatory Policies from the Mid-1990s," in Terao, Tadanori and Otsuka Kenji, eds., "*Policy Process and Dynamism in "Development and Environment": Japan's Experience and East Asia's Challenge*, Institute of Developing Economies, 2002 (both in Japanese).
31. *Chosun Ilbo*, July 8, 2002 (English-language edition).
32. Kojima, Michikazu, "Pollution Abatement by Small and Medium-sized Enterprises and in the Informal Sector: Japan's Experience and Other Asian Countries' Tentative Efforts in Lead Recycling," in Terao and Otsuka, eds.
33. See: http://www.jeas.or.jp/ecomark/english/index.html.
34. See the Global Ecolabelling Network at http://www.gen.gr.jp/.
35. Tachibana, Satoshi, Nemoto Akihiko, and Minowa Yasushi, "The Potentials of Forest Certification Systems: Trends in International Forest Certification and Trends in Indonesia and Malaysia," in Inoue, Makoto, ed., *Loss and Conservation of Forests in Asia*, Chuo Hoki, 2003, pp. 272–291 (in Japanese).
36. Shiraishi, Norihiko, "Progress of the Forest Certification Systems in the World and Japan," *Policy Trend Report 2002*, IGES Forest Conservation Project, 2002, pp. 95–104.
37. See, for example: Tasaka, Koa, *Contaminated Food Imports in Asia*, Ienohikari Kyokai, 1991 (in Japanese).
38. <www.ifat.org/>.
39. References: *Bonito and Tuna Yearbook 2001*, Suisan Shinchousha, 2001; Miyahara, Masanori, "International Management of Tuna Resources and Flag-of-convenience (FOC) Fishing Vessels," *International Fisheries Research*, vol 5, no. 1, 2002; and Yokemoto, Masafumi, "Japan's Marine Product Imports and Fishery Resource Management, with Sashimi Tuna as an Example," *The Journal of Tokyo Keizai University*, no. 234, 2003 (all in Japanese).
40. Basel Action Network and Silicon Valley Toxics Coalition, *Exporting Harm: The High-Tech Trashing of Asia*, 2002.
41. UNEP, "Illegal Trade in Ozone Depleting Substances: Is There a Hole in the Montreal Protocol?" *Ozone Action Newsletter Special Supplement*, No. 6, 2001.

*Chapter 3*

1. "Marginal land" in this chapter means land that is unsuited to agricultural production due to temperature, fertility, gradient, availability of irrigation water, or other conditions.
2. Calculated from FAO's FAOSTAT using the averages of the three-year periods 1964–1966 and 1998–2000, with food demand equaling the sum of animal feed, seed grain, grain for processing, amount of decrease, and gross food. Rice is converted to milled weight. FAO statistics include 47 countries in "Asia," including some Middle Eastern countries. The eight countries added in 1992 have been excluded here to make a chronological comparison possible. Statistics used in sections 2 and 3 have been prepared this way unless otherwise specified.
3. Calculated from FAOSTAT.
4. Otsuka, Shigeru, "Agribusiness and the Transformation of Asia's Food Market," in Nakano, Isshin and Sugiyama Michio, eds., *Globalization and the International Agricultural Market*, Tsukuba Shobo, 2001, pp. 197–223 (in Japanese).
5. Calculated from FAOSTAT.
6. Economic growth alone does not determine diet because traditional dietary customs such as the prohibition on beef in Hinduism or that on pork in Islam, influence the rate and type of dietary change.
7. World Bank, *World Development Report 1999/2000*.
8. Calculated from FAOSTAT.
9. Ministry of Agriculture, Forestry and Fisheries, *Agriculture, Forestry, and Fisheries Trade Report 2001*, 2001, p. 13 (in Japanese).
10. Unless otherwise specified, this example is based on: Green Earth Network, *Clean Development Mechanism Feasibility Studies for Global Warming Projects: Report on a Study of Possibilities for Greening China's Loess Plateau*, 2001 (in Japanese).
11. http://wwww.riceweb.org/riceprodasia.htm
12. Calculated from FAOSTAT.
13. Kyuma, Kazutaka, *Food Production and the Environment: Thoughts on Sustainable Agriculture*, Kagaku Dojin, 1997, pp. 34–35 (in Japanese).
14. Umali, Dina L., *Irrigation-Induced Salinity: A Growing Problem for Development and the Environment*, World Bank Technical Paper, no. 215, 1992, pp. 12–15.
15. Mase, Toru, "Irrigated Agriculture," in Tanaka, Akira, ed., *Introduction to Tropical Agriculture*, Tsukiji Shokan, 1997, p. 441 (in Japanese).
16. Umali, p. 18.
17. Ibid., p. 44.
18. Mase, p. 434.
19. Government of India, *Indian Agriculture in Brief* (27th ed.), 2000, pp. 13, 33.
20. Ibid.
21. Dhawan, B. D., *Studies in Minor Irrigation: With Special Reference to Ground Water*, Commonwealth Publishers, New Delhi, 1997, p. 35.
22. Government of India, Ministry of Water Resources, *Annual Report 1999–2000*, 2000, p. 92.
23. Singh, Joginder, et al., *Changing Scenario of Punjab Agriculture: An Ecological Perspective*, Indian Ecological Society and Centre for Research in Rural and Industrial Development, 1997, p. 27.
24. Government of India, Ministry of Water Resources.
25. Government of India, *Cultivation Practices in India*, NSS 54th Round, Report no. 451, 1999, pp. 33–35.

26. Suda, Toshihiko, "The High Price of Increasing Food Production in Developing Countries: Pesticide Contamination in Bangladesh," *Agriculture and Forestry Financing*, vol. 45, no. 11, November 1992 (in Japanese).

27. Bhatnagar, V. K., "Pesticide Pollution: Trends and Perspective," *ICMR Bulletin*, vol. 31, no. 9, September 2001, p. 4.

28. Uegaki, Takao, "Southeast Asian Pesticide Regulations and Their Problems," *International Cooperation in Agriculture and Forestry*, vol. 18, no. 3, 1995, p. 17 (in Japanese).

29. In 1996/97 India used 2,236 tons of methyl parathion. Government of India, *Indian Agriculture in Brief* (27th ed.), 2000, pp. 13, 60. Dieldrin and lindane are listed in a Punjab state agricultural guidebook as limited-use pesticides. Punjab Agricultural University, *Package of Practices for Crops of Punjab, Rabi, 2000–2001*, 2000.

30. Uegaki, p. 19.

31. Smith, Carl, "Pesticide Exports from U.S. Ports, 1997–2000," *International Journal of Occupational and Environmental Health*, vol. 7, no. 4, October–December 2001, <www.fasenet.org/pesticide_report97-00.pdf>. Does not include air and land shipments.

32. For a discussion of this, see: McMichael, P., *Development and Social Change*, Pine Forge Press, 1996, and Iwasa, Kazuyuki, "International Agricultural Development Projects and Agribusiness in Developing Countries," in Nakano and Sugiyama, eds. (in Japanese).

33. China and Taiwan had the first- and third-highest fresh and frozen fish exports of all developing economies in 1998 and 1999 (accounting for 14.8% and 11.8% of all developing economy fish exports). They are followed by South Korea in fourth place, Indonesia in sixth, Thailand in seventh, Singapore in eighth, and India in ninth.

34. Unless otherwise specified, production and trade data are calculated from: FAO, *Yearbook of Fishery Statistics: Aquaculture Production 1999, Yearbook of Fishery Statistics: Capture Production 1999*, and *Yearbook of Fishery Statistics: Commodities 1999*.

35. CP Group is Thailand's largest Chinese agribusiness group. It has grown into a representative Southeast Asian multinational engaging in the processing and export of various foods.

36. Unless otherwise noted, this section borrows much from these sources: Kawabe, Midori, "The Development of Shrimp Farming in Asia and the Emergence of Externalities," *Japanese Journal of Fisheries Economics*, vol. 46, no. 2, October 2001 (in Japanese); "Focus on Mangroves and Shrimp Farming," *WRM Bulletin*, no. 51, October 2001; and Hagler, M., *Shrimp: The Devastating Delicacy*, Greenpeace, 1997. On eel I have referred to: Qian, Hongu, *The Development of Eel Production for Export in China*, M.A. Thesis, Graduate School of Economics, Kyoto University, January 2001 (in Japanese).

37. These actions were taken against shrimp from China, Vietnam, Thailand, and Myanmar. *JETRO Food & Agriculture*, no. 2381, April 1, 2002 (in Japanese).

38. Goss, J., Burch, D., and Rickson, R., "Shrimp Aquaculture and the Third World," in Burch, D., et al. eds., *Australasian Food and Farming in a Globalised Economy*, Dept. of Geography and Environmental Science, Monash University, 1998, pp. 152–154.

39. On the relationship between agribusiness and haphazard development in Thailand, see: Skladany, M., and Harris, C., "On Global Pond: International Development and Commodity Chains in the Shrimp Industry," McMichael, P., ed., *Food and Agrarian Orders in the World Economy*, Praeger, 1995, pp. 182–185.

40. In July 2003 the National Research Institute of Aquaculture of Japan's Fisheries Research Agency announced that it was the first in the world to raise eels through their entire life cycle, but deploying the technology will still take time.

41. See *The State of the Environment in Asia* 2002/2003, pp. 185–186.

42. For details, see: Iwasa; and Casson, A., *The Hesitant Boom: Indonesia's Oil Palm Subsector in an Era of Economic Crisis and Political Change*, Center for International Forestry Research, 2000.

43. In both countries not only private enterprise, but also government-led small farmer settlement programs, play important roles. In Malaysia the Federal Land Development Authority boasts the largest scale in terms of production, and Indonesia has a Nucleus Estates and Smallholders program which places small farmers on plantation peripheries. Plantation development as discussed here takes these into consideration.

44. Iwasa, Kazuyuki, "Formation of the Palm Oil Market and Agribusiness in Malaysia," *Agricultural Marketing Journal of Japan*, vol. 9, no. 2, April 2001 (in Japanese); Chidley, L., *Forests, People and Rights*, International Campaign for Ecological Justice in Indonesia, 2002, Part I; Casson, A., pp. 13–14, 64.

45. Tenaganita and Pesticide Action Network Asia and the Pacific, *Poisoned and Silenced: A Study of Pesticide Poisoning in the Plantations*, 2002.

46. See Chidley, parts I and II.

47. *JETRO Food & Agriculture*, no. 2341, June 11, 2001 (in Japanese).

48. Chidley.

49. Fujishima, Koji, *Report: When Vegetable Imports Attain 3 Million Tons*, Ienohikari Kyokai, 1997, pp. 49–60; *Mainichi Shimbun*, August 11, 2002; Chen, Yongfu, *The Growing Vegetable Trade and Food Supply Capacity*, Norin Tokei Kyokai, 2001, pp. 23–37, 50–85 (all in Japanese). Investment in China by Taiwanese companies is for Chinese exports to Japan. They therefore serve as a bridge between China and Japanese companies.

50. Chen, pp. 68, 70.

51. Between 1997 and 2000 import declarations increased from 1,180,000 to 1,550,000, while the number of imported produce inspectors in Japanese ports of entry remained unchanged at 264 between 1997 and 2001, and the inspection rate declined from 13.4% to 7.2% between 1995 and 2000.

52. Pesticide contamination is an issue in China as well. Sampling inspections conducted in the third quarter of 2001 detected residual pesticides exceeding the standard in 47.5% of vegetables. In Hong Kong, the second-largest importer from China, Chinese-produced vegetables are called "poison vegetables."

53. Chen, p. 108; Vegetable Supply Stabilization Fund, *Trends in Chinese Producing Areas of Garlic, Ginger, and Other Produce: Mainly in Shandong Province*, 1996, pp. 27, 44, 49 (both in Japanese).

54. Only 20 to 30% of the garlic produced can be exported to Japan.

55. In 2001 commercial cultivation of GMO crops in China amounted to 1.5 million ha (a three-fold increase over the previous year), or 4% of the world total, and fourth behind the US, Argentina, and Canada.

56. On these trends, see: *The State of the Environment in Asia 1999/2000*, pp. 30–33; FAO, *The State of Food and Agriculture*, 1998, pp. 89–103; and Iwasaki, Misako and Ono, Kazuoki, *Rediscovering Asian Small-Scale Farming*, Ryokufu Shuppan, 1998, pp. 124–204 (in Japanese).

57. Some organizations handle clothing and furniture in addition to food and beverages. In Asia 54 organizations currently belong to the International Federation for Alternative Trade, and various other organizations are also active.

## Chapter 4

1. FAO, *State of the World's Forests 2001*, 2001.
2. FAO, *Forest Resources Assessment 1990: Global Synthesis*. This is the 1980--1990 average for the Asia-Pacific region.
3. IGES Forest Conservation Project, *Research Report on Phase I Strategy*, The Institute for Global Environmental Strategies, 2001, p. 95 (in Japanese).
4. Yamane, M. and Chantirath, Khampha, "Lao Cypress Forests: Causes of Degradation and the Present State of Conservation in Lao, P. D. R.," *International Review for Environmental Strategies*, vol. 1, no. 1, 2000, pp. 119–133.
5. The July 2000 G8 Summit in Okinawa had illegal logging on its official agenda, and delegates confirmed the issue's importance and the need for quick action on it. The Forest Law Enforcement and Governance East Asia Ministerial Conference held in Indonesia confirmed the importance of initiatives by both importing and exporting countries. In March 2002 the Second Session of the UN Forum on Forests issued a ministerial message that included initiatives for action against the illegal international trade in forest products.
6. For details, see: Verolme, H. J. H. and Moussa, J., *Addressing the Underlying Causes of Deforestation and Forest Degradation, Case Studies, Analysis, and Policy Recommendations*, Biodiversity Action Network, 1999, Washington, DC.
7. Japan's rice paddies have a storage capacity of about 5.1 billion tons of water, or about twice that of flood-control dams. See: Tabuchi, Toshio, *Rice Paddies of the World and Japan*, Noson Gyoson Bunka Kyoukai, 1999, p. 160 (in Japanese).
8. In areas with steep topography and heavy rainfall, the dikes surrounding terraced paddies and fields prevent soil erosion. Ibid., p. 166.
9. Moriyama, Hiroshi, *The Value of Preserving Rice Paddies*, Noson Gyoson Bunka Kyoukai, 1997, p. 49 (in Japanese).
10. Rieley, J. O., Susan, P. E. and Notohadiprawiro, T., *The Mega Rice Land Conservation Project in Central Kalimantan, Indonesia: Unsustainable Development*, 2000.
11. Suzuki, Kenji and Goto, Akira, "Environmental Conditions of Rainwater-fed Rice Agriculture in Northeast Thailand and the Anatomy of Agricultural Projection," Fujita, Kazuko, ed., *Water and Social Milieu in Monsoon Asia*, Sekai Shisousha, 2002, p. 124 (in Japanese); Sansanee, Choowaew, "Ricefield Agro-ecosystem and Biodiversity Conservation in the Lower Central Plain of Thailand," presentation at INTECOL VI, 2000.
12. Tsujii, Tatsuichi, "Paddy Field of Hokkaido: The Transition of Rice Agro-ecosystem in Northern Limit of Japan and its Impact on the Environment," presentation at INTECOL VI, 2000.
13. Moriyama, Hiroshi and Sprague, David S., *Traditional Paddy Field Landscape Acted to Preserve Japanese Dragonfly Fauna*, 2000.
14. Nakamura, Reiko and Hironaka, Kazumi, "Role of Rice Paddy as Wintering Habitat for Grus Monacha: A Case of Yashiro Valley, Japan"; Yamamoto, Hironobu, "Management and Conservation Strategies of Katano-Kamoike which is the Smallest Ramsar Site in Japan," both presentations at INTECOL VI, 2000.

15. Harada, Yuriko, "The Domestic Laws on the Conservation of Biodiversity of Rice Agro-ecosystem in Japan," presentation at INTECOL VI, 2000.
16. Report by the Bangladesh Rice Research Institute.
17. Hossain, Sanowar, "Rice Agro-ecosystem of Bangladesh: Sustainability through Integrated Farming Approach," presentation at INTECOL VI, 2000; Hossain, MZ, "Sustainable Use of Wetland Flood Plain Paddy Fields in Bangladesh," Asian Wetland Symposium Summary, 2001.
18. Ali, Ahyaudin Bin, "Fish sauna biodiversity in the Malaysian rice agro-ecosystem," 2000; Manor, Mashor, "Biodiversity of aquatic flora in the rice agro-ecosystem of Malaysia: the impact of human intervention," both presentations at INTECOL VI, 2000.
19. Salmah, Che, "Biodiversity of dragonfly in the rice agro-ecosystem of Malaysia as an indicator of ecosystem robustness," presentation at INTECOL VI, 2000.
20. Kuwabara, Ren, and Satrawaha, R., "Agro-Ecosystem of Rice: Fish Culture and the Annexed Ponds in Northeast Thailand and Laos," presentation at INTECOL VI, 2000.

## Part II

### Chapter 1
1. See: Shimazaki, Miyoko and Nihei, Takeo, "The Development of Economic Cooperation in Northeast Asia," in: Institute of Eurasian Studies, ed., *Eurasian Studies*, no. 27, Toyo Shoten, 2002, p. 22 (in Japanese). Some definitions of "Northeast Asia" cover broader areas, but this chapter will leave inland Asia to other chapters.
2. Koreans insist that it be called the "East Sea." "Northeast Asia" in this chapter's context is usually called the "Sea of Japan Rim Region" in Japanese.
3. A similar situation exists in the Sanjiang (Three-River) Plain located southwest of Vladivostok and north of Lake Khanka because Russia is more interetest in nature protection, and China in development. While the Russian side of the plain is a nature reserve, China has plans for developing the plain into the breadbasket of Northeast Asia by draining wetlands and growing crops including soy beans, rice, and barley. Already agricultural technologies have been introduced with Japanese assistance.
4. Previous volumes of this book and other chapters of this volume cover air pollution. Hence this chapter will describe only the activities of the Acid Deposition and Oxidant Research Center (founded in 1998 in Niigata City) and latest trends in its system for international cooperation.
5. See: *The State of the Environment in Asia 2002/2003*, pp. 76–79.
6. http://www.erina.or.jp/
7. Enviroasia, http://www.enviroasia.info/
8. Liu, Yongmao, Wang, Renhua and Zhai, Pingyang, *Controls and Standards on Methylmercury Pollution of Songhua River in China*, Kexue Chubanshe, 1998 (in Chinese).

### Chatper 2
1. Upper Mekong Associated survey Team of China, Laos, Myanmar and Thailand, Report on an Investigation of Waterway Transportation along the upper Mekong River of China, Laos, Myanmar, and Thailand, May 19, 1993.
2. The Joint Experts Group on EIA of China, Laos, Myanmar, and Thailand, Report on Environmental Impact Assessment: The Navigation Channel Improvement Project of the Lancang-Mekong River from China-Myanmar Boundary Marker 243 to Ban Houei Sai of Laos, September 2001.

3. McDowall, R. M., "Evaluation of 'Report on environmental impact assessment: The navigation channel improvement project of the Lancang-Mekong River from China-Myanmar Boundary Marker 243 to Ban Houei Sai of Laos' prepared by the Joint Expert Group on EIA of China, Laos, Myanmar and Thailand, September 2001; The Fisheries Impacts Reviewed, prepared for Mekong River Commission," January 2002; Finlayson, Brian, "Report to the Mekong River Commission on the 'Report on Environmental Assessment: The Navigation Improvement Project of the Lancang-Mekong River from China-Myanmar Boundary Marker 243 to Ban Houei Sai of Laos' prepared by the Joint Experts Group on EIA of China, Myanmar and Thailand September 2001," February 2002; Cocklin, Chris and Hain, Monique, "Evaluation of the EIA for the Proposed Upper Mekong Navigation Improvement Project Report, prepared for the Mekong River Commission: Environment Programme," December 2001.
4. Southeast Asia Rivers Network, et al., Mekong Rapids Under Fire, October 2002.
5. International Herald Tribune, July 31, 2001.
6. Pujaggan, March 23–24, 2002 (in Thai).
7. Bangkok Post, July 19, 2002.
8. The Nation, August 1, 2002.
9. People's Daily, January 20, 2002 (in Chinese).
10. Associated Press, January 24, 2002.
11. Bangkok Post, March 16, 2002.
12. According to interviews conducted by author Matsumoto, September 10, 2002.
13. AFP, March 4, 2000.
14. The Fisheries Office, A Study of the Downstream Impacts of the Yali Falls Dam in the Se San River Basin in Ratanakiri Province, Northeast Cambodia, May 20, 2000.
15. Phnom Penh Post, vol. 9, no. 9, April 28 to May 11, 2000.
16. Baird, Ian, An Update on the Situation in Communities Located Along the Se San River Impacted by the Yali Falls Dam in Northeast Cambodia, and Consultations with Local People Regarding Establishing a Network of Se San Communities, NTFP Project, July 2001.
17. Vietnam News Agency, June 15, 2002.
18. Vietnam News Agency, Asia Pulse, August 29, 2002.
19. This section makes use of achievements by the World Resources Institute's Mekong Regional Environmental Governance Project. Badenoch, Nathan, Transboundary Environmental Governance, World Resources Institute, 2002.
20. Interview by author (Matsumoto) conducted on March 9, 2003.
21. http://www.unece.org/env/eia/

## Chapter 3

1. Takaya, Yoshikazu, The World as Seen from "World Units": A Perspective for Regional Research, New Edition, Kyoto University Academic Publishing, 2001, pp. 36–122 (in Japanese).
2. Data for the 1950s to mid-1970s, and for mid-1970s to mid-1980s are quoted from Liu, Mingxing, et al., "Trend in Sandy Desertification of China's Land and New Ideas for Its Control," Regional Research and Development, 1996, no. 1; 1985 to 1995 data are quoted from Qie, Jianrong, "Total Effort at Controlling Sand Desertification of Land," Law Daily, December 27, 2001. The latest figure of 3,436 km²/year is the average for 1995 to 1999, and the sandy desertified area of 174.31 km² is for 1999: Lin, Xuan, "Second Announcement of State Forestry Bureau on Sandy Desertification and

Results of Observations on Sandy Desertification," *People's Daily*, January 31, 2002 (all in Chinese).

3. For data and information on dust storms, government afforestation projects, the causes of sandy desertification, and other areas, see Shen, Xiaohui, "Chapter 9: Sandy Desertification," Zheng, Isheng and Wang Shiwen, eds., *Critique of Environment and Development in China*, Social Sciences Publishing Company, 2001 (in Chinese).

4. See: Namjim, T., *Mongolia Then and Now* (two volumes), Japan Mongolia Folk Museum, 1998 (in Japanese); National Statistical Office of Mongolia, *Mongolian Statistical Yearbook* (annual), Ulaanbaatar; State Statistical Office of Mongolia, *Agriculture in Mongolia 1971–1995*, Ulaanbaatar, 1996; Yasuda, Yasushi, *Introduction to the Mongolian Economy*, Nihon Hyoronsha, 1996 (in Japanese); UN in Mongolia Website, <www.un-mongolia.mn/news/disaster/>.

5. Disaster caused by severe natural conditions, especially heavy snowfall.

6. National Statistical Office of Mongolia.

7. Humphrey, Caroline and Sneath, David, eds., *Culture and Environment in Inner Asia (1–2)*, The White Horse Press, 1996; Humphrey, Caroline, and Sneath, David, *The End of Nomadism?*, The White Horse Press, Cambridge, 1999; Oniki, Shunji, "The Nomadic Pastoralist Economy of the Mongolian Highlands and the Desertification of the Steppe," in: Society for Environmental Economics and Policy Studies, ed., *Environmental Problems in Asia*, Toyo Keizai Inc., 1998, pp. 279–291 (in Japanese).

8. *Ödriin sonin*, December 8, 2000, no. 289 (542), Ulaanbaatar (in Mongolian).

9. For a study of livestock overcrowding in this area, see: Müller, Franz-Volker and Bold, Bat-Očir, "On the Necessity of New Regulations for Pastoral Land Use in Mongolia," *Applied Geography and Development*, vol. 48, pp. 29–51, 1996.

10. Komiyayama, Hiroshi, "Mongolia, a Country of Nomadic Pastoralism," *Agriculture, Forestry, and Fisheries around the World*, January 2001, pp. 37–42 (in Japanese).

11. Buyantsogt Bat and Noel P. Russel, "Environmental Impacts of the Mongolian Economic Transition," in Nixon, Frederic, Bat Suvd, Luvsandorj, Puntsagdash, and Walters, eds., *The Mongolian Economy: A Manual of Applied Economics for an Economy in Transition*, Edward Elgar, 2000, Chapter 9, pp. 175–188; UNEP, *Nature and Environment in Mongolia*, UNEP, Ulaanbaatar, 1999.

12. Some works to see on this subject: Ad'yaasüren, Chimediin, Nyambuu, Khandyn, and Bat-Ireedüi, Jantsangiin, *Encyclopedia of Mongolian Customs (I–IV)*, Ulaanbaatar, 1999–2001 (in Mongolian); Batjargal, Z., "Sustainable Development: The Mongolian Experience," in Akiner, Shirin, Tideman, Sander, and Hay, Jon, eds., *Sustainable Development in Central Asia*, St. Martins's Press, 1998, Chapter 4, pp. 93–103; Means, Robin, "Environmental Entitlements: Pastoral Natural Resource Management in Mongolia," *Cahiers des Sciences Humaines*, vol. 32, no. 1, 1996, pp. 105–131; Means, Robin, "Community Collective Action and Common Grazing: The Case of Post-Socialist Mongolia," *The Journal of Development Studies*, vol. 32, no. 3, 1996, pp. 297–339; Tsubouchi, Toshinori, "Adaptation of Village Society to Harsh but Fragile Natural Environments: Symbiosis with Nature by Mongolian Nomadic Herders," in Kusano, Takahisa, ed., *Village Development and International Cooperation*, Kokin Shoten, 2002, Chapter 7, pp. 105–119 (in Japanese).

13. Mongol Bank, *Monthly Statistical Bulletin*, June 1995.

14. Estimates by the Kamo Research Office in the Biology Department of National University of Mongolia, 2002.

15. Although 44,000 were registered in Mongolia in 1990, that number doubled to 93,000 (of which 48,000 were registered in Ulaanbaatar) in 2001.

16. Mongolian Academy of Sciences, Institute of Geography.
17. Public Health Center of Ulaanbaatar City, 1995. However, respiratory damage stemming from indoor coal stove use is also found in Mandal Village.
18. Some initiatives have already been implemented, such as a thermal power plant improvement project by Japan (1992–1996) and a low-emission *ger* stove project by the World Bank (2000-), but systematic improvement of power plant efficiency is difficult because generating and emission treatment equipment is dated. The World Bank project aims to mitigate air pollution by building and selling improved stoves, but 982 of 2,000 stoves manufactured remain unsold (as of July 31, 2002) because they are very expensive to *ger* dwellers and because people who purchased them do not see much of a difference with their previous stoves.
19. Ulaanbaatar Office of Statistics.
20. Plants siting in and near the city include wool and cashmere processing plants, tanneries, shoe and clothing factories, and printing plants. Substances sometimes used at such plants include, for example, chrome and sulfides at tanneries, and chrome, cyanide, cadmium, copper, zinc, and manganese in the printing industry.
21. Ulaanbaatar Office of Statistics.
22. These landfill sites (one of which stopped accepting waste in 2001) merely use montane valleys adjoining Ulaanbaatar and *ger* villages. While all are managed by the national government, access is unrestricted by fences or walls, and one can often observe trash scavengers, and livestock that come to feed on kitchen waste.
23. This section is based primarily on work performed in the region by the Japan Research Association with Kazakhstan since 1991.
24. Purposively sampled subjects were 42 Aralsk residents, heads of local governments, and 33 others who were mainly people of comparatively high status and experts.
25. Kusumi, Ariyoshi, "Life and Opinions around the Aral Sea," Japan Research Association with Kazakhstan, *Impacts of Large-Scale Irrigated Agriculture in Central Asian Arid Regions on Ecosystems, Societies and Economies: 2001 Research Report*, August 2002, pp. 97–113 (in Japanese).
26. Baasan, Tudev, "Types of Desert in Mongolia," and Okada, Tomokazu, "Mongolia's Environment: Getting the Story on Desertification in Govi-Altay Province (Aimag)," *Overseas Environmental Cooperation Center, Japan Newsletter*, no. 34, 2001.

*Chapter 4*
Japan
1. A broad category that includes refuse, bulky waste, cinders, sludge, human waste, waste oil, waste acid, waste alkalis, and animal carcasses.

Republic of Korea
1. In 2001 NOx was 0.037 ppm and $PM_{10}$ was 70 $\mu g/m^3$.

People's Republic of China
1. *Washington Post*, September 9, 2001.
2. "Underground Elegy," *Nanfengchuang*, September 2001 (in Chinese).
3. *China Environment News*, May 2, May 30, July 11, and July 18, 2001 (in Chinese).
4. On the part of the Tenth Five-Year Plan pertaining to environmental protection, see: State Environmental Protection Administration, Policy and Legal Bureau, ed., *Comprehensive Collection of Chinese Environmental Policy on Transition to the Market Economy (2002)*, Huaxue Gongye Chubanshe (in Chinese).

Taiwan
1. Especially with regard to the "fourth nuclear power plant" whose construction was started under the Kuomintang government and scheduled for completion in 2004, former DPP leader Lin Yi-xiong declared, "If the new government does not stop construction of the fourth nuclear plant, it would violate the party's platform."
2. An October 2000 public opinion poll found 31% of the public supported the decision, while 50% did not.
3. See: Chiau, Wen-yan and Chen, Li-chun, "Environmental Issues and Challenges in Taiwan," *Research on Environmental Disruption*, vol. 32, no. 1 (in Japanese).

Republic of the Philippines
1. The council has issued a report setting forth the challenges to development and the environmental in the Philippines: Philippine Council for Sustainable Development (1997), *Philippine Agenda 21*, Republic of the Philippines.
2. A region comprising 16 cities and towns.
3. Section 9 of the Clean Air Act reads, in part: "Pursuant to Sec. 8 of this Act, the designation of airsheds shall be on the basis of, but not limited to, areas with similar climate, meteorology and topology which affect the interchange and diffusion of pollutants in the atmosphere, or areas which share common interest or face similar development programs, prospects or problems."

Socialist Republic of Vietnam
1. Nippon Koei Co., Ltd. and EX Corporation, The Study on Environmental Improvement for Hanoi City in the Socialist Republic of Vietnam Interim Report, vol. 3, March 1999, pp. 1-1 and 7–31.

Kingdom of Thailand
1. Later one of the power plants was moved to Ratchaburi Province and its fuel was switched from coal to natural gas (*Bangkok Post*, February 26, 2003).
2. *Bangkok Post*, May 23, 2002.
3. The Bo Nok power station EIS was approved before the power contract was signed.
4. *State of the Environment in Asia 2002/2003*, pp. 183–185.
5. The chief of the Mineral Resources Bureau claims that granting blanket concessions to vast areas is environmentally friendly because environmental damage is decreased by the slowed pace of development, and because promoting underground mine development lessens surface development, which has heavy environmental impacts (*Bangkok Post*, July 12, 2002).
6. *Bangkok Post*, November 6, 2001 and November 26, 2001.
7. The current national forestry policy prohibits not only large-scale logging, but also the gathering of branches for firewood and charcoal, and of non-wood items such as mushrooms and insects.
8. Specifically, the bill was changed so that a National Community Forest Committee to be established would centrally control the community forest standards and assume responsibility for monitoring provincial inspection committees, in these and other ways concentrating decision-making in the hands of the government. Additionally, a community forest could not be created by an organization or village, but required at least 50 persons age 18 or older from a traditional community that is native to the area, and which has over the past five years or more been actively engaged in forest preservation. And by linking requirements to permanent settlement and migration, the bill made it hard for hilltribes which have practiced traditional shifting cultivation to establish community forests.

9. *Bangkok Post*, November 9, 2001.
10. The Senate passed this changed bill in March 2002, but it must be returned to the House of Representatives for deliberation again before becoming law.
11. *Bangkok Post*, March 24, 2002.
12. See: Anan Ganjanapan, *Local Control of Land and Forest: Cultural Dimensions of Resource Management in Northern Thailand*, Chiangmai: Regional Center for Social Science and Sustainable Development, Faculty of Social Sciences, Chiangmai University, 2000; Anchalee Kongrut, "Changes worry activists," *Bangkok Post*, July 12, 2001; Kultida Samabuddhi, "Thaksin's First Year: Little Done for Environment," *Bangkok Post*, February 6, 2002; Saniysuda Ekachai, "Planned Changes Are a Slap in the Face," *Bangkok Post*, September 27, 2001; Saniysuda Ekachai, "Art for the Forest Sake," *Bangkok Post*, March 21, 2002; Sukran Rojanapaiwong (ed.), *State of the Thai Environment 1997–98*, Green World Foundation, 2000; and Tasaka, Toshio, *Eucalyptus Business: Thai Deforestation and Japan*, Shinnihon Shinsho, 1992 (in Japanese).

Malaysia
1. *Nihon Keizai Shimbun*, July 25, 2002.
2. This plan is to use high-speed optical fiber to link the Kuala Lumpur City Center with the new administrative city Putra Jaya to the south, the new information industry city Cyber Jaya, and the new international airport through a zone 50 km long, which would serve as a hub for the information industry and networks.
3. *State of the Environment in Asia 2002/2003*, pp. 190–191.
4. Aoki, Yuko, "Regional Waste Management Systems in Malaysia," Diss. Yokohama National University, March 2003 (in Japanese).
5. *Activities Report July 2000 Newsletter*, Malaysia Nature Society, 2000.
6. On sewage treatment privatization, see: *State of the Environment in Asia 2002/2003*, pp. 189–190.
7. *Eighth Malaysia Plan (2001–05)*, Economic Planning Unit, Prime Minister Department, Malaysia, 2001, pp. 283–284.
8. *The Star*, August 7, 2002.
9. http://www.rjc.edu.sg/studproj/sterling/1_perfect-blue/treaties.html (accessed September 15, 2004).
10. Department of Environment, Ministry of Science, Technology and the Environment, Malaysia/University of Putra Malaysia (DOE/UPM), *Investigation and Assessment of Municipal Landfill Sites in the Federal Territory of Kuala Lumpur*, vol. 1, vol. 2, 2000.
11. On the pollution and health damage caused by the IT industry, and related problems in Malaysia and other Asian countries, see: Yoshida, Fumikazu, *IT Pollution*, Iwanami Shinsho, 2001 (in Japanese). Published in English as: (1) Yoshida, Fumikazu, "IT Pollution Problems in Asia," *Economic Journal of Hokkaido University*, vol. 32, 2003. (2) Yoshida, Fumikazu, "Information technology waste problems in Japan," *Environmental Economics and Policy Studies*, vol. 5, no. 3, pp. 249–261, Springer, 2002.

Republic of Indonesia
1. As with the passage of the Decentralization Law in 1903 during the Dutch colonial era, and the 1974 Local Government Law and 1979 Rural Governance Law during the Suharto era.
2. This article does not chronicle the most recent changes in local autonomy in Indonesia, specifically the revision of two basic laws on local autonomy which took place in 2004. Laws No. 22 and No. 25 of 1999 were replaced by, respectively, Law No. 32 (Concerning Regional Administration) and Law No. 33 (Concerning Fiscal Balance

between the Central Government and the Regional Governments). Under Law No. 32, autonomous regional governments at all levels were given clear status as authoritative governments in environmental matters, and environmental conservation was given higher priority as one mandatory area to which the central government must tend in view of national interests.

3. ICG Group, "Indonesia: Natural Resources and Law Enforcement," *ICG Asia Report*, no. 29, 2001.
4. Department of Forestry, "Overcoming Illegal Logging and the Distribution of Illegal Forest Products," *ICG Asia Report*, no. 29, 2001.
5. Ibid.
6. *Jakarta Post*, April 5, 2002.
7. In June and July of 2002 a UN legal advisor reported that Indonesia's courts were highly corrupt. Suharto's son Hutomo "Tommy" Mandala Putra was found guilty of hiring hit men to assassinate a Supreme Court justice who had convicted him of corruption. Corruption of the judiciary, including the Supreme Court, is a major social issue.
8. World Bank, *Indonesia: Environment and Natural Resource Management in a Time of Transition*, 2001.
9. Suwondo, Kutut, *Decentralization in Indonesia*, INFID Annual Lobby, 2000; Aden, Jean, *Decentralization of Natural Resource Sectors in Indonesia*, East Asian Environment and Social Development Unit, 2001; Kuswanto, SA, "Antara Sentralisasi dan Desentralisasi Penglolaan Sumber Daya Air" (in Indonesian); Baktiar, Irfan, "Desentralisasi Penglolaan Sumber Daya Hutan di Kabupaten Wonosobo" <http://www.arupa.or.id/papers/02.htm> (in Indonesian); Potter and Suimon Badcoc, "Reformasi and Riau's Forests" <http://www.insideindonesia.org/edit65/potter.htm>; "Shrimp Farmers Charged with Destroying Mangrove Forests," *Jakarta Post*, August 12, 2002; "Illegal Sand Mining Runs along Cianjur Southern Coast," *Jakarta Post*, August 12, 2002; "Fishermen Complain about Trawlers, Dynamite Fishing," *Jakarta Post*, March 19, 2002; "Deforestation Leaves Dieng Plateau Barren," *Jakarta Post*, March 19, 2002; and "Agar Tidak Tumbuh Raja-Raja Kecil," *Media Transparensi* <http://www.transparensi.or.id/majalah/edisi6/6berita_5.html>.
10. World Bank, op. cit., p. 121.
11. Ferrazzi, Gabe, *Legal Standings and Models of Local Government Functions in Selected Countries: Implications for Indonesia (SfDM Report 2002-1)*, Department Dalam Negeri, March 2002, p. 11.

India
1. Nomura, Yoshihiro and Endo, Takako, "India's Environmental Laws and Administrative System," Nomura, Yoshihiro and Sakumoto, Naoyuki, eds., *Environmental Laws in Developing Countries: Southeast and South Asia*, Institute of Developing Economies, 1994 (in Japanese).
2. World Bank, *India's Environment: Taking Stock of Plans, Programs and Priorities*, India Nepal Bhutan Country Department, South Asia Regional Office, World Bank, 1994.
3. Anil Agarwal, Sunita Narain and Srabani Sen, *State of India's Environment: The Citizen's Fifth Report, Centre for Science and Environment*, 1999, p. 376.
4. The Delhi Metro subway began operating over part of its length in late 2002, and it is scheduled to operate in the city center in September 2005. Aimed at reducing motor vehicle air pollution and traffic congestion, the project is being built with funding from

Japan and other countries. But Dunu Roy, representative of the Hazard Center, points out that setting appropriate fare levels, securing a stable supply of electric power, and other actions are needed before the Delhi Metro can become eco-friendly transportation for the citizens.

5. On Supreme Court decisions, compliance with them, and the costs and benefits of factory relocation since 1996, see: Vinish Kathuria, "Delhi: Relocating Polluting Units: Parochialism vs Right to Live?" *Economic and Political Weekly*, January 20, 2001.

6. According to a hearing at the Tata Energy Institute (New Delhi) in December 2001. However, *The Hindustan Times* of June 22, 2002 reported that all observed pollution values had improved at seven locations in Delhi according to Central Pollution Control Board observations.

7. <http://www.cseindia.org/html/cmp/air/cng/cng´cngorder´note8.htm>

8. Tsujita, Yuko, "India: Emergence of a New Market and Attempts at Reviving Traditional Insights," *IDE World Trends*, no. 73, 2001, pp. 14–15 (in Japanese).

9. Centre for Science and Environment, *State of India's Environment: The Citizen's Fourth Report: Dying Wisdom*, 1997.

10. Centre for Science and Environment, *Tanks of South India*, 2001.

11. Centre for Science and Environment, *Making Water Everybody's Business: Practice and Policy of Water Harvesting*, 2001.

12. Shiva, V., *Biopiracy: The Plunder of Nature and Knowledge*, South End Press, 1997.

13. Yamana, Mika, "Legal Protection of Biological Resources and Traditional Knowledge: India's proposal on 'Biopiracy,'" *Kyoto Women's University Contemporary Study Society*, no. 119, 2000, pp. 119–132 (in Japanese).

## Part III

1. Economic Inequality, Poverty and Human Development
 1. UNDP. Human Development Report 1990, Oxford University Press, 1990.
 2. UNDP. Human Development Report 1997. See the 2001 edition for the calculation method.
 3. Ibid.

2. Population and Gender: Reproductive Health and Rights
 1. For a more detailed treatment, see pp. 228–231 of the *State of the Environment in Asia 2002/2003*.
 2. The law's Japanese name literally means 'Basic Law for a Society with the Joint Participation of Men and Women.'

5. Disputes over Regulating GM Crops and Food
 1. See, for example: Watanabe, Mikihiko and Futamura, Satoshi, eds., *Biological Resource Access: The Bioindustry and Asia*, compiled by the Japan Bioindustry Association, Toyo Keizai, 2002 (in Japanese).
 2. USDA National Agricultural Statistics Service, 'Acreage,' a supplement report to *Crop Production*, June 28, 2002.
 3. Hisano, S. and Altoe, S. M., 'Brazilian Farmers at a Crossroads: Biotech Industrialization and Agriculture or New Alternatives for Family Farmers?' a paper presented at the conference of CEISAL, July 3–6, in Amsterdam, The Netherlands.
 4. James, C., 'Global Status of Commercialized Transgenic Crops,' *ISAAA Briefs*, no. 30, ISAAA, Ithaca, NY, 2003.

5. Threshold means the rate of adulteration with GM ingredients, which is the cutoff point determining whether food is labeled GM (required) or non-GM (voluntary).
6. USDA Foreign Agricultural Service (FAS), 'Republic of Korea, Biotechnology: A Summary of Korean Regulations on Agro-Biotechnology Products 2002,' *GAIN Report*, no. KS2034, July 31, 2002.
7. FAS, 'Taiwan, Biotechnology: Bioengineered Food Regulation Update,' *GAIN Report*, no. TW1041, October 12, 2001.
8. FAS, 'People's Republic of China, Biotechnology: GMO Administration Regulations 2001,' *GAIN Report* no. CH1024, June 15, 2001; 'People's Republic of China, Biotechnology: Interim Regulations Extended,' *GAIN Report*, no. CH3113, July 18, 2003.
9. Sakarindr Bhumiratana, 'Report on Biosafety Policy Options and Capacity Building Related to Genetically Modified Organisms in the Food Processing Industry of ASEAN,' A report prepared for UNIDO, June 2002.
10. *Korea Times*, 'Activists oppose promotion of genetically modified products from U.S.,' January 25, 2002.
11. Reuters, 'U.S. says more talks needed on China's GMO rules,' February 7, 2002; 'U.S. senators ask China to delay biotech food rules,' February 19, 2002.

6. Forest Resources on the Decline
1. FAO, *State of the World's Forests 2001*, 2001. See also http://www.fao.org/forestry.
2. Because some forests should be protected and others cannot be logged, actual consumption must be far smaller. An estimated 12% of the world's forested area is in protected areas.
3. FAO, *State of the World's Forests 1999*, 1999.

7. Local Perspectives in Protected Area Management
1. FAO, *State of the World's Forests*, 2001.
2. IUCN, *1997 United Nations List of Protected Areas*, 1997.
3. World Resources Institute, et al., *World Resources 2000–2001*.

8. Illegal Wildlife Trade Flourishes
1. IUCN, *2000 IUCN Red List of Threatened Species*, 2001.
2. Primack, R.B. 1995. *A Primer of Conservation Biology*, Sinauer Associates, Sunderland, Massachusetts.
3. Japan is the main Asian importer of parrots.
4. Ministry of Finance data.

9. Coastal and Marine Environments in Crisis
1. ESCAP and ADB, *State of the Environment in Asia and the Pacific 2000*, United Nations, 2000, p. 100.
2. Pp. 70–88. Since that chapter discusses tanker accidents, the author will omit such discussion here.
3. ESCAP and ADB.
4. Chia, L. S. and H. Kirkman, *Overview of Land-Based Sources and Activities Affecting the Marine Environment in the East Asian Seas*, Regional Seas Report and Studies Series, UNEP/GPA Coordination Office and EAS/RCU, 2000, p. 35.
5. Ibid., p. 39.
6. ESCAP and ADB, p. 102.
7. Ibid.
8. Ibid., pp. 112–120.

9. See: *State of the Environment in Asia 2002/2003*, pp. 70–88.
10. For a discussion of such trends in Thailand, see: Yamao, Masahiro, *Development and Cooperatives: Advancement of Farming and Fishing Village Cooperatives in Thailand*, Taga Publishing, 1999, and Yamao, Masahiro, "Current State and Challenges of Coastal Marine Resources Management in Thailand," Proceedings of 50th Conference of the Japanese Society of Fisheries Economics, May 31, 2003 (both in Japanese).

10. Deepening Water Crisis

1. *Sustaining Water, Easing Scarcity: A Second Update*, 1997.
2. Falkenmark, Malin, Jan Lundquist, and Carl Widstrand, "Macro-scale Water Scarcity Requires Micro-scale Approaches: Aspects of Vulnerability in Semi-arid Development," *Natural Resources Forum*, vol. 13, no. 4, 1989. However, a subsequently modified index is used here in place of the original index. Gleick, Peter H., *The World's Water 2002–2003: The Biennial Report on Freshwater Resources*, 2002, Island Press, Chapter 4. This book is useful as a survey of water scarcity indexes.
3. Annual per capita water availability is what the FAO terms "total actual renewable water resources." Theoretically this is the maximum amount of water that can be used in one's own country, and is the total of water in the country plus that flowing in from outside. But because this value also takes into account the amount of water drawn by other countries pursuant to official and unofficial agreements, it changes due to circumstances at any given time (FAO, *Review of World Water Resources by Country*, 2003).
4. Gleick, Peter H., "Basic Water Requirements for Human Activities: Meeting Basic Needs," *Water International*, no. 21, 1996.
5. Singapore purchases about half its household and industrial water from Malaysia and obtains it through a pipeline, but recently there is disagreement between them on price. *Asahi Shimbun* 07/13/2004.
6. These results are quite different from those of Gleick. For example, Gleick has 28.8 liters/person/day for Vietnam. It is possible the table's figures are too large because withdrawal data were used.

13. Urbanization Proceeds Rapidly

1. United Nations, *World Population Prospects*, 2002.

14. Motor Vehicles and Air Pollution

1. As of July 2004, there is little or no discussion about road pricing in the Tokyo Metropolitan Assembly, and the practice is not about to be instituted.
2. The following works contain comprehensive and detailed treatments of options for dealing with environmental problems stemming from motor vehicle use in Asia: Economic and Social Commission for Asia and the Pacific, *Road Transport and the Environment: Areas of Concern for the Asia and Pacific Region*, United Nations, 1997; Kojima, M., C. Brandon and J. Shah, *Improving Urban Air Quality in South Asia by Reducing Emissions from Two-Stroke Engine Vehicles*, Working Paper 21911, The World Bank, 2000.

15. Burgeoning Wastes and New Management Policies

1. United Nations Economic and Social Commission for Asia and the Pacific, *State of the Environment in Asia and the Pacific 2000*, 2000.
2. *State of the Environment in Asia 1999/2000*, pp. 164–166.

16. The Internationalization of Recycling
 1. *State of the Environment in Asia 2002/2003*, Part I Chapter 3, "Wandering Wastes," and Part III-17, "Paper Recycling: The North-South Connection."
 2. On the handling of data and the analysis method employed in this section, see: Yamashita, Hidetoshi, "Using Data to Track Cross-Border Recycling," *Frontiers of Environment and Development*, no. 2, pp. 115–125, 2002, published by the Course of International Studies, Graduate School of Frontier Sciences, the University of Tokyo (in Japanese).

18. Nuclear Power Development and KEDO
 1. "Long-Term Energy Supply and Demand Outlook," 1999.

19. International Contact among Environmental NGOs Grows More Vigorous
 1. Membe rship organizations that assume various forms and number over 200,000 nationally.
 2. Currently estimated to number about 1 million nationally, these include welfare facilities such as private schools, schools for the handicapped, and private environmental organizations.

20. Trends in Environmental ODA, and the Need for Post-Project Evaluation
 1. *State of the Environment in Asia 2002/2003*, pp. 187–188.
 2. Ibid., pp. 183–185, 200–201.
 3. *State of the Environment in Asia 1999/2000*, p. 16.

# Afterword

Today the Bon festival begins. All the elementary school children in our village have gathered in our yard and are busy with preparations to welcome the spirits of the dead, who come to visit us on this occasion. It is common throughout Japan to welcome the spirits with a bonfire, but here in Kamioi, a rural village on the southern extremity of Tsukuba City, the time-honored practice is the "Bon rope."

Yesterday, the 12th, the children went around to all the approximately 50 homes here gathering rice straw, which adults used to make thin cords. This morning everyone braided the cords into a rope 20 centimeters thick and about 10 meters long. The rope is finished by adding to its end a dragon's face made of straw and about 40 centimeters across, which is completed by adding eggplant eyes and a mouth of red peppers. That finishes the preparations, which are performed under the guidance of the children's grandfathers.

Every year these preparations are made next to a bamboo grove on a part of our property, yet I have been able to participate only once. This year I took some of my vacation time to help, but I ended up in my study writing this afterword instead.

Today's Bon rope main event takes place from late afternoon and into the night. To start with, the children pull the rope from the village to the cemetery while loudly chanting, "We're going to meet the spirits," and after reaching the cemetery's center they make three circuits. By this time the spirits of the ancestors have gotten on the rope, so they visit all the homes in the village while loudly chanting, "The Bon rope has come." When people hear the children approaching their homes, they open their front doors and await the Bon rope's arrival. As the children chant and pull the rope around a house once, that family's ancestral spirits enter the house from the front door. The children receive a little money, and move on to the next house. After visiting all the homes, the children total up the money and split it evenly.

Such events, handed down from one generation to the next, give different generations opportunities to work together, and the moment you notice that, it gives you a sense of spiritual gratification. What's more, it confirms that you're "involved" with the people around you, and that provides a sense of reassurance. Though it takes me two hours each way between my home and work in Tokyo, living here gives me a number of opportunities to recharge my batteries, so to speak. Along with field work in Kalimantan and visiting my family in Yamanashi, the annual events here are literally healing to me, even though work obligations often prevent my participation.

I hope to always maintain this feeling of experiencing these little happinesses. But doing so makes it necessary to keep the world in a state of peace, because the little happinesses we experience in everyday life readily crumble owing to anti-peace acts such as war and environmental damage. For that reason we should not simply withdraw into our own lives, nor should we have no concern for things happening in society and the world. There are in fact many victims suffering every day from pollution and wars. We have to put ourselves in their shoes and take action.

Unfortunately, the August 6, 2003 peace declaration by the mayor of Hiroshima accurately describes the situation: "A world free of nuclear weapons and war is receding from our grasp, and dark clouds hover wherever we look." If that is so, we must deal with the situation, but nothing can be done if we are indifferent to events at home and abroad. The "involvement" discussed in the Introduction is a concept meant to extricate us from such an impasse. One expects that concrete action based on this "involvement principle" can stave off the environmental damage arising throughout Asia and at the same time foster the little happinesses that have taken root in various places.

Since its start in 1979, the Japan Environmental Council (JEC) has grappled with a variety of problems including military issues, poverty, development, and trade, all from the perspective of the environment. In other words, the environment has been our approach to bringing about a society of peace. JEC is characterized by people of various specialized disciplines but adhering to no certain factions, who have carried on voluntary activities (both research and practical action). The organization has a number of accomplishments in contributing to better environmental policy in Japan, and in the future members intend to leave a solid record of achievement in the progress of environment policy not only in Japan, but the rest of Asia as well.

Doing so demands that people of my age and the younger generation make ourselves more capable. JEC's senior members are well-known scholars representing various disciplines, but the rest of us will achieve nothing by just revering them for their capabilities. Instead, we must look for new ways to press forward while succeeding to their critical awareness and achievements. Though we younger members cannot compare individually to the senior members in terms of capabilities, we can build immense capabilities using a network of individuals who share a critical awareness. It is increasingly crucial that we expand our network of researchers and NGO people across national borders.

Fortunately, many leading and younger researchers and activists were involved in the publication of this book as editorial committee members, writers, or collaborators. Just as with the Bon rope that I described above, it is highly significant that this book was published as a joint effort of several generations, from young members to old hands.

As mentioned in the Preface, I have taken over the important task of editor-in-chief from Professor Teranishi Shun'ichi (Hitotsubashi University). Our editorial office comprised four people including me. I had a truly enjoyable time working with the other three, very capable people: Kojima Michikazu (Institute of Developing Economies), who handled the planning, Oshima Ken'ichi (Ritsumeikan University), who did the accounting, and Yamashita Hidetoshi (Hitotsubashi University), who assisted me. I also think we did well in dividing up the work of editing and proofreading the manuscripts; I did Part I, Mr. Kojima did Part II, and Mr. Oshima did Part III.

But here I must be honest in confessing the immense contribution of Professor Teranishi, who served as one of the chief editors and also took on the job of editorial supervision for this book. Because serving as editor-in-chief for the previous two volumes had totally familiarized him with all the tasks involved, he was always there with appropriate advice when needed. What's more, he offered incisive comments on content at the editing stage, and pointed out things I failed to notice at the proofreading stage. The new editorial office was like a just-hatched chick taking its first steps as the mother watches.

I wish to express my gratitude to the editorial committee members, writers, collaborators, and everyone else who worked on this project for cheerfully cooperating when our inexperienced editorial staff made requests. Our thanks also go to Toyo Keizai Inc. and Mr. Kojima Shinichi of its Book Publications Bureau for publishing the Japanese versions of the series, as well as to Mr. Sato Takashi, who was directly responsible for this volume in print.

It is our hope that people working on environmental conservation in Asia and globally, as well as those who plan to take such action, will read this book and put it to good use in pursuing their initiatives. We are already at work on the fourth volume in this series, and we hope that this endeavor will continue to benefit from the help and cooperation of so many people.

August 13, 2003

At home in Tsukuba and feeling I should be helping with the Bon rope ...

INOUE Makoto
Editor-in-chief for the Japanese-language version

# Translator's Afterword

With two previous volumes of this series under my belt, one would think that it would be getting easier, and that I'd be feeling like an old hand while doing the third. But on the contrary, the broad variety of topics, fields, countries, and regions covered in this series sometimes makes me feel even more overwhelmed than before. What's more, the lack of an editor means that the job of editing "on the fly" is also thrown into my lap. My snap, and largely arbitrary, judgments made while translating on what needs cutting, enhancement, rearrangement, and clarification are partly behind the product, so I must accept part of the responsibility for what you read. Needless to say, helpful comments from readers are welcome.

Normally, such circumstances would make it impossible to get this translation out, so the fact that you are reading this now is in a large part due to the generous cooperation and valuable help of many people whose names are found in the list of "Editors, Writers, Collaborators, and Assistants." In fact, this series would not exist if it were not for a great deal of dedicated volunteer effort on the part of all these people.

But I want to use the rest of my space to bring up a timely topic of critical importance: Oil.

Since this series specializes in Asia, let's begin with Japan's situation. Modern Japan and its economy simply would not exist without cheap oil. With at least 60% of its food imported and a population of over 100 million, Japan is highly dependent on food imports to feed its people, yet those imports would be far more expensive than now without cheap fossil fuels to manufacture, transport, and apply pesticides and fertilizers, and to produce, harvest, and transport the crops long distances. The inexorable rise in oil price will make food imports more and more expensive, and eventually uneconomical. And with the heavy toll that expensive oil will exact on Japan's economy, dark clouds are gathering on the horizon.

Yet, Japan is not the only Asian country that will be severely affected. Industrialization and development in other Asian economies—especially soaring energy demand in the populous countries of China and India—are putting an extremely heavy strain on oil supplies. Although millions of Asians live in want and count on development to lift them out of poverty, heavily oil-dependent development must inevitably be aborted at some point.

And while globalization has been touted as the antidote to poverty throughout the world, it too is a child of low-priced energy. As long as oil is cheap, moving raw materials, goods, and even labor around the world seems to make sense as a good way to earn money and create jobs, but at a certain point on the oil-price scale, this too will become uneconomical, limiting "globalization" to tasks that can be done over the Internet or telephone lines. It is not an encouraging picture.

Yet, every cloud has a silver lining, and peak oil is no exception. Just like stricter environmental standards or the Kyoto Protocol, expensive oil is an opportunity to make badly needed changes that will include rebuilding local economies (which were devastated in the first place by cheap oil), moving quickly to secure and develop renewable energy sources, and generally "powering down" national economies to adjust to a low-energy world. Lower consumption of fossil fuels will also have environmental and health benefits.

Because much of Asia is still undeveloped and therefore dependent on local economies, development in such regions should be oriented toward building on those local economies instead of making people dependent on an energy supply that would lead to the collapse of those economies, only to soon recede out of their reach again. In this sense, being undeveloped could actually give some regions a head start in development for a better life that is not heavily dependent on oil. And it goes without saying that special attention is needed to ensure the equitable distribution of energy and resources, something which has eluded humankind throughout history.

With the reality of peak oil near at hand, one would expect more open discussion on the momentous—and potentially cataclysmic—changes that are inevitably going to happen as fossil fuels become increasingly costly, yet the world's politicians are strangely silent on this issue. It is up to NGOs and the activist community to get this topic on the official agenda as soon as possible.

Energy wars or a better world—it is our choice, but we must make it today.

Rick DAVIS
Ashigawa, Japan

January 2005

# Index